BENZENE

POLLUTION ENGINEERING AND TECHNOLOGY

A Series of Reference Books and Textbooks

EDITORS

RICHARD A. YOUNG

Editor, Pollution Engineering
Technical Publishing Company
Barrington, Illinois

PAUL N. CHEREMISINOFF

Associate Professor
of Environmental Engineering
New Jersey Institute of Technology
Newark, New Jersey

1. Energy from Solid Wastes, *Paul N. Cheremisinoff and Angelo C. Morresi*

2. Air Pollution Control and Design Handbook (in two parts), *edited by Paul N. Cheremisinoff and Richard A. Young*

3. Wastewater Renovation and Reuse, *edited by Frank M. D'Itri*

4. Water and Wastewater Treatment: Calculations for Chemical and Physical Processes, *Michael J. Humenick, Jr.*

5. Bifouling Control Procedures, *edited by Loren D. Jensen*

6. Managing the Heavy Metals on the Land, *G. W. Leeper*

7. Combustion and Incineration Processes: Applications in Environmental Engineering, *Walter R. Niessen*

8. Electrostatic Precipitation, *Sabert Oglesby, Jr. and Grady B. Nichols*

9. Benzene: Basic and Hazardous Properties, *Paul N. Cheremisinoff and Angelo C. Morresi*

Additional volumes in preparation

BENZENE

Basic and Hazardous Properties

**PAUL N. CHEREMISINOFF
ANGELO C. MORRESI**

MARCEL DEKKER, INC. New York and Basel

Library of Congress Cataloging in Publication Data

Cheremisinoff, Paul N
    Benzene, basic and hazardous properties.

    (Pollution engineering and technology ; 9)
    Includes bibliographical references and index.
    1. Benzene. I. Morresi, Angelo C., joint
author. II. Title. III. Series.
TP248.B4C47   662'.669   79-17512
ISBN 0-8247-6860-4

COPYRIGHT © 1979 BY MARCEL DEKKER, INC. ALL RIGHTS RESERVED.

Neither this book nor any part may be reproduced or transmitted in any form or by any means, electronic or mechanical, including photocopying, microfilming, and recording, or by any information storage and retrieval system, without permission in writing from the publisher.

MARCEL DEKKER, INC.
270 Madison Avenue, New York, New York 10016

Current printing (last digit):
10 9 8 7 6 5 4 3 2 1

PRINTED IN THE UNITED STATES OF AMERICA

Preface

Science, technology, and legislation regulations regarding the environment are rapidly advancing and changing. Environmental backgrounds (the area external to the plant), as well as worker exposure (occupational safety and health), the handling, transportation, and disposal of materials and wastes have all come under close scrutiny by regulatory agencies, environmental groups, the public, and the worker, who have all become increasingly concerned with the effects of certain materials which were once widely accepted but which, with the advance of science and technology, have demonstrated properties that may be detrimental to health, life, and property if not handled properly.

Benzene is one material which has been widely accepted as an article of commerce. Its value in our modern technological society is great and the curtailment of its use would undoubtedly have an impact on our way of life. The purpose of this study was to pull together the data dealing with benzene and present in one source a reference dealing with the basic properties, production, uses, handling, and transportation requirements and controls for safe, effective utilization that is within the state of the art.

The intent of this book is to provide a source for quick, timely, evaluation/analysis data in recognition of the great level of interest and importance for benzene. An additional objective was to produce a reference of reasonable size, and in this effort the authors have attempted to distill and present only that information which was without repetition and which was felt to be of significant importance to those concerned with the basic and hazardous properties of this commodity.

Our acknowledgments and thanks go to the many individuals whose valuable suggestions and assistance made this book useful and informative—in particular, to Dr. Nicholas Cheremisinoff, for his assistance and to Dr. Yen-Hsiung Kiang for his expert preparation of Chapter 8, to which he gave his valuable time and knowledge.

Paul N. Cheremisinoff
Angelo C. Morresi

Contents

Preface		iii
Chapter 1	Properties	1
	General Description	1
	Benzene Reactions	1
	Benzene Physical Properties and Data Base	4
Chapter 2	Methods of Manufacture	9
	Petroleum-Derived Benzene	9
	Coal-Derived Benzene	11
	Production Capacity	12
	Production Plants	12
Chapter 3	Uses	20
	Rubber-Tire Manufacturing	27
	Rubber Products	27
	Adhesives, Gravure Printing Inks, Printing and Publishing, Paints, and Miscellaneous Chemicals Industry	30
	Paint Removers	30
	Gasoline	30
	Laboratories	32
	Production of Major Benzene Derivatives	32
Chapter 4	Safe Use, Handling, and Storage	46
	Fire Safety and Storage	46
	Spills	47
	Waste Disposal	48
	Protective Clothing and Equipment	48
	Respirators	48
	Good Handling Practices	49

	Emergency and First Aid	49
	Employee Training and Information Training	51
	Hazardous Materials and Shipping Document Requirements	62
	Typical Hazardous Materials Shipment Violations	66
Chapter 5	Monitoring	70
	Monitoring Requirements	70
	Medical Monitoring	71
	Biological Monitoring	71
	Biological Monitoring in Practice	77
	Air Sampling and Analysis Methods	82
Chapter 6	Health Effects	94
	Entrance to the Body	94
	Acute Exposure Effects	94
	Chronic Exposure Effects	95
	Benzene Toxicity Data	98
	Benzene Substitutes' Toxicity	112
Chapter 7	Benzene in the Environment	117
	Benzene Fate	117
	Sources Emitting Benzene into the Environment	117
	Benzene Exposures	119
Chapter 8	Controlling Benzene Emissions by Yen-Hsiung Kiang	129
	Characteristics of Benzene Emissions	129
	Carbon Adsorption	130
	Wash Oil Absorption	139
	Thermal Oxidation	143
	Catalytic Oxidation	151
	Combination Systems	154
References		155
Appendix A:	Exposure to Benzene; Liquid Mixtures: Occupational Safety and Health Standards	161
Appendix B:	Occupational Exposure to Benzene: Occupational Safety and Health Standards	173
Appendix C:	Benzene: Regulation as an Intentional Ingredient or as a Contaminant; Products Requiring Special Labeling; and Substances Requiring Special Packaging	227
Index		245

BENZENE

1
Properties

GENERAL DESCRIPTION

Benzene is a clear, colorless compound. It is a highly flammable liquid with a low boiling point and high vapor pressure, effecting vapors that are three times heavier than air—thus, in a work environment resultant vapors/fumes are found to accumulate at ground levels.

Benzene's solvent properties and its solubility with most organic solvents and compounds have made it an article of commerce and a significant factor in industry. It is soluble with acetone, carbon tetrachloride, ether, chloroform, alcohol, carbon disulfide, glacial acetic acid, and numerous other organic substances.

Benzene is a cyclic aromatic hydrocarbon with a ring containing six carbon atoms. It is in the shape of a hexagon with ahydrogen atom attached to each carbon. The structural formula of benzene is given in Fig. 1.1. Studies [1] in X-ray diffraction and electron diffraction show the benzene molecule to be flat and symmetrical with all bond angles at 120° and with equal bond lengths (Fig. 1.2).

Like other liquid aromatic hydrocarbons, benzene is highly refractive, non-polar, and nonreactive with low viscosity and low surface tension. Benzene has a sweet, pleasant odor that is characteristic of the aromatics. Its taste is rather sharp effecting a warm tingling sensation [2,3,4,5].

BENZENE REACTIONS

Basically, as a result of its structure, benzene is a stable compound. In chemical reactions it normally undergoes hydrogen atom substitution instead of addition. Reactions such as sulfonation, hydrogenation, nitration, chlorination, and bromination are common. Figure 1.3 exhibits some typical benzene reactions. Many benzene reactions require high temperature and pressures in the presence of catalysts in order to effect completion. For example, water and benzene are nonreactive unless high temperatures and pressures are applied.

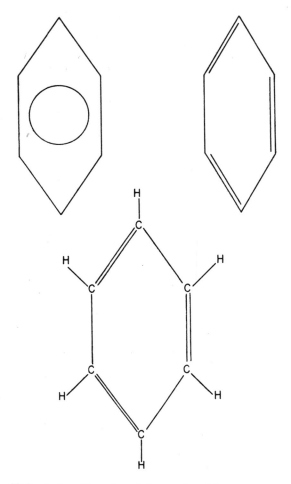

FIG. 1.1 Structural formula of benzene.

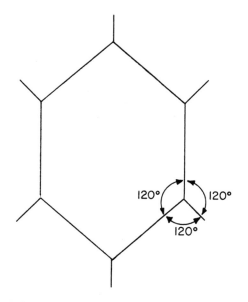

FIG. 1.2 Shape of a benzene molecule.

Properties

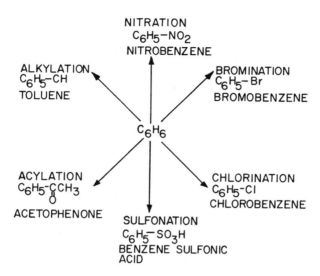

FIG. 1.3 Typical benzene reactions.

The following benzene reactions are of commercial importance and exemplify these nonreactive characteristics of benzene [6,7].

1. Phenol production: Chlorination/hydrolysis

$$C_6H_6 + Cl_2 \xrightarrow[\text{Catalyst}]{\text{Metallic iron}} C_6H_5Cl + HCl$$

$$C_6H_5Cl + NaOH \text{ (6.8\% solution)} \xrightarrow{360°C \ 4500 \text{ lb/in}^2}$$

$$C_6H_5Na + HCl \longrightarrow C_6H_5OH + NaCl$$
$$\text{phenol}$$

2. Cyclohexane production: Hydrogenation

$$C_6H_6 + 3H_2 \xrightarrow[\text{150-200°C, 25 atm}]{\text{Metallic nickel catalyst}} C_6H_{12}$$
$$\text{cyclohexane}$$

3. Styrene production: Benzene-ethylene/dehydrogenation

$$C_6H_6 + C_2H_4 \xrightarrow[\text{Catalyst}]{\text{Phosphoric acid}} C_6H_5C_2H_5$$
$$\text{ethyl benzene}$$

$$C_6H_5C_2H_5 \xrightarrow[\text{Catalyst (600°C)}]{\text{Cr}_2O_3 - Al_2O_3} C_6H_5CH=CH_2 + H_2$$
$$\text{styrene}$$

BENZENE PHYSICAL PROPERTIES [3,4,5,6,8,9,10,11,12,13] AND DATA BASE (See also Tables 1.1-1.6.)

CAS Number	000071432
Molecular Formula	C_6H_6
Line Notation (WLN)	R
Common Name	Benzene
Synonyms	Benzol
	Benzole
	Benzolene
	Bicarburet of hydrogen
	Carbon oil
	Coal naphtha
	Coal tar naphtha
	Cyclohexatriene
	Mineral naphtha
	Motor benzol
	Nitration benzene
	Phene
	Phenyl hydride
	Pyrobenzol
	Pyrobenzole
	Benzolo (Italian)
	Benzeen (Dutch)
	Benzen (Polish)
	Fenzin (Czech)

Physical State (at T and P): Liquid at standard conditions

Molecular Weight: 78.11

Melting (Freezing) Point: 5.53°C (42°F)

Boiling Point (at 760 nnHg): 80.1°C

Vapor Density (Air = 1): 4.0 at 90°C

Vapor Pressure at 20°C: 75 nnHg

Flash Point (Closed Cup): -11.1°C

Autoignition Point: 580°C (1076°F)

Surface Tension (at 25°C): 28.18 dyn/cm

Viscosity (Absolute): 0.6468 at 20°C, centipoise

Properties

Flammability Limits (in Air)
  Lower (lel): 1.4%
  Upper (uel): 7.9%

Specific Gravity (Water = 1): 0.879 at 20/20°C

Critical Density: 0.300 g/nl

Critical Temperature: 289.45°C

Critical Pressure: 48.6 atm

Heat of Combustion (at 25°C and Pressure = Constant): 9.999 kcal/g

Heat of Fusion (at Melting Point): 30.1 cal/g

Heat of Vaporization (at 100°C): 8.09 kcal/mol
    (at 25°C): 103.57 cal/g

Solubility in Water (% by Weight at 20°C): 0.17
    (at 20°C): 0.082 g/100 ml

Evaporation Rate (Ether = 1): 2.8
    (Butyl Acetate = 1): 5.6

Air Saturated Percent Benzene (at 760 mmHg and 26°C): 13.15
    (at 20°C): 319 g/m^3 air
    (at 30°C): 485 g/m^3 air

Refractive Index (at 29°C): 1.50

Partition Coefficient (Octanol/Water): 2.14 average

Density Saturated Air/Benzene Mixture (at 760 mmHg; 26°C; air = 1): 1.22

TABLE 1.1 Thermal Conductivity of Benzene

Temperature (°F)	Thermal conductivity K = Btu/(hr)(ft^2)(°F/ft)
Liquid 86	0.092
Gas[a] 32	0.0052
115	0.0073
212	0.0103
363	0.0152
413	0.0176

[a]Thermal conductivity of air @ 32°F = 0.0140 Btu/(hr)(ft^2)(°F/ft).

TABLE 1.2   Vapor Pressure of Benzene

Vapor pressure (mmHg)	Temperature (°C)
1	−36.7
5	−19.6
10	−11.5
20	−2.6
40	+7.6
60	15.4
100	26.1
200	42.2
400	60.6
760	80.1
1520	103.8
3800	142.5
7600	178.8
15200	221.5
22800	249.5
30400	272.3
38000	290.3

TABLE 1.3 Compressibility of Benzene

Temperature (°C)	Pressure (Mbar)	Compressibility (B) per Mbar (B × 10^6)
17	5	89
20	200	77
20	400	67

TABLE 1.4 Specific Heat

Temperature (°C)	Specific heat (cal/g °C)
65	0.482
60	0.419
10	0.340
-50	0.299
-100	0.227
-200	0.124
-250	0.0399

TABLE 1.5 Heat of Combustion (at 25°C and Constant Pressure)

Kcal/mol	Cal/g	Btu/lb
	<u>Benzene stage: gas</u>	
	[To form H_2O (liquid) and CO_2 (gas)]	
789.08	10,102.4	18,172
	[To form H_2O (gas) and CO_2 (gas)]	
757.52	9698.4	17,446
	<u>Benzene state: liquid</u>	
	[To form H_2O (liquid) and CO_2 (gas)]	
780.98	9998.7	17,986
	[To form H_2O (gas) and CO_2 (gas)]	
749.42	9594.7	17,259

TABLE 1.6 Combustion of Benzene

Heat of vaporization (Hy) in Btu/lb	169
HB gas-gas (Btu/lb)	17,446
Stoichiometric mixture (% volume)	2.71
Flammability limit %	
Stoichiometric, lean	43
rich	336
Autoignition temp. (°F)	1097
Maximum flame speed in ft/sec (U_f)	1.3
Adiabatic temp. (°F at U_f)	4150

2
Methods of Manufacture

The production of benzene is principally effected by the petroleum and petrochemical industries in which 95% of the benzene in the United States is produced. Nearly 50 plants produce petroleum-derived benzene from processes such as catalytic reforming, hydrogenation of pyrolysis gasoline, and hydrodealkylation of toluene.

The other 5% of the benzene produced is coal-derived. Until 1959 coal was the primary source of benzene, a by-product of steel mill coking processes. Coal-derived benzene production is not expected to increase in the near future [14, 15].

PETROLEUM-DERIVED BENZENE

A process flowchart for the production of petroleum-derived benzene is given in Fig. 2.1. The benzene content in crude oil is small, making its extraction uneconomical. Therefore, several processes have evolved to increase benzene yield from crude oil.

Catalytic Reformation

Catalytic reformation is a process which transforms lower-octane hydrocarbons into a high-octane yield with a high aromatics content. After various cleaning and fractionating steps the crude oil is converted to petroleum naphtha. This naphtha, in the presence of a catalyst, chemically reacts to form a petroleum fraction (the reformate) with a high content of aromatics (toluene, xylene, benzene, etc.). The benzene content of this reformate is between 5% and 10%.

The reformate is then fractionally distilled to recover the benzene. Silica gel adsorption, extraction distillation, and solvent extraction are processes used in the benzene recovery operation. Chemical reactions can include:

$$\text{Hexane} \xrightarrow{\text{catalyst}} \text{benzene} + \text{hydrogen}$$

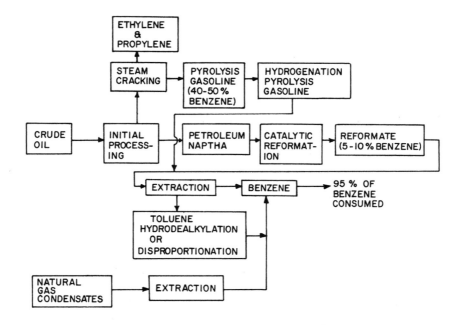

FIG. 2.1 Block flow diagram for petroleum-derived benzene.

$$\text{Substituted cyclohexane} \xrightarrow{\text{catalyst}} \text{benzene} + \text{hydrogen} + R_1 + R_2$$

$$\text{Methyl cyclopentane} \xrightarrow{\text{catalyst}} \text{benzene} + \text{hydrogen}$$

Figure 2.2 shows a block flow diagram of the catalytic reformation process for benzene. Nearly 67% of benzene is produced by this process [16,17].

Pyrolysis Gasoline

This is a high-benzene content coproduct of the naphtha and gas oil cracking to ethylene and propylene process. Pyrolysis gasoline is about 40-50% benzene with other aromatics present. Hydrogenation of the pyrolysis gasoline is the first process step before the aromatics can be extracted. Benzene is then selectively recovered from this aromatic stream. Nearly 10% of the benzene consumed is manufactured by this process.

Hydrodealkylation of Toluene

This is the process whereby benzene and methane are effected from the catalytic reaction of toluene and hydrogen. The resultant stream is fractionated producing a high-quality benzene. Almost 10% of the benzene demand is satisfied by this process.

Methods of Manufacture

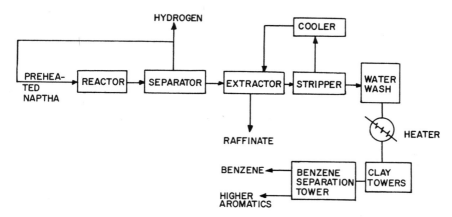

FIG. 2.2 Catalytic reformation process for the production of petroleum-derived benzene. (From Ref. 16.)

Toluene Disproportionation

This is the catalytic production of benzene and xylene from toluene and other methylbenzenes. Presently there is no benzene produced in the United States by this method [16,17,18,19,20].

COAL-DERIVED BENZENE

A process flowchart for the production of coal-derived benzene is given in Fig. 2.3. There are approximately 2 gal benzene per ton coal and it is recovered as a coke production by-product. Benzene comprises about 55-70% of the coke by-product intermediate, light oil, which also contains toluene, xylene, and other hydrocarbons [19,20].

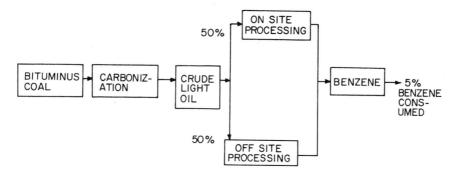

FIG. 2.3 Block flow diagram for coal-derived benzene.

Light oil recovery of benzene initially involves the production of light oil from the coal carbonization to coke process. In the coal-to-coke process, high temperatures ($\approx 2000°$ F) are used to derive the volatile gases from the coal. After several cleaning and cooling operations, the condensate formed is light oil. Steam distillation is then utilized to recover the benzene from the light oil. The distillation process often involves sulfuric acid washing separating the light oil into toluene, xylene, benzene, and residues.

PRODUCTION CAPACITY

Figure 2.4 shows U.S. benzene production and inventory rates for the years 1968 through 1975. The demand for benzene through 1974 was increasing at levels of 9% per year. However, the consumption of benzene is expected to increase only at a level of 5.6% per year over the period 1976 through 1985. Between 1976 and 1981 the benzene growth rate is predicted to increase 7.5% per year. Table 2.1 shows the present benzene production rates and expected 1981 production rates in the United States. Growth rates are also given in Table 2.1.

Coal-derived benzene production will not increase. Therefore, all the growth will be in the petroleum-derived benzene sector. Catalytic reformation capacity should increase by 180 million gallons, toluene hydrodealkylation by 250 million gallons, and pyrolysis gasoline production should more than double by 1981.

PRODUCTION PLANTS

The U.S. plants producing petroleum-derived benzene are listed in Table 2.2. 48 plants presently have the capability of producing petroleum-derived benzene. More than half of these plants (25) utilize benzene as an intermediate. As noted in Table 2.2, many of these plants can produce benzene by three different

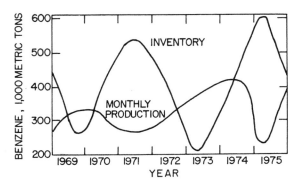

FIG. 2.4 Monthly plot of U.S. benzene production from 1968-1975. (From Ref. 21.)

Methods of Manufacture

processes. These plants are concentrated in the South and Southwest of the country, Louisiana and Texas.

Coal-derived benzene, being a by-product of coke, is initially processed in coke plants which produce the intermediate light oil. Table 2.3 lists the coke-producing (or light oil) plants in the United States. Those plants that process the light oil into benzene are listed in Table 2.4. Some of those plants listed in Table 2.3 are also listed in Table 2.4 because they are coke plants with on-site benzene production facilities. The other facilities purchase their light oil for benzene production.

TABLE 2.1 Benzene Production Capacity in the United States

Benzene source	Benzene production (million gallons)		Annual percentage
	1976	1981	
Coal-derived benzene	150	150	0
Petroleum-derived benzene			
Catalytic reformation	860	1040	3.9
Pyrolysis gasoline	250	620	19.9
Toluene hydrodealkylation	560	810	7.7
Total benzene production	1820	2620	7.6

Source: From Ref. 17.

TABLE 2.2 U.S. Producers of Petroleum-Derived Benzene

Company	Location	Production method
Allied Chemical Corp. Union Texas Petroleum Div.	Winnie, Texas	Catalytic reformate
Amerada Hess Corp. Hess Oil Virgin Islands Corp. (subsid.)	St. Croix, Virgin Islands	Catalytic reformate
American Petrofina, Inc. American Petrofina Co. of Texas (subsid.)	Port Arthur, Texas	Not available
Cosden Oil & Chemical Co. (subsid.)	Big Spring, Texas	Catalytic reformate Toluene dealkylation

TABLE 2.2 (continued)

Company	Location	Production method
Ashland Oil, Inc. Leigh Valley Chemical Co. (div.) Petrochems. Div.	Ashland, Kentucky	Catalytic reformate Toluene dealkylation Extraction from light oil (coal-derived benzene)
	North Tonawanda, N.Y.	Catalytic reformate Extraction from light oil (coal-derived benzene)
Atlantic Richfield Co. Arco Chem. Co.	Houston, Texas	Catalytic reformate Toluene disproportionation
	Wilmington, Calif.	Catalytic reformate
Beckman Industries, Inc. Biological & Fine Chems. Div. Bioproducts Dept.	Palo Alto, Calif.	Not available

CF&I Steel Corp.
  The Charter Co. [Charter Oil Co. (subsid.)]
  Charter International Oil Co. (subsid.)
	Houston, Texas	Catalytic reformate

Cities Service Co., Inc. Northamerican Petroleum Group	Lake Charles, Louisiana	Catalytic reformate
Coastal States Gas Corp. Coastal States Marketing Inc. (subsid.)	Corpus Christi, Texas	Catalytic reformate
Commonwealth Oil Refining Co., Inc. Commonwealth Petrochems. Inc. (subsid.)	Penuelas, Puerto Rico	Catalytic reformate Toluene dealkylation
Crown Central Petroleum Corp.	Pasadena, Texas	Catalytic reformate
Dow Chemical U.S.A.	Bay City, Michigan	Fractionation of benzene/toluene stream

Methods of Manufacture

TABLE 2.2 (continued)

Company	Location	Production method
Dow Chemical U.S.A. (continued)		Toluene dealkylation Extraction from light oil (coal-derived benzene)
	Freeport, Texas	Toluene dealkylation
	Plaquemino, Louisiana	Pyrolysis gasoline
Exxon Corp. Exxon Chem. Co. (div. Exxon Chem. Co. U.S.A.)	Baton Rouge, Louisiana Baytown, Texas	Not available Not available
Gulf Oil Corp. Gulf Oil Chems. Co. (div. Petrochems. Div.)	Alliance, Louisiana	Catalytic reformate
	Philadelphia, Pennsylvania	Catalytic reformate Toluene dealkylation
	Port Arthur, Texas	Catalytic reformate Pyrolysis gasoline
Kerr-McGee Corp. Southwestern Refining Co. Inc. (subsid.)	Corpus Christi, Texas	Not available
Marathon Oil Co.	Texas City, Texas	Catalytic reformate
Mobil Oil Corp. Mobil Chem. Co. (div. Petrochems. Div.)	Beaumont, Texas	Catalytic reformate Pyrolysis gasoline
Monsanto Co. Monsanto Polymers and Petrochems. Co.	Chocolate Bayou, Texas	Pyrolysis gasoline Toluene dealkylation
Penzoil Co. Atlas Processing Co. (subsid.)	Shreveport, Louisiana	Pyrolysis gasoline Extraction from condensate containing naturally recurring benzene
Phillips Petroleum Co. Phillips Puerto Rico Core, Inc. (subsid.)	Guayama, Puerto Rico	Toluene dealkylation Catalytic reformate
Shell Chem. Co. Base Chems.	Deer Park, Texas	Catalytic reformate Pyrolysis gasoline

TABLE 2.2. (continued)

Company	Location	Production method
Shell Chem. Co. (continued)	Odessa, Texas	Catalytic reformate
	Wood River, Illinois	Catalytic reformate Extraction from light oil (coal-derived benzene)
Shelly Oil Co.	El Dorado, Kansas	Catalytic reformate
Standard Oil Co. (Indiana) Amoco Oil Co. (subsid.)	Texas City, Texas	Catalytic reformate
The Standard Oil Co. (Ohio) BP Oil Inc. (subsid.)	Marcus Hook, Pennsylvania	Not available
Sun Oil Co. Sun Oil Co. of Pennsylvania (subsid.)	Marcus Hook, Pennsylvania	Catalytic reformate
	Toledo, Ohio	Not available
	Tulsa, Oklahoma	Catalytic reformate
Suntide Refining Co. (subsid.)	Corpus Christi, Texas	Catalytic reformate
Texaco, Inc.	Port Arthur, Texas	Catalytic reformate
	Westville, New Jersey	Catalytic reformate
	Convent, Louisiana	Not available
Union Carbide Corp. Chems. & Plastics Div.	Taft, Louisiana	Pyrolysis gasoline
Union Oil Co. of Calif.	Beaumont, Texas	Not available
	Lemont, Illinois	Catalytic reformate Extraction for light oil (coal-derived benzene)
Union Pacific Corp. Chemplin Petroleum Co. (subsid.)	Corpus Christi, Texas	Catalytic reformate

Source: From Refs. 16 and 22.

Methods of Manufacture

TABLE 2.3 Coke-Producing Plants in the United States [16, 22]

Company	Location	Production (tons/day)
Alan Wood Steel	Conshohocken, Pennsylvania	1640
Armco Steel Co.	Hamilton, Ohio	1820
	Houston, Texas	1079
	Middletown, Ohio	1155
Bethlehem Steel Corp.	Bethlehem, Pennsylvania	4580
	Sparrows Point, Maryland	8950
	Lackawanna, New York	7360
	Johnstown, Pennsylvania	4525
	Burns Harbor, Indiana	5310
CF&I Steel Corp.	Pueblo, Colorado	2370
Crucible, Inc.	Midland, Pennsylvania	1280
Cyclops Corp.	Portsmouth, Ohio	1200
Donner-Hanna Coke	Buffalo, New York	2810
Ford Motor Co.	Dearborn, Michigan	4000
Inland Steel Co.	Indiana Harbor Wks, Indiana	8475
Interlake, Inc.	Chicago, Illinois	1925
	Erie, Pennsylvania	575
	Toledo, Ohio	890
International Harvester	S. Chicago, Illinois	1980
Jones & Laughlin, Inc.	Aliquippa, Pennsylvania	4300
	Pittsburgh, Pennsylvania	5585
Kaiser Steel Corp.	Fontana, California	3900
Lone Star Steel	Lone Star, Texas	956
National Steel	River Rouge, Michigan	5400
	Weirton, West Virginia	8770
	Granite City, Illinois	2350
Republic Steel	Birmingham, Alabama	988
	Cleveland, Ohio	5381
	Gadsden, Alabama	1976

TABLE 2.3 (continued)

Company	Location	Production (tons/day)
Republic Steel (continued)	Massillon, Ohio	471
	S. Chicago, Illinois	1260
	Warren, Ohio	1216
	Youngstown, Ohio	2462
Sharon Steel Corp.	Fairmont, West Virginia	550
Shenango, Inc.	Neville Island, Pennsylvania	1580
U.S. Steel Corp.	Clairton, Pennsylvania	22,007
	Duluth, Minnesota	1748
	Fairfield, Alabama	7433
	Fairless Hills, Pennsylvania	2645
	Gary, Indiana	12,601
	Geneva, Utah	3830
	Lorain, Ohio	3587
Wheeling-Pitts.	E. Steubenville, West Virginia	4773
	Monessen, Pennsylvania	1600
Youngstown Sheet	Campbell, Ohio	3130
	E. Chicago, Illinois	3275
	Youngstown, Ohio	Not operating

TABLE 2.4 U.S. Producers of Benzene from Light Oil

Company	Location	Source of light oil
Armco Steel Corp.	Middletown, Ohio	On site
Ashland Oil, Inc.	Ashland, Kentucky	Purchased
Bethlehem Steel Corp.	Bethlehem, Pennsylvania	On site
	Lackawanna, New York	On site
	Sparrows Point, Maryland	On site

Methods of Manufacture

TABLE 2.4 (continued)

Company	Location	Source of light oil
CF&I Steel Corp.	Pueblo, Colorado	On site
Dow Chemical, U.S.A.	Bay City, Michigan	Purchased
Interlake, Inc.	Toledo, Ohio	On site
Jones & Laughlin, Inc.	Aliquippa, Pennsylvania	On site
Northwest Industries Inc. Lone Star Steel Co. (subsid.)	Lone Star, Texas	On site
Shell Oil Company	Wilmington, California	Purchased
	Wood River, Illinois	Purchased
Skelly Oil Company	El Dorado, Kansas	Purchased
Sun Oil Company	Corpus Christi, Texas	Purchased
Tenneco, Inc.	Chalmette, Louisiana	Purchased
Union Oil Company of California	Lemont, Illinois	Purchased
U.S. Steel Corp.	Clairton, Pennsylvania	On site
	Geneva, Utah	On site

Source: From Ref. 16.

3
Uses

Prior to World War II benzene was primarily used in the rubber industry. During the war toluene was required in large quantities in the manufacture of explosives. At that time the benzene content of toluene was relatively high, thus, as more toluene was used in products, the use of benzene also expanded [22].

The solvent properties of benzene are significant, as it mixes well with a large number of organics. Benzene is presently utilized primarily as an intermediate in the manufacture of other chemicals. Further, about 90% of domestically produced benzene is consumed in the production of a half-dozen chemicals [14, 23].

1. Ethylbenzene
2. Cumene
3. Cyclohexane
4. Nitrobenzene
5. Maleic anhydride
6. Dodecylbenzene

At the end of this chapter the production techniques for these compounds are discussed.

Benzene is used to increase the octane rating of unleaded gasoline. It is utilized in laboratories, and in the manufacture of pesticides, paint removers, rubber cements, and detergents. In fact, the usage of benzene and its derivatives has become so widespread that there really is no facet of our lives today that is not touched by them.

Figures 3.1 through 3.19 outline the commercial uses of benzene and its derivatives. Some of their uses include: rubbers, plastics, dyes, deodorants, insecticides, herbicides, fungicides, explosives, nylon, film, food products, drugs, lube oil additives, perfumes, pharmaceuticals, and paint and varnish removers.

Although benzene is apparent in numerous applications, it is generally used as a raw material to produce several common intermediates. In Fig. 3.20 the flow of benzene in the United States is given.

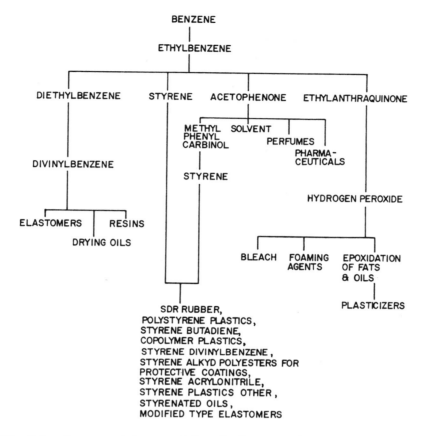

FIG. 3.1 Commercial uses of the ethylbenzene derivative of benzene.

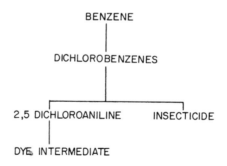

FIG. 3.2 Commercial uses of the dichlorobenzene derivatives of benzene.

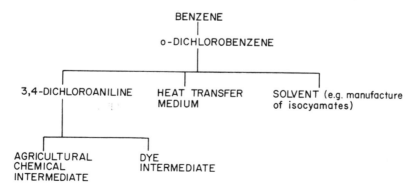

FIG. 3.3 Commercial uses of the orthodichlorbenzene derivative of benzene.

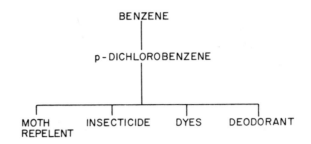

FIG. 3.4   Commercial uses of the paradichlorbenzene derivative of benzene.

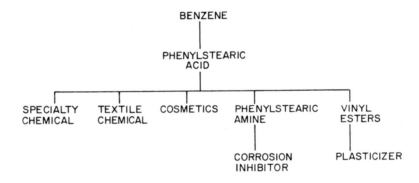

FIG. 3.5   Commercial uses of the phenylstearic acid derivative of benzene.

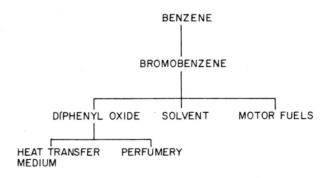

FIG. 3.6   Commercial uses of the bromobenzene derivative of benzene.

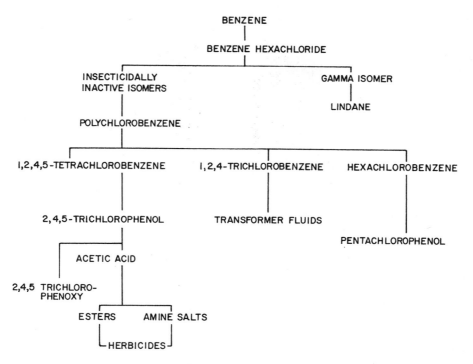

FIG. 3.7 Commercial uses of the benzene hexachloride derivative of benzene.

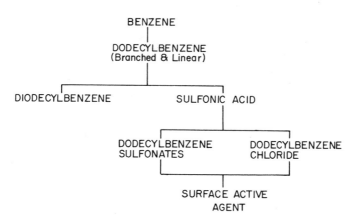

FIG. 3.8 Commercial uses of the dodecylbenzene derivative of benzene.

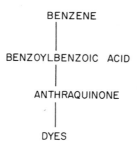

FIG. 3.9 Commercial uses of the benzoylbenzoic acid derivative of benzene.

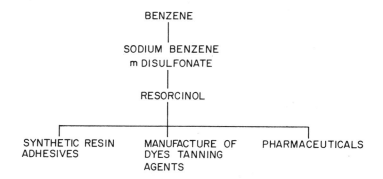

FIG. 3.10 Commercial uses of some meta-derivatives of benzene.

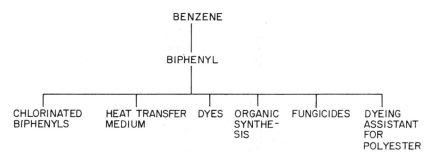

FIG. 3.11 Commercial uses of the biphenyl derivative of benzene.

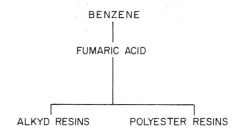

FIG. 3.12 Commercial uses of the fumaric acid derivative of benzene.

Normally, the production of these benzene derivatives is accomplished in closed systems. Further, the final products are produced with no appreciable benzene content. Table 3.1 lists the common benzene derivatives and their estimated benzene content. Only the nitrobenzenes contain relatively significant amounts of benzene. In Table 3.2 several petroleum products are listed along with their estimated volumetric benzene content [24].

As the government regulations concerning benzene become more restrictive and the evidence of the detrimental health effects mount, industry is beginning to phase out its use of benzene and instituting more rigid controls. Final OSHA

Uses

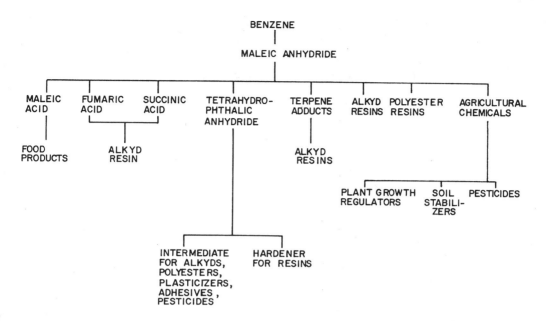

FIG. 3.13 Commercial uses of the maleic anhydride derivative of benzene.

TABLE 3.1 Benzene Content of Major Benzene Derivatives

Benzene derivative	Estimated % volume benzene
Ethylbenzene	0-0.3
Styrene	Trace
Cumeme	Trace
Phenol	0-Trace
Maleic anhydride	Trace
Nitrobenzenes	0-3.0
Cyclohexane	0-0.5
Alkylbenzenes	0-Trace
Aniline	0-Trace
Chlorobenzene	0-Trace

Source: From Ref. 24.

TABLE 3.2 Several Petroleum Products and Their Volumetric Benzene Content

Benzene-containing products	Estimated volume percent benzene
Aviation gasolines	0.4- < 3
Automotive gasolines	0.4-2
Farm tractor fuels	0-Trace
Diesel fuel oils	0-Trace
Aviation turbine fuels	0-3
Gas turbine fuel oils	0-Trace
Liquefied petroleum gases	~ 0
Fuel oils and kerosene	0-Trace
<u>Solvent naphthas</u>	
Coal tar	
Aromatic petroleum	0-3
Stoddard	0-Trace
VM&P	0-3
Others	0-3
<u>Aromatic solvents</u>	
Toluene	0-4.0
Xylenes	0-0.1
<u>Other products</u>	
Hexane	0-1.5
Coke-oven tar	0-0.3
Lubricating oils	0-Trace
Paint removers	< 0.5
Paint thinners	< 0.5-2.3

<u>Source</u>: From Ref. 24.

Uses

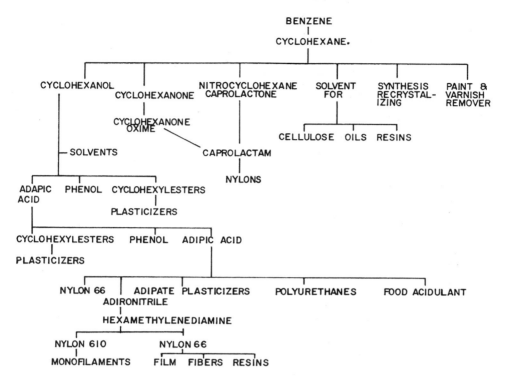

FIG. 3.14   Commercial uses of the cyclohexane derivative of benzene.

regulations published on February 10, 1978 on the occupational exposure limit to benzene established an 8-hr time-weighted average (TWA) limit of employee exposure to benzene of 1 ppm. This rule is only the beginning in government efforts to restrict the commercial use of benzene.

RUBBER-TIRE MANUFACTURING

Benzene or benzene contaminated hydrocarbons are utilized in several rubber-tire manufacturing processes. Substitutes are being found for benzene in the industry. Benzene usage is significant in rubber adhesives, synthetic rubbers, phenolic antioxidants, and hydrophilic-polymers manufacturing. There are 207 plants utilizing benzene in the rubber tire industry [22, 23, 24].

RUBBER PRODUCTS

Benzene utilization in the rubber products industry is inherent in many processes including spreading and coating operations. The synthetic rubbers and adhesives also require benzene. There are 197 plants utilizing benzene in the

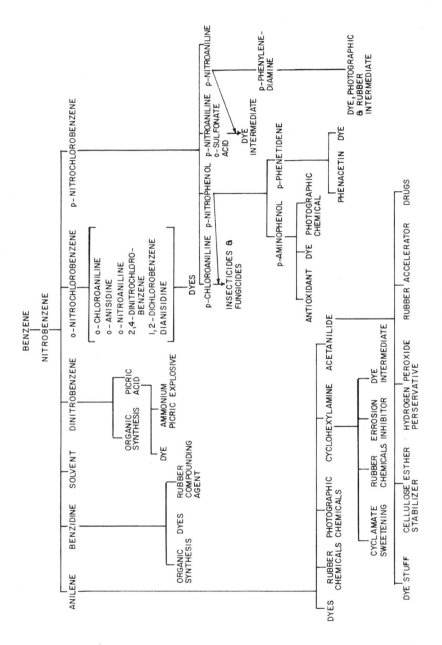

FIG. 3.15 Commercial uses of the nitrobenzene derivative of benzene.

Uses

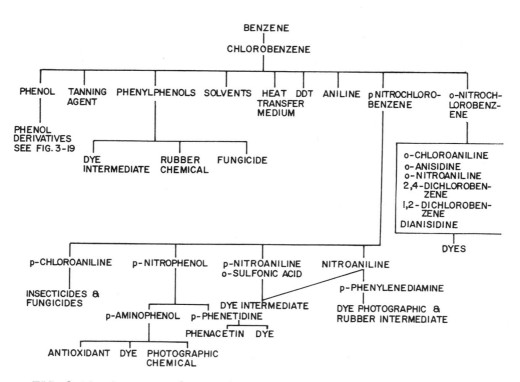

FIG. 3.16 Commercial uses of the chlorobenzene derivative of benzene.

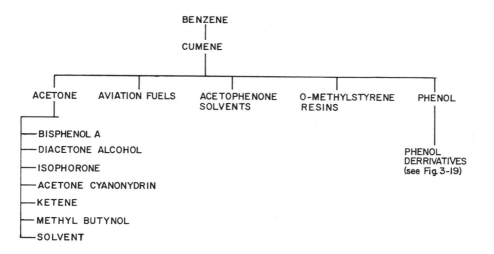

FIG. 3.17 Commercial uses of the cumene derivative of benzene.

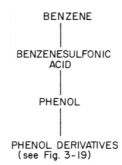

FIG. 3.18 Commercial uses of the benzenesulfonic acid derivative of benzene.

ADHESIVES, GRAVURE PRINTING INKS, PRINTING AND PUBLISHING, PAINTS, AND MISCELLANEOUS CHEMICALS INDUSTRY

Benzene utilization in these industries has been minimized if not eliminated through the use of alternates. The benzene base of adhesives and inks has been replaced with toluene, xylene, heptane, petroleum naphtha, or trichloroethylene. These solvents as well as methylene chloride have been used to replace benzene in the organic chemical synthesis industry.

The only problem with substitution is that the replacement solvent might be more dangerous than the one for which the substitution is being made. Or, as with the case of replacing benzene with toluene, the benzene contamination of toluene could be significant when considering exposure. However, the benzene content of toluene in most cases is small and heretofore considered insignificant. The benzene contamination of rubber cement is a maximum of 1% and only results from hydrocarbon solvent contamination [23, 24, 25].

PAINT REMOVERS

Benzene is used in the formulation of paint removers. However, the trend in the industry has been reformulation with other solvents such as methylene chloride, toluene, chlorinated hydrocarbons, and mineral spirits.

In an effort to eliminate benzene exposure, many firms have discontinued whole product lines. Methylene chloride is acknowledged as a superior paint remover as compared to benzene except its cost is 15% higher [23, 25].

GASOLINE

Benzene is a normal component of unleaded gasoline, used to increase its octane rating. Although the benzene content is normally below 2%, its gasoline content ranges between 0.5 and 4.3% due to crude fuel grade and seasonal variations [25].

Uses

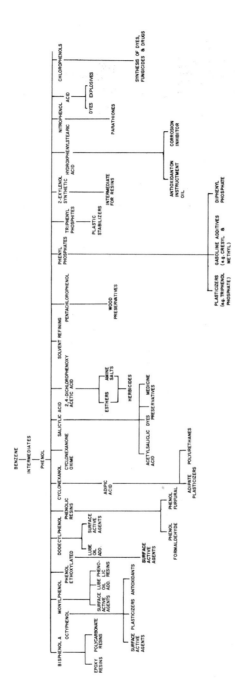

FIG. 3.19 Commercial uses of benzene through its phenol derivative.

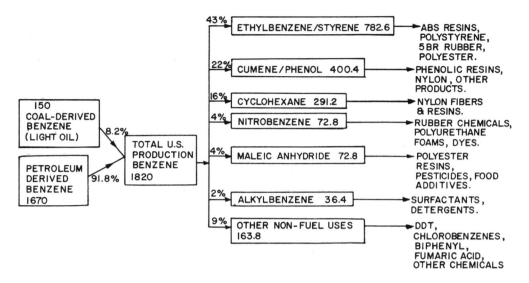

FIG. 3.20 Flow of benzene through the United States and benzene consumption (estimated usage in million gallons).

LABORATORIES

Benzene is used in more than 12,000 laboratories in the United States. It is used as a reagent in sample collection, preparation, and extraction. Laboratory hoods are used to minimize laboratory personnel exposure.

PRODUCTION OF MAJOR BENZENE DERIVATIVES
[3, 16, 17, 20, 24, 25, 26]

Ethylbenzene/Styrene

A major portion of U.S. consumption of benzene goes into the production of ethylbenzene, 95% of which in turn is utilized in the manufacture of styrene.

There are two processes for the manufacture of ethylbenzene. The first process accounts for only about 5% of the ethylbenzene production. It involves the extraction of ethylbenzene from mixed xylenes.

The second process accounts for 95% of the ethylbenzene production. It is the catalytic alkylation of benzene with ethylene and proceeds as follows:

$$\text{Benzene + ethylene} \xrightarrow{\text{catalyst}} \text{ethylbenzene}$$

The final product contains only trace amounts of benzene.

The ethylbenzene is then dehydrogenated in the presence of a catalyst to produce styrene. The reaction is as follows:

Uses

Ethylbenzene $\xrightarrow{\text{catalyst}}$ styrene + hydrogen

Table 3.3 gives the U.S. plants manufacturing ethylbenzene and styrene and production capacities. Figure 3.21 is a usage flowchart of ethylbenzene in the United States.

Cumene

About one-fifth of the U.S. benzene production goes into the manufacture of cumene, 95% of which is a phenol intermediate. The production of cumene is a process of catalytic alkylation of benzene with propylene as follows:

Benzene + propylene $\xrightarrow{\text{catalyst}}$ cumene

Only trace amounts of benzene are available in cumene. The cumene is then processed into phenol.

Approximately 5% of the phenol domestically produced is effected directly from benzene via chlorination and sulfonation processes.

Table 3.4 gives the U.S. plants producing cumene and/or phenol and production capacities for 1976. Figure 3.22 is a usage flowchart of the benzene-to-cumene-to-phenol process.

Cyclohexane

Cyclohexane production accounts for approximately 16% of U.S. benzene consumption. Benzene is hydrogenated in the presence of a catalyst to form cyclohexane. The process is as follows:

Benzene + hydrogen $\xrightarrow{\text{catalyst}}$ cyclohexane

Only trace amounts of benzene contaminate the cyclohexane.

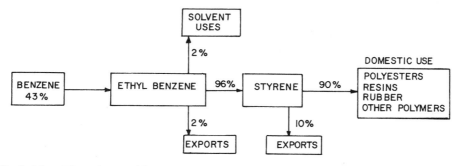

FIG. 3.21 Flowchart of benzene utilization in the production of ethylbenzene and styrene in the United States.

TABLE 3.3 U.S. Plants Manufacturing Ethylbenzene and Styrene and Their Production Capacities (MM lb/yr)

Company	Ethylbenzene capacity	Full capacity benzene requirements	Capacity for styrene production at plant
American Petrofina, Inc. Cosden Oil & Chemical Co. subsidiary, Big Spring, Texas	45	27	90
Arco/Polymers, Inc., Houston, Texas,	100	--	100
Port Arthur, Texas,	440	330	0
Beaver Valley, Pennsylvania	0	NP[a]	440
The Charter Company Charter Oil Co. subsidiary, Houston, Texas	35	--	0
Commonwealth Oil Refining Co., Inc. Styrochem Corp. subsidiary, Penuelas, Puerto Rico	160	--	0
Cos-Mar, Inc. Carville, Louisiana	1520	1160	1300
Dow Chemical U.S.A., Freeport, Texas,	1865	1420	1430
Midland, Michigan	550	420	400
El Paso Natural Gas Co., El Paso Products Co. subsidiary, Odessa, Texas	275	210	150
Foster Grant Baton Rouge, Louisiana	970	740	820
Gulf Oil Corp. Gulf Oil Chemicals Div., Petrochemicals Div., Welcome, Louisiana	550	420	525
Monsanto Company Monsanto Polymers & Petrochemicals Co., Aluin, Texas,	50	Combined	0
Texas City, Texas	1450	1130	
Oxirane Chemicals Company Channelview, Texas	1200	910	1000

TABLE 3.3 (continued)

Company	Ethylbenzene capacity	Full capacity benzene requirements	Capacity for styrene production at plant
Phillips Petroleum Company Petrochemical & Supply division, Phillips, Texas	NA[b]	--	0
Standard Oil Company (Indiana) Amoco Chemicals Corp. subsidiary, Texas City, Texas	945	720	840
Sun Oil Company Suntide Refining Company subsidiary, Corpus Christi, Texas	95	62	80
Tenneco, Inc. Tenneco Oil Company division, Chalmette, Louisiana	26	--	0
Union Carbide Corp. Seadrift, Texas	340	260	300
Totals	10,616	7809	8775

[a]NP = not pertinent.
[b]NA = not available.

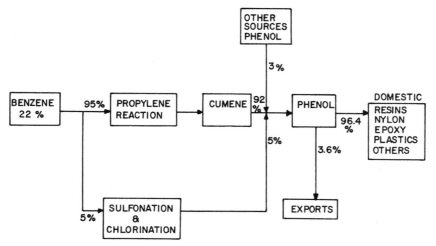

FIG. 3.22 Flowchart of benzene utilization in the production of cumene and phenol in the United States.

TABLE 3.4 U.S. Plants Producing Cumene and/or Phenol and Their Production Capacities (MM lb/yr)

Company	Phenol capacity	Cumene capacity	Full capacity benzene requirements
Allied Chemical Corp. Specialty Chemicals Div., Philadelphia, Pennsylvania	550	0	0
Ashland Oil, Inc. Lehigh Valley Chemical Co. Div., Ashland, Kentucky	0	350	235
Clark Oil & Refining Corp. Clark Chemical Corp. subsidiary, Blue Island, Illinois	88	110	74
Coastal States Gas Corp. Coastal States Marketing Inc. subsidiary, Corpus Christi, Texas	0	140	94
Dow Chemical U.S.A. Midland, Michigan, Oyster Creek, Texas	48 400	10 0	49 0
Ferro Corporation Productol Chemical Div., Sante Fe Springs, California	NA[a]	0	0
Georgia Pacific Corp. Chemical Division, Plaquemine, Louisiana	265	0	0
Gulf Oil Corporation Gulf Oil Chemicals Div., Philadelphia, Pennsylvania Port Arthur, Texas	0 0	450 450	300 300
Kalama Chemical, Inc. Kalama, Washington		0	0
Koppers Company, Inc. Organic Materials Div., Follansbee, West Virginia	NA	0	0
Marathon Oil Company Texas City, Texas	0	190	130
The Merichem Company Houston, Texas	NA	0	0
Monsanto Company Monsanto Polymers & Petrochemicals Co., Chocolate Bayou, Texas	500	650	435

TABLE 3.4 (continued)

Company	Phenol capacity	Cumene capacity	Full capacity benzene requirements
Reichhold Chemicals, Inc. Tuscaloosa, Alabama	150	0	165
Shell Chemicals, Inc. Deer Park, Texas	500	700	470
Skelly Oil Company El Dorado, Kansas	95	135	90
Standard Oil Company (Indiana) Amoco Chemicals Corp. subsidiary, Texas City, Texas	0	50	50
Standard Oil Company of California Chevron Chemical Co. subsidiary, Richmond, California	55	0	0
El Segundo, California	0	100	67
Stimson Lumber Northwest Petrochemical Corp. Div., Ana Cortes, Washington	NA	0	0
Sun Oil Company Sun Oil Co. of Pennsylvania subsidiary, Suntide Refining Co. subsidiary, Corpus Christi, Texas	0	250	170
Texaco, Inc. Westville, New Jersey	0	260	175
Union Carbide Corp. Chemicals & Plastics Division, Bound Brook, New Jersey	150	0	0
Union Carbide Corp. Union Carbide Corporation division, Penvelas, Puerto Rico	200	640	430
U.S. Steel Corp. U.S.S. Chemicals Division, Clairton, Pennsylvania	NA	0	0
Haverhill, Ohio	315	0	0
Totals	3371	4485	3217

[a]NA = not available.
Source: From Refs. 21 and 24.

Table 3.5 gives the U.S. plants producing cyclohexane from benzene and their production capacities. Figure 3.23 is a flowchart of cyclohexane in the United States.

Nitrobenzene

Nitrobenzene production consumes about 4% of the benzene in the United States. The process involves the nitration of benzene with nitric and sulfuric acids. The process is as follows:

$$\text{Benzene} + \text{nitric acid} \xrightarrow{\text{sulfuric acid}} \text{nitrobenzene} + \text{water}$$

Table 3.6 gives the U.S. plants producing nitrobenzene and their production capacities. Figure 3.24 is a flowchart of nitrobenzene in the United States.

Maleic Anhydride

Less than 4% of the benzene in the United States goes into the production of maleic anhydride. Maleic anhydride production involves the oxidation of benzene in the presence of a catalyst as follows:

$$\text{Benzene} + \text{oxygen} \xrightarrow{\text{catalyst}} \text{maleic anhydride} + \text{carbon dioxide} + \text{water}$$

Table 3.7 gives the U.S. plants producing maleic anhydride from benzene and their production capacities. Figure 3.25 is a flowchart of maleic anhydride in the United States.

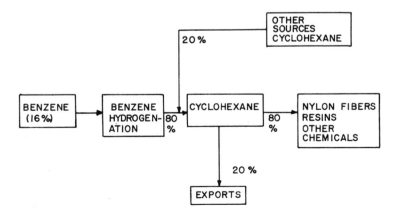

FIG. 3.23   Flowchart of benzene utilization in the production of cyclohexane in the United States.

TABLE 3.5 U.S. Plants Producing Cyclohexane from Benzene and Their Production Capacities (MM lb/yr)

Company	Cyclohexane capacity	Full capacity benzene requirements
American Petrofina, Inc. Cosden Oil & Chemicals Co. subsidiary, Big Springs, Texas	12	10
Commonwealth Oil Refining Co., Inc. Cotco Cyclohexane, Inc. subsidiary, Penuelas, Puerto Rico	40	33
Exxon Corporation Exxon Chemical Co. division, Exxon Chemical Co. U.S.A., Baytown, Texas	40	33
Gulf Oil Corporation Gulf Oil Chemicals Co. division, Petrochemicals Division, Port Arthur, Texas	33	27
Phillips Petroleum Company Borger, Texas, Sweeny, Texas	40 85	0 NA[a]
Phillips Petroleum Company Phillips Puerto Rico Corp., Inc. subsidiary, Guayama, Puerto Rico	70	58
Sun Oil Company Sun Oil Company of Pennsylvania subsidiary, Tulsa, Oklahoma	20	17
Texaco, Inc. Port Arthur, Texas	40	33
Union Oil Company of California Beaumont, Texas	34	28
Union Pacific Corporation Champlin Petroleum Company subsidiary, Corpus Christi, Texas	22	18
1976 Totals	436	257

[a]NA = not available.

TABLE 3.6 U.S. Plants Producing Nitrobenzene from Benzene and Their Production Capacities (MM lb/yr)

Company	Nitrobenzene capacity	Full capacity benzene requirements
Allied Chemical Corp. Specialty Chemicals Division, Moundsville, West Virginia	55	36
American Cyanamid Company Organic Chemicals Division, Bound Brook, New Jersey, Willow Island, West Virginia	85 60	56 39
E. I. DuPont deNemours Corp., Inc. Elastomer Chemicals Department Industrial Chemicals Department, Beaumont, Texas Gibbstown, New Jersey	310 200	203 131
First Mississippi Corporation First Chemical Corp. subsidiary, Pascagoula, Mississippi	135	88
Mobay Chemical Corporation Industrial Chemicals Division, New Martinsville, West Virginia	135	88
Monsanto Company Monsanto Industrial Chemicals Co., Sauget, Illinois	10	6.5
Rubicon Chemicals, Inc. Geismar, Louisiana	275	180
1976 Totals	1265	827.5

Source: From Refs. 21 and 24.

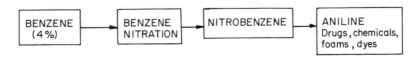

FIG. 3.24 Flowchart of benzene utilization in the production of nitrobenzene in the United States.

TABLE 3.7 U.S. Plants Producing Maleic Anhydride and Their Production Capacities (MM lb/yr)

Company	Maleic anhydride capacity	Full capacity benzene requirements
Allied Chemical Corporation Specialty Chemicals Divisionk Moundsville, West Virginia	20	27
Ashland Oil, Inc. Ashland Chemical Division, Neal, West Virginia	60	80
Koppers Company, Inc. Organic Materials Division, Bridgeville, Pennsylvania	34	45
Cicero, Illinois	10	0
Monsanto Company Monsanto Industrial Chemicals Co., St. Louis, Missouri	105	110
Petro-Tex Chemical Corporation Petro-Tex Chemical Company subsidiary, Houston, Texas	50	66
Reichhold Chemicals, Inc., Elizabeth, New Jersey,	30	40
Morris, Illinois	60	80
Standard Oil Company (Indiana) Amoco Chemicals Corp. subsidiary, Joliet, Illinois	60	0
Tenneco Inc. Tenneco Chemicals Inc., Organic & Polymers Division, Fords, New Jersey	26	35
U.S. Steel Corp. U.S.S. Chemicals Division, Neville Island, Pennsylvania	80	106
1976 Totals	535	589

Source: From Refs. 21 and 24.

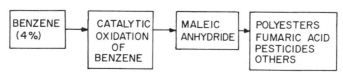

FIG. 3.25 Flowchart of benzene utilization in the production of maleic anhydride in the United States.

Dodecylbenzene

Less than 2% of the benzene in the United States is consumed by dodecylbenzene production. Synonyms for dodecylbenzene are alkylbenzene and detergent alkylate. There are two processes utilized to convert benzene to dodecylbenzene.

The first process involves the alkylation of benzene in the presence of a sulfuric acid or hydrogen fluoride catalyst which proceeds as follows:

$$\text{Benzene} + \text{dodecene} \xrightarrow{\text{catalyst}} \text{dodecylbenzene}$$
$$\text{(branched polymer)}$$

This process results in hard dodecylbenzene.

The second process involves the alkylation of benzene with dodecyl chloride in the presence of a catalyst which proceeds as follows:

$$\text{Benzene} + \text{dodecyl chloride} \xrightarrow{\text{catalyst}} \text{dodecylbenzene} + \text{hydrogen chloride}$$

This process results in soft dodecylbenzene.

Table 3.8 gives the U.S. plants producing dodecylbenzene and their production capacities. Figure 3.26 gives the flowchart of dodecylbenzene in the United States.

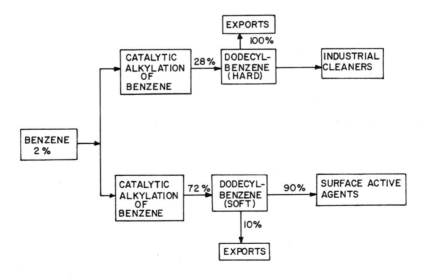

FIG. 3.26 Flowchart of benzene utilization in the production of dodecylbenzene.

Uses

TABLE 3.8 U.S. Plants Producing Dodecylbenzene and Their Production Capacities (MM lb/yr)

Company	Dodecylbenzene capacity			Full capacity benzene requirements
	Branched	Linear	Total	
Continental Oil Company Baltimore, Maryland	25	190	215	80
Monsanto Company Monsanto Industrial Chemicals Co., Chocolate Bayou, Texas	--	225	225	84
Standard Oil Co. of California Chevron Chemical Co. subsidiary, Richmond, California	220	--	220	82
Union Carbide Corporation Chemicals & Plastics Division, Institute, West Virginia	--	150	150	56
Witco Chemical Corporation Witfield Chemical Division, Carson, California	--	55	55	21
1976 Totals	245	620	865	323

Source: From Refs. 21 and 24.

Chlorobenzene

Less than 1% of the U.S. benzene production is utilized in chlorobenzenes. Three dichlorobenzene isomers are produced from chlorobenzene. Initially, benzene is chlorinated in the presence of a catalyst to produce chlorobenzene as follows:

$$\text{Benzene} + \text{chlorine} \xrightarrow{\text{catalyst}} \text{chlorobenzene} + \text{hydrogen chloride}$$

The chlorobenzene is then also catalytically chlorinated to produce ortho- and paradichlorobenzene.

$$\text{Chlorobenzene} + \text{chlorine} \xrightarrow{\text{catalyst}} \text{orthodichlorobenzene or paradichlorobenzene} + \text{hydrogen chloride}$$

Metadichlorobenzene is also produced but production information about it is not available.

Table 3.9 gives the producers of chlorobenzenes in the United States. Figure 3.27 is a flowchart of chlorobenzene and dichlorobenzene production.

TABLE 3.9 U.S. Plants Producing Dichlorobenzenes and Their Production Capacities

Company	Dichlorobenzene capacity (MM lb/yr)			Total
	Ortho	Meta	Para	
Allied Chemical Corporation Industrial Chemicals Division, Syracuse, New York	8	--[a]	12	20
Dow Chemicals, U.S.A. Midland, Michigan	25	--	38	63
Eastman Kodak Company Eastman Organic Chemicals, Rochester, New York	--	NA[b]	--	NA
Guardian Chemical Corporation Eastern Chemical Division, Hauppauge, New York	--	NA	--	NA
ICC Industries, Inc. Chemical Divisions, Niagara Falls, New York	NA	NA	NA	NA
Monsanto Company Monsanto Industrial Chemicals Co. Sauget, Illinois	16	--	12	28
Montrose Chemical Corp. of Calif. Henderson, Nevada	NA	--	NA	NA
PPG Industries, Inc. Chemical Division, Natrium, West Virginia	20	--	30	50
Solvent Chemical Co., Inc., Malden, Massachusetts, Niagara Falls, New York	3 20	-- --	NA 20	NA NA
Standard Chlorine Chemical Company, Inc., Delaware City, Delaware Kearny, New Jersey	40 16	0 --	60 NA	NA NA

[a] -- = not manufactured.
[b] NA = not available.
Source: From Refs. 21 and 24.

Uses

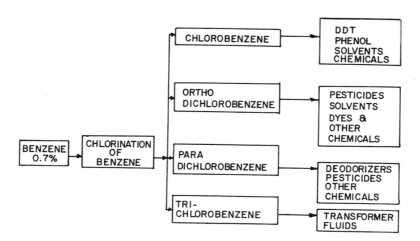

FIG. 3.27 Flowchart of benzene utilization in the production of chlorobenzenes in the United States.

4

Safe Use, Handling, and Storage

FIRE SAFETY AND STORAGE

In a regulatory sense the term benzene denotes solid, liquid, or gaseous benzene including liquid mixtures containing benzene and the vapors emanating from these liquids.

Benzene in the liquid state is classified as a Class IB flammable liquid because of its high flammability. Class IB flammable liquids are those with boiling points above 100°F and flashpoints below 73°F. Benzene vapors can form explosive mixtures in the ambient. These vapors are heavier than air and susceptible to ground level ignition sources, sometimes a relatively large distance from the point of release. A benzene vapor concentration greater than 3250 ppm can result in fire explosions. The presence of benzene in levels where fires or explosions are a potential results in such locations being designated by the Occupational Safety and Health Administration (OSHA) as Class I Group D. The following National Fire Prevention Association Codes apply:

Flammable and Combustible Liquid Code	NFPA No. 30
National Electric Code	NFPA No. 70
Static Electricity	NFPA No. 77
Lightning Protection Code	NFPA No. 78
Fire-Hazard Properties of Flammable Liquids, Gases and Volatile Solids	NFPA No. 325M

Storage areas for benzene should be cool and well ventilated. Storage containers should be tightly closed and grounded before opening. Electrical interconnection between storage containers and receiving vessels is required by OSHA and nonsparking tools used when opening and closing the containers.

Ignition sources must be removed from benzene storage areas or locations where benzene is used. Fire extinguishers should be provided in readily accessible locations. The extinguishing media for benzene are carbon dioxide, dry chemical, or foam. Large amounts of water can be used in fighting a benzene fire. However, a solid stream of water should not be used since this has the effect of spreading the fire. Water sprays should be used on the fire. Further water sprays should be used to disperse unignited vapors and to cool fire-exposed containers [5,15,27]. Storage area locations should be made available to local fire departments in case of problems.

Benzene is a basically stable compound. However, its storage and handling areas should be kept cool because the application of heat leads to its instability. It is also incompatible with oxidizing agents. Its decomposition products can be toxic gases and vapors such as carbon monoxide.

SPILLS

During transfer operations there is more likelihood of high exposure to benzene or spills. Handling areas should be well ventilated and in extreme cases respirators for employees provided (see pp. 48-49).

During transfer operations of closed systems, pipelines and receiving vessels should be pressure-tested prior to each transfer. The possibility of large spills during a closed-system transfer is greater because the bursting of a line during such a transfer may not be readily evident if pipelines are buried or out of sight.

In all situations good work practices should be established to avoid spills and to protect the environment once a spill has occurred. Curbing of drum storage areas and diking of storage tanks are essential. Drains from such storage areas normally have closed valves. These storage areas are not drained until their contents are checked.

In the event of a benzene spill all potential ignition sources should be eliminated. The benzene should be cleaned up immediately. For relatively small spills, absorbant pads or absorbant particles should be spread to absorb the benzene. For larger spills, the benzene should be pumped into containers using explosion-proof motors prior to the use of absorbants. Depending on the type of wastewater treatment available, the residue should be flushed with large volumes of water to remove the benzene contact potential.

Should the spill occur indoors, the space should be well ventilated to remove an explosive concentration and reduce worker exposure. Benzene should not be flushed into a confined space like a sewer because of the probability of vapor concentrations reaching the lower explosive limit [5,27,28]. Furthermore, most municipalities have ordinances against the discharge of volatile flammable substances to their sewer systems.

The Environmental Protection Agency (EPA) has issued regulations [29] dealing with spill control of hazardous substances. The EPA designated as hazardous 271 substances, including benzene. The Federal Water Pollution Control Act (1977) established harmful quantities for these substances. Penalties for releasing harmful quantities of these materials have also been established.

One thousand pounds of benzene is defined as a harmful quantity. Benzene is classified as category "C" for which a specific toxicity range is established. All categories and their toxicity ranges are given in Table 4.1.

TABLE 4.1 Units of Measurement, Categories, and Toxicity Ranges for Harmful Substances

Category	Unit of measurement (lb)	Toxicity range
X	1	LC^a < 0.1 mg/liter
A	10	0.1 mg/10 LC 50 < 1 mg/liter
B	100	1 mg/10 LC 50 < 10 mg/liter
C	1000	10 mg/10 LC 50 < 100 mg/liter
D	5000	100 mg/10 LC 50 < 500 mg/liter

^{a}LC = lethal concentration.
Source: From Ref. 29.

WASTE DISPOSAL

Rags, contaminated protective clothing, absorbant materials, and other materials with benzene residues should be stored in tightly closed containers. Their disposal in an approved chemical landfill may be adequate. However, combustion in an incinerator with multiple chambers or afterburners which insure complete reduction would be more suitable.

Liquid benzene for disposal is best incinerated to render it into its innocuous components. Again, proper incineration techniques must be employed. Otherwise, the whole concept of environmental protection is negated as uncombusted benzene vapors are spewn out the incinerator stack.

PROTECTIVE CLOTHING AND EQUIPMENT

Benzene contact with the skin should be prevented by wearing protective clothing. Gloves, uniforms, smocks, and laboratory coats are necessary to reduce contact due to spills and splashes. Of utmost importance is eye protection. Safety goggles are required where benzene spills and splashes to the eyes may occur, and those areas where facial sprays may occur full length face shields of an 8-in. minimum length are required.

When benzene is splashed on clothing that is permeable, the clothing should be removed immediately and laundered before reuse.

In those workplaces where benzene contact with the body is evident, impermeable clothing including boots, aprons, gloves, and sleeves should be worn.

RESPIRATORS

Under normal circumstances respirators are not allowed to bring workers exposure to benzene vapors within permissible exposure limits. Engineering

Safe Use, Handling, and Storage

controls must be used to reduce and maintain worker exposure to benzene; these controls include ventilation, hoods, leak prevention, improved maintenance of equipment, alternate work methods, and so on.

However, in those work spaces where engineering controls and worker scheduling to reduce benzene exposure are impractical, then respirators must be worn by those who will work in these high-benzene exposure areas.

Any respirators that are used must be approved by the National Institute for Occupational Safety and Health (NIOSH). Table 4.2 gives recommended respirator usage for various benzene concentrations greater than 1 ppm [5,15].

A respirator selected from Table 4.1, which is appropriate for a high concentration of benzene may be used in areas of lower vapor concentrations. Canisters or cartridges must be changed prior to the end of its service life or at the end of each work shift, whichever comes first. In those areas where a mixture of toxic substances is present, a respirator to control the various compounds in combination should be used.

Worker usage of respirators can cause problems in terms of comfort. Many workers develop frequent or continuous breathing difficulties when using a respirator. These persons should be examined by a doctor to ascertain their ability to wear this type of protective device. If the prognosis is negative, then a rotation of the worker out of the hazardous areas is necessary. Other persons, when working with respirators in hot atmospheres, develop skin irritations. These people should be allowed to wash and rinse their faces and face shields periodically.

GOOD HANDLING PRACTICES

In any areas where the ambient time-weighted-average benzene concentration is greater than 1 ppm, all edible and personal care products such as cosmetics should be prohibited. Tobacco products should also be prohibited in these locations, for the obvious reason that they present a potential fire hazard, but also for the not-too-obvious reason of avoiding benzene contamination.

Eyewash and shower stations should be available in easily accessible locations for the immediate flushing of the eyes in an emergency. Any clothing that has become contaminated with liquid benzene should be removed immediately and those contacted areas of the body rinsed and washed under showers. Clean change rooms should be provided to promote the practice of changing clothes at work to avoid contamination at home.

EMERGENCY AND FIRST AID

Plan of Attack

A plan of attack for emergencies involving benzene releases should be developed. The plan should establish procedures whereby respirators and protective clothing are provided at the scene to affected individuals and those workers correcting the problem. The area should be cordoned off to prevent onlookers and sightseers from entering the emergency area. The previously mentioned eyewash

TABLE 4.2 Recommended Respirator Usage for Various Benzene Concentrations Greater than 1 ppm

Concentration of benzene or condition of use	Respirator type
≤ 10 ppm	(1) Chemical cartridge respirator with organic vapor cartridges and half mask
	(2) Any supplied air respirator with half mask
≤ 50 ppm	(1) Chemical cartridge respirator with organic vapor cartridges and full facepiece
	(2) Any supplied air respirator with full facepiece
	(3) Any organic vapor gas mask
	(4) Any self-contained breathing apparatus with full facepiece
≤ 1000 ppm	(1) Supplied air respirator with half mask in positive pressure mode
≤ 2000 ppm	(1) Supplied air respirator with full facepiece helmet or hood in positive pressure mode
≤ 10,000 ppm	(1) Supplied air respirator and auxiliary self-contained breathing apparatus with full facepiece in positive pressure mode
	(2) Open-circuit self-contained breathing apparatus with full facepiece in positive pressure mode
Entry into unknown concentrations or fire fighting	(1) Open circuit self-contained breathing apparatus with full facepiece in positive pressure mode
Escape only	(1) Any organic vapor gas mask
	(2) Any self-contained breathing apparatus with full facepiece

Source: From Ref. 5.

Safe Use, Handling, and Storage

and shower stations should be part of the plan. An alarm system should be installed to alert personnel of the imminent danger. In order to avoid confusion at such times instructions should be written and available to all personnel. Individuals should be designated as emergency leaders (with alternates) who can assess the problem, inform personnel of the dangers involved, and take action—from handing out the respirators to getting first aid to the worker whose breathing has stopped after inhaling a high concentration of benzene.

First Aid

Eyes

Eyes that have contacted benzene should be washed immediately with large volumes of water. A physician who can determine the extent of damage and prescribe further remedies should be seen at once.

Skin

Skin that has contacted benzene should be washed at once with large volumes of water and soap. In many instances benzene contact with the skin is by way of contaminated clothing. When benzene is spilled on the clothing, the clothing should be removed immediately and the contacted skin tissue flushed with water and soap.

Inhalation

Inhalation of high concentrations of benzene is highly dangerous. Victims should be brought to fresh air or given oxygen immediately. If the victim's breathing has ceased, then artificial respiration should be applied. A physician should be consulted at once.

Ingestion

The ingestion of benzene is highly toxic. Vomiting should <u>not</u> be induced in victims who have swallowed it. Medical assistance should be obtained.

EMPLOYEE TRAINING AND INFORMATION TRAINING

In any workplace where benzene is handled, used, or stored, including production, reactions, releases, packaging and repackaging, and transportation, a training program must be instituted to ensure worker's cognizance of benzene's properties, toxicity, and safety procedures.

The training program should include descriptions of the following:

1. Benzene properties and health hazards.
2. Signs and symptoms of benzene exposure and action to be taken when they are evident.
3. Safe work and good housekeeping practices.
4. Emergency procedures for leaks, spills, and fires, including protective clothing to be worn in such instances.
5. The importance of protection from benzene contact; the proper clothing and cleaning requirements—to ensure worker compliance.
6. The importance of the time-weighted-average exposure to a benzene level of 1 ppm.
7. The care that must be taken whenever and wherever benzene or benzene products are used, handled, stored, and transported.
8. The importance of respirators, their effectiveness, and the health hazards effected by nonuse.
9. The need for and the requirements of the medical surveillance program.
10. The availability of written benzene usage, health hazard, and training program procedures.

The training program should be part of a worker's initial indoctrination courses and should be taken at least annually thereafter.

Labels and Signs

Areas within a workplace where benzene concentrations exceed the permissible airborne exposure are considered "regulated areas." These areas should be designated and have limited access by authorized personnel only. Furthermore, within 30 days after an area has been designated as a regulated area OSHA must be notified of its location, process operations affecting the benzene concentration, and employees and their benzene exposure in the area. Signs shall be posted in these designated areas and shall read as shown in Fig. 4.1.

All containers of benzene or products with any benzene content must be marked with caution labels. These caution labels should read as shown in Fig. 4.2.

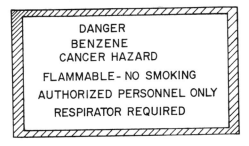

FIG. 4.1 Sign to be posted in benzene-regulated areas. (From Ref. 15.)

Safe Use, Handling, and Storage

FIG. 4.2 Caution labels to be affixed to all benzene and products with any benzene content containers. (From Ref. 15.)

The caution labels are to be permanently affixed to the containers. Pipelines and transport vehicles and vessels are specifically omitted from benzene caution-labeling requirements. However, in the next section the proper packaging, labeling, and identification of benzene are considered with regard to Department of Transportation requirements.

Department of Transportation (DOT) Requirements

Benzene is listed as a DOT hazardous material in CFR 172.101. These regulations specify the proper identification, packaging, and marking for hazardous materials which are offered for transportation on public roads.

The hazardous materials description and proper DOT shipping name for benzene is "BENZENE." It has a hazard class of "flammable liquid" and containers require flammable liquid labels. The DOT hazard classes are given below.

DOT Hazard Classes

Explosives A

Explosives B

Explosives C

Flammable liquid

Flammable solid

Oxidizer

Corrosive liquid

Nonflammable gas

Flammable gas

Poison A

Poison B

Radioactive material

Organic peroxide

The maximum quantity of benzene that may be shipped in one package on a passenger carrying aircraft or railcar is 1 qt. Ten gallons is the maximum quantity in one package that may be shipped on a cargo-only aircraft.

DOT Drums

Figure 4.4 illustrates a typical light gauge closed head drum, which, depending on the specifications, may be suitable for flammable liquids.

Shipment of Flammable Liquids

The requirements for packaging, marking, and labeling for the shipment of flammable liquids are prescribed in sections 173.118, 173.119, 173.132, and 173.144 of Title 49, Code of Federal Regulations (CFR). Sections 173.118 and 173.119 are regulations specifically pertaining to benzene [30,31]. Table 4.3 outlines the DOT requirements for packaging, marking, and labeling for the shipment of flammable liquids.

Flammable Liquid Label

Each flammable liquid label, affixed to or printed on a package, must be durable and weather resistant. Each diamond (square-on-point) label must be at least 4 in. on a side with each side having a black solid line border 1/4 in. from the edge [7]. It should appear as in Fig. 4.3. The flammable liquid label is red, the printing and symbol are black.

Requirements for Shippers of Hazardous Materials

This guide is intended to serve as an aid for in-house when reviewing shipping procedures. It does not include or refer to all applicable DOT requirements. Shippers of hazardous materials are legally obligated to comply with the applicable DOT regulations.

An available up-to-date copy of the regulations and use of it are essential. DOT regulations are minimum requirements. State and local agency requirements are frequently more stringent. Familiarity with and reference to such regulations when applicable is also essential.

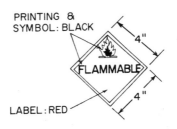

FIG. 4.3 Flammable liquid label.

Safe Use, Handling, and Storage 55

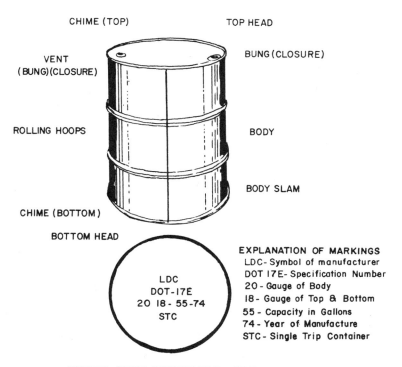

FIG. 4.4 Typical light gauge closed head drum.

TABLE 4.3 The Requirements for Packaging, Marking, and Labeling for the Shipment of Flammable Liquids as Prescribed in Section 173.118, 119, 128, 132, and 144 of Title 49, CFR. (This Does Not Include Flammable Liquids Indicated by the "No Exemption" Statement.)

Container type and capacity	Specific requirements for packaging, marking and labeling for the following flash point ranges:	
	Below 100°F but not less than 73°F	Below 73°F
1. Inside-packed in strong outside containers		
a. Metal—1 qt or less and others—1 pt or less	DOT specification packaging, marking, and labeling are not required, except that shipment via air carrier must be labeled, or DOT specification packaging, marking, and labeling are not required when the flash point range is marked on the outside package.	DOT specification packaging, marking, and labeling are not required, except shipments via air carrier must be labeled.
b. Glass or earthenware—1 qt or less. Special requirements for ink, paint, and	DOT specification packaging, marking, and labeling are not required, except that shipments via air carrier must be labeled, or DOT specification packaging, marking, and labeling are not required when the flash point range is marked on the outside package.	DOT specification packaging, marking, and labeling are not required, except shipments via air carrier must be labeled.
c. Metal—greater than 1 qt but not more than 1 gal and others—greater than 1 pt but not more than 1 gal	DOT specification packaging, marking, and labeling are not required when the flash point range is marked on the outside container, except that shipments via air carrier must be	DOT requirements for packaging, marking, and labeling are prescribed in Section 173. Specific section references are listed in Section 172.5.

Safe Use, Handling, and Storage

labeled and outside containers having inside containers of over 1-qt capacity each must be marked with the name of the contents and labeled.

d. Fiberboard body with metal tops and bottoms, leakproof—1 gal or less. Special requirements for cement and metal—5 gal or less. Special requirements for ink, paint, and cement

DOT specification packaging, marking, and labeling are not required, except that shipments via air carrier must be labeled and outside containers having inside containers of over 1-qt capacity each must be marked with the name of the contents and labeled.

DOT specification packaging, marking, and labeling are not required, except that shipments via air carrier must be labeled and outside containers having inside containers of over 1-qt capacity each must be marked with the contents and labeled.

e. Containers—greater than 1-gal

DOT specification packaging is not required when the flash point range is marked on the outside container. NOTE: The outside containers must be marked with the name of the contents and labeled.

DOT requirements for packaging, marking, and labeling are prescribed in Section 173. Specific references are listed in Section 172.5.

2. Outside:

a. Containers—110 gal or less

DOT specification packaging is not required when the flash point range is marked on the outside container. NOTE: The outside containers must be marked with the name of the contents and labeled.

DOT requirements for packaging, marking, and labeling are prescribed in Section 173. Specific section references are listed in Section 172.5.

b. Containers—greater than 110 gal

DOT requirements for packaging, marking, and labeling are prescribed in Section 173. Specific section references are listed in Section 172.5.

DOT requirements for packaging, marking, and labeling are prescribed in Section 173. Specific section references are listed in Section 172.5.

Technical Personnel:

1. Need to be familiar with the definitions of the hazardous material classes as defined in the DOT Hazardous Materials Regulations
2. Must determine or obtain the proper <u>DOT classification</u> and <u>shipping name</u> for each material shipped falling within any of the hazardous material classifications
3. Must determine or obtain complete information as to the DOT authorized containers that may be used for the quantities and chemical characteristics of the hazardous materials to be shipped
4. Must prepare any specific instructions needed by carrier personnel to handle transportation emergencies involving the hazardous material being shipped

Packaging and Shipping Personnel:

1. Need to be familiar with, and have available for reference, the packaging, labeling, and marking regulations for hazardous materials
2. Must have accurate information indicating what DOT containers must be used for all hazardous material shipped
3. Must determine the proper DOT shipping name, class, label, and required container marking
4. When repacking or breaking down shipments received from others for further distribution, must not assume that the original shipments were correct and must follow the same checking procedure as they would for a shipment originating in your plant
5. When calling a carrier for pick up of hazardous materials, must give him specific information as to the DOT shipping name, class, and quantity involved.

Before hazardous materials shipments are tendered to carriers, the following must be checked:

1. That DOT authorized containers have been used
2. That proper closures have been used and no leaks are evident
3. That outside packages are properly labeled and have required outside marking
4. That shipping documents include proper DOT shipping name, hazardous material class, signed certificate, proper count, and weight
5. That incompatible materials are not tendered in the same shipment

When tendering a hazardous material shipment to a carrier, the driver should be made aware that it contains a hazardous material.

When hazardous materials are loaded by shipper personnel, a carrier representative should have an opportunity to approve the placement, securing, and blocking of the material before closing out and sealing the load, if seals are used.

Safe Use, Handling, and Storage

If placards are required, a carrier should not be permitted to transport hazardous material shipment unless such placards are properly displayed.

Requirements for Carriers of Hazardous Material

This guide is intended to serve as an aid for in-house use when reviewing carrier hazardous materials procedures. It does not include or refer to all applicable DOT requirements. Both carriers and shippers of hazardous materials are legally obligated to comply with the applicable DOT regulations.

An available up-to-date copy of the regulations and use of it are essential. DOT regulations are minimum requirements. State and local agency requirements are frequently more stringent. Familiarity with and reference to such regulations when applicable is also essential.

Sales Personnel:

1. Need to be familiar with classes of hazardous materials, proper DOT shipping names, packaging requirements, labels, marking, and documentation requirements.
2. When calling on shippers, must determine whether they may ship hazardous materials and, if so, which materials
3. Must work with the shippers concerning shipping requirements and act as liaison with shipper when discrepancies are discovered
4. Must keep the dispatcher and/or traffic department advised as to what hazardous materials can be expected from individual shippers

Dispatchers and Order Takers:

1. Must have a broad hazardous material knowledge involving proper DOT shipping names, classes, required packaging, compatibility, labeling, placarding, and shipping documents
2. As phone pick-up orders are received, must check with shippers to determine what, if any, hazardous materials are included in the shipment
3. Must maintain a current list of regular hazardous material shippers including classes, quantities, and special problems that may exist
4. Must make certain that when pickup orders are given to driver (written or verbal) that hazardous material information is included and understood
5. Must make certain that drivers call for specific instructions when suspect shipments are encountered
6. Must maintain a current list of sources of help and information to be used when contaminations and other emergencies occur and when unrecognizable chemicals are encountered
7. Must have specific procedures prepared for use when hazardous materials incidents occur
8. Must make certain that drivers have adequate placards and means of securing hazardous material shipments in vehicles

Pickup and Delivery Drivers:

1. Must have a broad hazardous material knowledge involving proper DOT shipping names, classes, required packaging, compatibility, labeling, marking requirements, placarding, and shipping documents
2. Must be able to recognize discrepancies in documents, packaging, labeling, and compatibility
3. Must inspect all hazardous material shipments prior to loading
4. Must refuse to accept hazardous material shipments from shippers unless in compliance with the regulations (shippers must certify that articles are packaged, etc., in compliance with DOT regulations)
5. Must contact carrier management for instructions concerning any suspect shipments offered to him
6. Must have available and must use proper placards when required
7. Must make sure that all hazardous material is properly blocked and secured in his vehicle before moving
8. Must make sure that hazardous material containers will not be damaged by other freight, or by nails or rough truck floors and sides
9. Must make certain that class A or B poisons are not loaded with foodstuffs or other contaminable cargo
10. Must understand the proper procedures for handling, disposal, and/or decontamination in case of accident or incidents involving hazardous materials
11. Must be able to report full details concerning any hazardous material incident, including detailed information as to cause, container damage, specific container identification, and corrective action taken
12. Must refuse to accept hazardous material freight from shippers or interline carriers if the shipping documents are improperly prepared or do not check out with the freight involved or if the containers are leaking, damaged or otherwise improper
13. Must see that understand their responsibilities as to attendance requirements when transporting a hazardous material

Road Drivers:

1. Must have a broad hazardous material knowledge involving proper DOT shipping names, classes, required packaging, compatibility, labeling, marking requirements, placarding, and shipping documents
2. Must have in their possession, and available for immediate use, proper shipping papers covering all hazardous materials loaded on their vehicle
3. Must have specific instructions (preferably written) as to handling procedures in case of hazardous material incidents
4. Must know what to do and what information to pass on to firemen, policemen, and others should an emergency arise
5. Must report all discrepancies and irregularities observed during trip (including such things as leaking containers or defective tank truck valves)

Safe Use, Handling, and Storage

6. Must understand their responsibilities as to attendance requirements when transporting a hazardous material
7. Must refuse to accept hazardous material freight from shippers or interline carriers if the shipping documents are improperly prepared or do not check out with the freight involved or if the containers are leaking, damaged, or otherwise improper

Traffic and Freight Billing Personnel:

1. Must be familiar with hazardous material classes, proper DOT shipping names, and other hazardous material shipping document requirements.
2. Must "flag" all improper documents and/or questionable hazardous shipments for verification, and not prepare bills or process suspect shipments.
3. Must make certain that all required information is reproduced and not abbreviated on freight bills.
4. If the hazardous material is moving under a DOT Special Permit, must make sure that the Special Permit is included on the freight bill. If the Special Permit contains any carrier conditions, etc., then the shipper must give the carrier a copy of that Special Permit.
5. If special emergency instructions including telephone numbers have been supplied by the shipper, must make sure that such information included on the freight bill.

Claims and OS&D Personnel:

1. Must be familiar with general hazardous materials regulations and particularly the marking and container requirements.
2. Must obtain complete container identification information for all damaged hazardous material containers, including specifications as shown in 49 CFR Part 173 (and required marking for such shown in Part 178).
3. Must determine which container failed when inside shipping containers are used and whether the inside container is an <u>authorized</u> container.
4. Must determine whether a non "Spec" container used for hazardous material is in fact an <u>authorized</u> non spec container. (If so, any written report to DOT, such as the hazardous materials incident report, should then indicate "Authorized Non-Spec Container.")
5. Must obtain accurate information describing the container damage or failure in detail, including cause.
6. Must prepare or be able to assist in the preparation of the report to the Department of Transportation.
7. The above information is frequently very helpful in determining shipper liability.

Dock Foreman and Freight Handlers

1. Must have a broad hazardous material knowledge involving proper DOT shipping names, classes, required packaging, compatibility, labeling, marking requirements, placarding, and shipping documents.

2. Must check hazardous material freight against documents. If they do not check out, they must make certain that the discrepancies are resolved before freight is allowed to move any further.
3. Must refuse to accept hazardous material freight from shippers or interline carriers if the shipping documents are improperly prepared or do not check out with the freight involved or if the containers are leaking, damaged, or otherwise improper.
4. Must inspect all hazardous material freight for leakage or damage each time it is handled.
5. When damaged containers are discovered, must isolate and make certain they are not moved until they are in proper condition for further transportation; must make certain that all container information is obtained for use in preparing the required report to the Department of Transportation (NOTE—in some instances an immediate telephone notification to DOT is also required).
6. When contamination occurs or when it is necessary to dispose of hazardous materials or containers, they must make certain that a qualified individual supervises such activities.
7. Must make certain that incompatible hazardous materials are not loaded into the same vehicle.
8. Must make certain that proper placards are placed on vehicles when required and that placards are removed or covered when not required.
9. Must make certain that hazardous material containers will not be damaged by other freight or by nails or rough sides and flooring within the vehicle.
10. Must make certain that all hazardous material is properly blocked and secured before closing out vehicle.
11. Must make certain that class A or B poisons are not loaded with foodstuffs or other contaminable cargo.
12. Must be familiar with appropriate loading and storage regulations.

HAZARDOUS MATERIALS AND SHIPPING DOCUMENT REQUIREMENTS

The following information has been abstracted from the Code of Federal Regulations, Title 49, CFR, Parts 100-199. In addition, you will note that CFR, Title 14, Part 103 and CFR, Title 46, Parts 140-149 are also referenced. Use the applicable Code(s) for the mode of transport.

Requirements for all Shippers

Shipping Papers (Title 49, CFR, Sec. 173.427)

1. Each shipper offering for transportation any hazardous material subject to the regulations in this chapter, shall describe that article on the shipping

Safe Use, Handling, and Storage

    paper by the shipping name prescribed in Sec. 172.5 of this chapter, and by the classification prescribed in Sec. 172.4 of this chapter, and may add a further description not inconsistent therewith. Abbreviations must not be used. The total quantity by weight, volume, or as otherwise appropriate must be shown.

2. Where the regulations (except Sec. 173.402) exempt the packages from labeling the exemption must be indicated by the words "No Label Required" immediately following the description on the shipping paper.

Certification (Title 49, CFR, Sec. 173.430)

1. Each shipper offering for transportation any hazardous material subject to the regulations in this chapter must show on the shipping paper the following certificate which must be signed by the shipper:

This is to certify that the above named articles are properly classified, described, packaged, marked and labeled, and are in proper condition for transportation, according to the applicable regulations of the Department of Transportation.

2. Shipping papers for <u>air shipments</u> in foreign commerce must be made out in duplicate and the shipper's certificate must be executed on both copies.
   a. For shipments on <u>passenger-carrying aircraft</u>, the shipper must also add the words:

This shipment is within the limitations prescribed for passenger-carrying aircraft.

    b. The shipper may also add the words "... and to the IATA Restricted Articles Regulations."

3. Shipper certification is not required for shipments to be transported by the shipper except for shipments which are to be reshipped or transferred from one carrier to another, or for bulk shipments in cargo tanks supplied by the carrier.

<u>Rail Shipments</u>

Shipping Papers (Title 49, CFR, Sec. 174.510)

1. The conductor of each train transporting hazardous materials must have in his possession a copy of the shipping paper showing the information required in paragraph (1) of this section.

Switching Ticket (Title 49, CFR, Sec. 174.431)

1. When the initial movement is a switching operation, the switching order, switching receipt, or switching ticket and copies thereof, prepared by the

shipper, must bear the placard endorsement and the shipper's certificate prescribed by Sec. 173.430 and 174.574(a) and (b) of this Chapter.

Public Highway (Common, Contract, or Private Carriers

Shipping Papers (Title 49, CFR, Sec. 177.817)

1. The driver of each motor vehicle transporting hazardous materials must have in his possession a copy of the shipping paper showing the information required in paragraphs (a) and (b) of this section.
    a. Each carrier offering or delivering for rail transportation any loaded motor vehicle, trailer, semitrailer, or container containing hazardous materials must show on the shipping paper the information required in paragraphs (a) and (b) of this section, the description of the vehicle or container, and the kind of placards applied.

Water Shipments

Shipping Papers (Title 46, CFR, Subpart 146.05)

Shipping paper requirements for shipments by water are almost identical to the requirements for shipments by rail or highway. The regulations are divided into requirements for domestic shipments (Subpart 146.05-12) export shipments (Subpart 146.05-13), and import shipments (Subpart 146.05-14). The two significant differences are as follows: First, for other than domestic shipments, when the proper shipping name of a hazardous material is an "NOS" (Not Otherwise Specified) entry in the list of hazardous materials (Subpart 146.05-5), this description must be qualified by the "chemical name of the commodity" in parentheses, e.g., "Corrosive liquid, NOS" (caprylyl chloride). Second, in connection with the entry (on the shipping documents) of each hazardous material, the description is required to include the kind and color of label applied to the package when the label is required [Subpart 146.05-12(f)(5) and (6)].

Air Shipments

Certification (Title 14, CFR, Part 103, Sec. 103.3)

1. No shipper may offer, and no person operating an aircraft may knowingly accept, any dangerous articles for shipment in an aircraft unless there is accompanying the shipment a clear and visible statement that the shipment complies with the requirements of this Part. In the case of shipments in passenger-carrying aircraft, the shipper shall also state the shipment complies with the requirements in this Part for carrying dangerous articles in passenger-carrying aircraft. The shipper or his authorized agent shall sign the statement or stamp it with a facsimile of his signature. The person operating an aircraft may rely on the shipper's

Safe Use, Handling, and Storage

statement as prima facie evidence that the shipment complies with the requirements of this Part.
2. The shipper shall execute the required certificates in duplicate. One signed copy accompanies the shipment and the originating air carrier retains the other signed copy.

Notification of Pilot in Command
(Title 14, CFR, Part 103, Sec. 103.25)

Whenever articles subject to the provisions of this Part are carried in an aircraft, the operator of the aircraft shall inform the pilot in command, before takeoff, in writing, of the proper shipping name, classification of each hazardous article, quantity in terms of weight, volume or as otherwise appropriate, and the location of the dangerous article in the aircraft. The person marking the cargo load manifest shall mark it conspicuously to indicate the dangerous articles.

All Modes of Transport

Exemptions—Standard Conditions Applicable to Packages,
Containers, and Shipments (Title 49, HM-127)

The outside of each package must be plainly and durably marked "DOT-E" followed by the number assigned. On portable tanks, cargo tanks and tank cars, the markings must be in letters at least two inches high on a contrasting background.

Each shipping paper issued in connection with any shipment made under an exemption must bear the notation "DOT-E" followed by the number assigned and the entries required by Sec. 173.427 of Title 49.

When an exemption issued to a shipper contains special carrier requirements, the shipper shall furnish a copy of the exemption to the carrier before or at the time a shipment is tendered.

Foreign Shipments

Imports and Export Shipments (Title 49, CFR, Sec. 173.9)

Import shipments of hazardous materials offered in the United States in original packages for transportation by any mode of transport must comply with all requirements of Title 49, CFR, Parts 100-199.

The <u>forwarding agent</u> (shipper) must file with the initial carrier in the United States a <u>properly certified</u> shipping order or other shipping papers as prescribed in Title 49, CFR, Parts 100-199.

NOTE: For detailed regulatory requirements refer to Sec. 173.9.

TYPICAL HAZARDOUS MATERIALS SHIPMENT VIOLATIONS

This partial checklist is intended to be used in checking shipments of commodities which are classified as hazardous by the Hazardous Materials Regulations (Title 49, Code of Federal Regulations) and which are shipped on carriers subject to the regulations. NOTE: DOT exemptions may permit deviations from the normal packaging regulations.

Shipping Papers

1. No proper shipping name and/or classification
2. Proper shipping name and/or classification abbreviated
3. Lack of wording "No Label Required" on shipments exempt from specification packaging, marking, and labeling
4. No exemption number on shipments moving under DOT exemptions
5. Color of label used in lieu of classification

Marking of Containers

1. No commodity description on container not of "exempt" size (commodity description per shipping name in Title 49, CFR, Sec. 172.5)
2. No exemption number on container shipped under a DOT exemption
3. No name and address of consignee on package which exceeds "exempt" size

Labeling

1. No label on container of hazardous materials which exceeds "exempt" size
2. Label on container not compatible with classification on shipping papers
3. Surface label on containers destined for air shipment
4. Less than two Radioactive Material labels on container, other than "small quantities."

Containers

Steel:

1. Labeled containers of 5 gal and under with no DOT specification marking (without further overpack)
   a. 5-gal, 29-gauge metal pails
   b. 5-gal rectangular can
2. Labeled containers of over 5-gal size with no DOT specification marking
   a. "Rule 40" containers
   b. Foreign containers

Safe Use, Handling, and Storage

3. Labeled containers with temporary repairs
   a. Damage sealed with tape, putty, or screws
   b. Shipped upside down
4. Labeled containers in improper condition
   a. Dented
   b. Rusted or corroded
5. Labeled containers on which specification markings are illegible
6. Labeled reused containers marked "NRC"—(Look for old date of manufacture and evidence of reuse such as dents, rust, and paint layers)
7. Labeled reused containers marked "STC" and 17-C, 17-E and 17-H with no reconditioner's marking
8. Labeled 55-gal open-head drums with two rolling hoops and/or less than 5/8 in ring bolt, nondrop forged-ring lugs, and/or "lever lock" ring closures (good possibility of non-DOT specification)

Corrugated Fiber Board:

1. Labeled box with no DOT specification marking when inside containers are larger than "exempt" size
2. Labeled box with DOT specification marking constructed of less than 175 lb-test singlewall fiber board (check boxmaker's certificate)
3. Labeled box with DOT specification marking which is poorly constructed (i.e., gaps and uneven closures, seams, and joint separating)
4. Water-damaged boxes
5. Improperly closed boxes (i.e., masking tape, scotch tape, string)

<u>Placarding</u>

Placarding for benzene shipments under DOT would follow the prescribed requirements for flammable liquids.

Motor Vehicles and Freight Containers

Motor vehicles and freight containers containing flammable liquids of 1000 lb or more must be placarded. Motor vehicles should be placarded on all four sides and must be visible from the facing direction. Figures 4.5 and 4.6 show

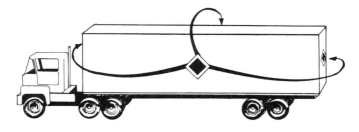

FIG. 4.5   Proper placarding for motor vehicles.

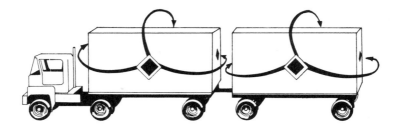

FIG. 4.6 Proper placarding for vehicles in combination.

the proper placarding for motor vehicles and vehicles-in-combination. Placards must be located and maintained such that they are clearly visible.

On freight containers with volumes greater than 640 cubic feet placarding is required on all four sides. Freight containers less than 640 cubic feet require one placard if shipped by air and one placard or label if shipped by land or water. Figure 4.7 shows the proper placarding.

Rail Cars

These require placarding for the shipment of any quantity of flammable liquids. The exception is when 1000 lb is shipped in trailer on flat car (TOFC) or containers on flat car (COFC). Placards must be attached to all four sides of the rail car as shown in Fig. 4.8.

Cargo Tanks and Portable Tanks

Cargo tanks must be placarded if they contain any quantity of flammable liquids. They must remain placarded until they have been emptied and residues flushed. Portable tanks 1000 gal or greater must be placarded on all four sides and those less than 1000 gal may be placarded on two sides. Before removal of flammable liquid placard the portable tank must be cleaned [31,32]. The previous discussion does not include all DOT requirements.

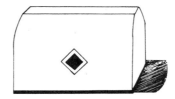

FIG. 4.7 Proper placarding for freight containers.

Safe Use, Handling, and Storage

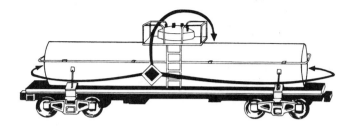

FIG. 4.8 Proper placarding for rail cars.

5
Monitoring

MONITORING REQUIREMENTS

Air samples which are representative of the air breathed by employees over an 8-hr period shall be taken and levels of benzene exposure determined. This representative sample shall not include air breathed through a respirator.

Sampling shall take place in those work environments where benzene is utilized, handled, transported, stored, packaged and repackaged, and produced, reacted, or released. Workplaces where employees are exposed to a benzene concentration at or above 1 ppm (TWA) or who come in dermal contact with benzene must be identified. These areas should be restricted to limit their occupancy [5].

Employee exposure to benzene can be indicated through observations, measurement techniques, employee complaints, and process and operational procedures.

The frequency of monitoring is dependent on the initial monitoring findings. Should the initial monitoring show benzene levels to be below the "action" level (0.5 ppm) then no further monitoring is required. However, changes in process, equipment, personnel, or additional benzene exposures such as leaks, spills, and breakdowns which may lead to benzene exposure above the 1 ppm limit may require further monitoring [5].

Should the initial monitoring show benzene concentrations to be greater than the 0.5 ppm (8-hr average) action level, but below the 1 ppm standard, then monitoring is to be conducted at least quarterly. Monitoring is continued until two consecutive measurements are below the action level. These two consecutive measurements must be at least 7 days apart. At this time monitoring can be discontinued.

Should the initial monitoring show benzene concentrations to be greater than the 1 ppm standard then monitoring is to be conducted at least monthly. When two consecutive monthly levels are below this permissible level, then the monitoring frequency can be reduced to quarterly. Again, when two consecutive quarterly measurements indicate benzene levels to be below the action level, then monitoring can be discontinued [5,22].

MEDICAL MONITORING

In those workplaces where air monitoring as shown above indicates benzene concentrations to be above permissible levels, medical examinations must be provided. The establishment of the medical surveillance program is a requirement and should include the availability of a licensed physician and cost examinations during working hours (after hours depending on physician availability).

This medical examination will involve a medical history consisting of previous benzene exposures, family blood dyscrasias (i.e., genetically related hemoglobin alterations, abnormal bleeding, dysfunctioning blood constituents), records of liver or renal dysfunctions, previous drug medication and alcohol and drug usage, systemic infections, and previous exposures to other blood toxins or bone marrow toxins inside or outside of the work environment.

Also included in the examination is a complete blood count, serum bilirubin, and urinary phenol evaluation. Laboratory analysis of blood samples are all encompassing. Other tests which would appear pertinent to the examining physician should be included. These medical examinations shall be provided to employees under the following circumstances:

1. Prior to assignment in an area with airborne benzene concentrations above the permissible level.
2. Twice a year for exposure to atmospheres containing benzene above the permissible level during the previous 6 months.
3. Symptoms of benzene exposure.
4. Examinations should be continued for those tested and showing blood alterations. The examinations should continue monthly until blood conditions return to normal [5].

The examining physician should be made aware of the reason for testing, employment activities of the individual, protective equipment utilized, exposure levels or anticipated exposure levels, and if necessary, previous medical information.

Needless to say, the requirements of benzene monitoring are stringent. Many companies have found these requirements to be so strict that they have been forced to phase out the use of benzene in their processes and operations. With the added costs of a benzene monitoring program more expensive solvents and raw materials become cost effective. Furthermore, programs to develop alternate processes that do not utilize benzene can be considered with the costliness of the benzene monitoring program being the impetus.

BIOLOGICAL MONITORING

Normal hemological values are given in Table 5.1. Biological monitoring is an important aspect of the benzene monitoring program. It should be conducted simultaneously with the air monitoring to determine benzene sources of entry if not already known [6]. Figure 5.1 illustrates human metabolic action on benzene.

TABLE 5.1   Normal Hematologic Values

<u>Cell counts</u>		
Erythrocytes	Male	4.6-6.2 million/mm^3
	Female	4.2-5.4 million/mm^3
Leukocytes	Total	5,000-10,000/mm^3
	Differential	
	Myelocytes	0%
	Immature polymorphonuclears	3-5%
	Segmented neutrophils	54-62%
	Lymphocytes	25-33%
	Monocytes	3-7%
	Eosinophils	1-3%
	Basophils	0-0.75%
Platelets		150,000-350,000/mm^3
Reticulocytes		0.5-1.5% of erythrocytes
<u>Corpuscular values for erythrocytes</u>		
Mean corpuscular hemoglobin		27-31 picograms
Mean corpuscular volume		82-92 cu micra
Mean corpuscular hemoglobin concentration		32-36%
Hematocrit	Male	40-54%
	Female	37-47%
Hemoglobin	Male	14.0-18.0 g%
	Female	12.0-16.0 g%
<u>Serum bilirubin concentration</u>		
Total		0.3-1.1 mg%
Direct		0.1-0.4 mg%
Indirect		0.2-0.7 mg%

<u>Source</u>: From Ref. 33.

Monitoring

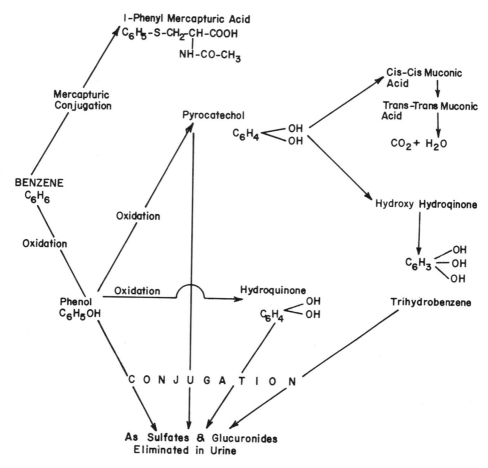

FIG. 5.1 Human metabolic action on benzene. (From Ref. 33.)

The purpose of biological monitoring is to determine whether or not benzene has been introduced to an individual and the extent of uptake. The levels indicated by biological monitoring are associated with the intoxication risks of benzene.

After benzene vapor is breathed, its rate of uptake by the body is about 60% [35], the rest is expired. The benzene is quickly metabolized, concentrating in fatty tissues. A great portion of the benzene is oxidized to phenols which passes from the body via urine [6,36]. Thus, benzene in expired air, urinary indicators, and benzene in the blood are strong considerations in benzene toxicology.

Blood

Because of the rapid metabolic action which is performed on benzene, blood indications for the presence of benzene are not good. Its residence time in the blood is very short-lived [6]. The correlation between airborne benzene concentrations and blood levels is undocumented [37,38]. Table 5.2 gives a summary of the examination of workers for benzene in their blood. They had been exposed to airborne benzene at their workplaces. Note that only about 30% of the workers were determined to have benzene in their blood.

Table 5.3 gives a detailed blood analysis performed on samples taken from 13 workers exposed to benzene. Two normal persons are also included as the control. Benzene is therefore measured as an effect on blood components, producing abnormalities. As shown in Table 5.3, individual blood tests by themselves may not be strong indicators of benzene poisoning. However, Table 5.4 illustrates that various combinations of tests can reveal benzene poisoning with a high certainty.

Breathanalysis [39]

Studies have shown a correlation between benzene exposure and benzene content in the breath. Benzene residues have been detected as long as 24 hr after exposure [40]. However, the use of breathanalysis as a biological monitor for benzene is limited to a secondary method to supplement other monitoring techniques. This is because of the rapid uptake of benzene by the body, unknowns with regard to rate of uptake, and the questionable effects of other compounds on the rate of uptake [41].

Urinary Sulfate Analysis

The urinary sulfate analysis is a measurement of the inorganic-to-total-sulfate ratio. This ratio is affected by the reaction between metabolized benzene and sulfate radicals and was once considered a good indicator of benzene exposure. However, this method has not proved to be as acceptable as urinary phenol analysis [43].

Urinary Phenol Analysis [6,22,44,45]

Studies have shown that a urinary phenol assay performed at the end of the standard workday which results in urinary phenol greater than 75 mg/liter indicates an exposure to benzene of greater than 10 ppm. The results of the urinary phenol assay taken at the end of the workshift should be compared to a similar test taken prior to beginning the workday. Such a comparison can show the rate of benzene absorption.

There are two basic analytic techniques for urinary phenol analysis. Colorimetric methods [14,17-20] and gas chromatograph techniques [44,50] are available for urine phenol determinations.

Table 5.5 gives the correlation of urinary phenol levels with equivalent benzene exposure concentrations. Table 5.6 illustrates the accuracy of the urinary phenol analysis [6] with regard to predicted and actual benzene exposure levels.

TABLE 5.2 Examinations of Workers Breathing Airborne Concentrations of Benzene and Their Blood Findings

Group	Room	Local ventilation	Average benzene in air (ppm)		Blood findings	
			Summer	Winter	Number of persons examined	Number positive
I-A Small amount of benzene; no local ventilation; low benzene content in air	150B 60 27A	— — —	100 150 110		9 1 2	2 0 1
I-B Small amount of benzene; no local ventilation; high benzene content in air	27B 59 61A 61B	— — — —	700 150 130 1360	210 210 580	2 9 12 1	0 1 6 1
II-A Large amount of benzene; local ventilation; low benzene content in air	78A 150A 75B	+ + +	70 90 100	90	0 1 3	1 1
II-B Large amount of benzene; local ventilation; high benzene content in air	91 50B 50A 75A	+ + + +	180 130	400 430 500 330	5 3 4 10	*0 1 1 1
III Large amount of benzene; no local ventilation; high benzene content in air	78B 23 83 95	— — — —	340 620 1800		1 6 9 3	0 2 6 2
Total					81	26

Source: From Ref. 35.

TABLE 5.3 Blood Analysis Performed for 13 Workers Exposed to Benzene

Plant code no.	Hemoglobin (Hb)	Red blood count (RBC)	White blood count (WBC)	Poly (%)	Lymphocytes (%)	Large mononuclears (%)	Eosin (%)	Trans (%)
23	65	4,376,000	5300	58	36	3.5	1.5	0.5
23	75	4,400,000	5200	55	39	3.5	2.0	0.5
23	85		4100					
23	50		4800					
27	55	4,304,000	4667	55	36	5.0	1.0	2.0
59	70	5,424,000	6140	47	47	3.5	0.5	1.0
61	85		4450					
61	50		4000					
61	40	1,736,000	3000					
61	75	1,736,000	2850					
61	80	800,000	4200					
83	23	1,055,000	3000					
	27		1450	58	36	5.0	1.0	0.0
	41							
	30	2,100,000	2100					
	29	1,365,000	2200	44	49	6.0	1.0	0.0
95	55	3,193,000	3100	50	39	1.5	7.0	1.5
95	70	4,968,000	3600	47	41	0.5	8.0	3.0
Normal male	90–110	5,500,000	7500	65–70	30	1–2	1–2	2–4
Normal female	50–100	4,500,000 5,000,000	7500	65–70	30	1–2	1–2	2–4

Source: From Ref. 37.

TABLE 5.4 Benzene Poisoning as Revealed by Combination of Various Blood Tests

Combined tests[a]	Cases of poisoning revealed by given test combinations	
	Number	%
MCV + RBC	61	82.4
MCV + WBC	59	79.7
MCV + Hb	59	79.7
MCV + platelets	57	77.0
RBC + platelets	56	75.7
RBC + WBC	54	73.0
RBC + Hb	51	68.9
MCV + RBC + WBC + platelets	72	97.3
MCV + RBC + WBC	69	93.2
MCV + RBC + platelets	66	89.2
MCV + RBC + Hb	65	87.8
Single tests		
MCV	48	64.9
RBC	47	63.5
Platelets	31	41.9
WBC	30	40.5
Hb	30	40.5
Total positive cases having complete blood studies	74	100.0

[a]MCV = mean corpuscular volume; RBC = red blood count; WBC = white blood count; Hb = hemoglobin
Source: From Ref. 42.

It is seen that from the content of the urinary phenol the benzene exposure level can be estimated to within a couple of factors.

BIOLOGICAL MONITORING IN PRACTICE

The various biological methods described previously by themselves may not be strong indicators of benzene exposure. However, used in conjunction with each

TABLE 5.5 Correlation Between Urinary Phenol Levels and Equivalent Benzene Exposure Concentrations

Urine phenol (mg/liter)	Approx. av. equiv. benzene air level (ppm)
100	10
120	13
140	16
160	19
180	22
200	25
220	27
240	29
260	31
280	33
300	35
320	38
340	41
360	44
380	47
400	50
420	53
440	56
460	59
480	62
500	65
520	68
540	71
560	74
580	77
600	80

Source: From Ref. 6.

TABLE 5.6 Comparison Between Urinary Phenol Content Estimated Airborne Benzene Exposure Concentrations and Actual Benzene Sampling Data

| Occupation | Urine phenol (mg/liter) | Benzene in air (ppm) ||
		Estimated from urine phenols	Air sampling data (TWA)
Agitator operator	105	10	1.3
Agitator operator	107	10	10.7
Benzol loader	<65	<5	1.7
Benzol still operator	<65	<5	6.7
Benzol oil still operator	<65	<5	0.8
Naphthalene operator	115	12	8.5
Analyst	105	10	2.4
Chemical observer	68	5	12.0
Foreman	<65	<5	none
Repairman	<65	<5	2.6
Chemical observer	65	5	17.1
Chemical observer	112	11	12.2
Chemical observer	66	5	6.5
Control tester	66	5	14.6
Stillman	212	24	39.2
Chemist	157	17	8.8
Pumpman helper	302	36	55
Pumpman helper	84	7	9.5

Source: From Ref. 6.

other benzene exposure can be accurately identified. This was shown in our blood analysis discussion in Table 5.4. The general blood tests of red cell count, white cell count, hematocrit, hemoglobin, mean corpuscular volume (MCV), mean corpuscular hemoglobin (MCH), and mean cell hemoglobin concentration (MCHC) normally don't show consistent correlations between benzene exposure and abnormalities. Yet an indication of benzene exposure can be obtained using a series of these tests. A deviation from the norm in one of the above tests could be the indicator that triggers a more extensive investigation [5,22].

Urinary phenol analysis is often used as a screening mechanism. The analysis can be performed on all personnel expected to have been exposed to benzene in high concentrations. Or it can be performed routinely on personnel to determine if benzene exposure has occurred. Should these screenings give positive indications, more elaborate blood analysis would then be performed by a hematologist.

The urinary phenol analysis is recommended by OSHA as the biological monitoring method to be used. Factors which could affect test results include foods or beverages consumed during the test that increase the production of phenol, drugs and medications, and other exposures to benzene and biphenyls [5].

In order to establish a urinary phenol analysis program, urine samples are taken from each person assigned to a workplace with a benzene concentration greater than 1 ppm TWA prior to his entry into the area. Obviously this is not possible for workers who have worked previously in the area. For these areas urine sampling and analysis should be completed every 3 months. Those persons with urinary phenol levels greater than 75 mg/liter should receive a report of their tests and be further tested on a daily basis for a before and after workplace benzene exposure comparison. Confirmation of benzene exposure should result in the taking of steps to limit future employee exposure. Further evaluation may be necessary. Employees with high urinary phenol levels (> 75 mg/liter) should be monitored every month until the condition disappears.

As a result of this testing, the implementation of personnel reassignment, process and operational changes, and/or protective controls may be warranted [5,22].

Biological Method for Sampling and Analysis of Benzene

The recommended biological method for urinalysis is taken from the NIOSH publication, "Criteria for a recommended standard... Occupational Exposure to Benzene." NIOSH derived this method from Ref. 15.

Collection of Urine Samples

"Spot" urine specimens of about 100 ml are collected as close to the end of the working day as possible. If any worker's urine phenol level exceeds 75 mg/liter, procedures are instituted immediately to determine the cause of the elevated urine phenol levels and to reduce benzene exposure to the worker. Weekly specimens are collected as described above until three consecutive weekly determinations indicate that urinary phenol levels are below 75 mg/liter.

After thoroughly washing their hands with soap and water, workers shall collect urine samples from single voidings in clean, dry specimen containers having tight closures and at least a 120-ml capacity. Collection containers may be made of glass, waxed paper, or other disposable types if desired. 1 ml of a 10% copper sulfate solution, as a preservative, is added to each sample and samples are immediately stored under refrigeration, preferably at 0-4°C. Refrigerated specimens will remain stable for approximately 90 days. If shipment of samples is necessary to perform analyses, the most rapid method available means shall be employed utilizing acceptable packing procedures as specified by the carrier. Proper identification of each specimen shall include as a minimum, the worker's name and the date and time of collection.

Analytical

<u>Principle of the method</u>   Urine samples are treated with perchloric acid at 95°C to hydrolyze the phenol conjugates, phenyl sulfate, and phenyl glucuronide, formed as detoxification products following benzene absorption. The total phenol is extracted with diisopropyl ether and the phenol concentration is determined by gas chromatography analysis of the diisopropyl ether extract.

<u>Apparatus</u>

1. Gas chromatograph with a flame ionization detector and equipped with a 5-ft × 3/16-in column packed with 2-w/w polyethylene glycol adipate on universal 'B' support. Operating conditions are as follows:

Column temperature	150°C
Detector temperature	200°C
Injection port temperature	200°C
Carrier gas	Nitrogen
Carrier gas flowrate	60 ml/min

2. Water bath
3. Glass-stoppered, 10-ml volumetric flasks
4. 1-ml, 2-ml, and 5-ml volumetric pipets
5. 5-μl syringe

<u>Reagents</u>

1. Phenol
2. Perchloric acid
3. Diisopropyl ether
4. Distilled water

<u>Procedure</u>

Hydrolysis of phenol conjugates: Pipet 5 ml of urine into a 10-ml, glass-stoppered, volumetric flask. Add perchloric acid, mix by swirling, and transfer the lightly stoppered flask to a water bath at 95°C. After 2 hr, remove the flask from the water bath and allow it to cool at room temperature.

Diisopropyl ether extraction of phenol and cresols: Pipet 1 ml of diisopropyl ether into the flask and adjust the volume to 10 ml with distilled water. Shake vigorously for 1 min to extract the phenol and cresols. Allow the aqueous and ether layers to separate.

Gas chromatographic analysis for phenol: Inject 5 μl of the diisopropyl ether layer into the gas chromatograph and record the attenuation and area of the phenol peak. Under the conditions described, phenol is eluted in 100 sec, orthocresol in 130 sec, and meta- and paracresols in 320 sec.

<u>Standards preparation</u> A 50-mg/liter standard aqueous solution of phenol is prepared. A 5-ml aliquot of the standard solution is then subjected to the hydrolysis, extraction, and gas chromatographic analysis procedures described under "Procedure" above.

<u>Calculations</u> Determine the phenol concentration in the urine by comparing the gas chromatographic peak area of the sample with that of the 50-mg/liter standard and adjust the value to a specific gravity of 1.024.

<u>Specific gravity correction</u> Due to the magnitude of correction which is required, samples having uncorrected specific gravities less than 1.010 shall be rejected and additional samples shall be obtained.

Based on a survey of a large population in the United States in connection with urinary lead excretion, Levine and Fahy found the mean specific gravity to be 1.024. Many investigators throughout the world now use this figure. Buchwald [45] in 1964 determined the mean specific gravity for residents in the United Kingdom to be 1.016, a value now frequently used for Northern Europeans. The importance of specific gravity adjustments can be seen in that a specific gravity of 1.016 will give results having two-thirds the value of those corrected to 1.024. It is important, therefore, that a value be chosen for standardization; since greater acceptance seems to be for 1.024, this value has been selected for adjustment of urinary concentrations of benzene recommended for biological monitoring.

$$\text{Corrected concentration} = \frac{\text{observed concentration} \times 24}{\text{last two digits of specific gravity (e.g., } 1.0\underline{21})}$$

AIR SAMPLING AND ANALYSIS METHODS

Air sampling methods for benzene are varied and dependent on the type of sample required and order of magnitude of concentration to be sampled.

For a simple screening study, Level 1 investigation [51], the grab sample technique, is easily used. An evacuated bomb is often used. Benzene laden air can also be pumped through a bomb for a time period to allow the internal atmosphere to become pure sampled air. A sampling bag can also be filled by a pump with the quantity of air needed for analysis. Polyethylene bags are often used, but cannot store the sample for any extended period of time because the benzene vapor can permeate the walls of the bag. Teflon or other "exotic" lined bags are sometimes used to keep a sample intact.

For the more extensive qualitative, quantitative time-weighted average study, Level 2 investigation [51], the charcoal tube method is utilized. The recommended NIOSH sampling and analytical procedures for the determination of benzene include the charcoal tube method (these are given at the end of this chapter). Benzene is adsorbed on charcoal as the air sample passes over it. After the desired sampling period the charcoal, which is contained in a small tubelike vessel, is transferred to a test tube and the benzene desorbed with carbon disulfide. This desorbed solution is then analyzed on a gas chromatograph for the presence and quantity of benzene.

Silica gel can also be used to absorb benzene. However, unlike benzene, silica gel has a strong affinity for moisture. The moisture adversely affects the silica gel's ability to absorb benzene. Prefilters for moisture removal are thus required [50].

Bubblers, vessels in which the air sample is passed through nitrating solutions, are utilized to collect benzene samples. The nitrating solutions are contained in a series of bottles (scrubbers) and the air sample bubbles through each until all the benzene has been removed [6,51]. The portability of a series of scrubbing bottles is limited, thus the use of this method for industrial hygiene applications is not preferred.

Although sampling the ambient workplace atmosphere for benzene at various locations may result in adequate results, the preferred sampling industrial hygiene technique is to obtain breathing-zone samples over the entire exposure period. To obtain this type of sample, portable self-contained units including charcoal tube, pump, and battery are carried by the worker being analyzed. The units clip onto his belt with a sampling tube and probe attaching to his collar. In this manner, the worker's breathing zone or actual exposure to benzene can be determined.

After obtaining a sample, analysis of that sample is required. The preferred NIOSH analytical method is the gas chromatograph, and this procedure is also given at the end of this chapter.

Additional instrumentation, including mass spectrometry, flame ionization, and electron capture detector are often used in conjunction with the gas chromatograph. These other techniques are utilized to supplement the gas chromatograph, depending on the type of sample and the detection limit required. Table 5.7 outlines some sample types and the analytical techniques used. Also given are detection limits.

Other analytical methods used to detect benzene include ultraviolet spectrophotometry [51,52,53], colorimetry, and colorimetric indicator tubes (detector tubes).

The first step away from laboratory analysis of samples is the portable instrument. Flame ionization and infrared spectrometry instruments are the most common kinds. Portable instruments provide direct reading of benzene concentrations and rapid and relatively inexpensive analysis. Atmospheric levels of benzene can be screened repeatedly to determine when hazardous conditions exist. At that time more elaborate laboratory techniques can be utilized [51].

A typical infrared portable analyzer is shown in Fig. 5.2. This instrument utilizes an activated charcoal filter cartridge and infrared filters. These variable infrared filters can be set at any wavelength within their range and absorbance measured in parts per million.

TABLE 5.7 Various Sample Types and Analytical Techniques

Sample	Extraction solvent	Analytical technique
Air absorbed on methanol	Methanol	UV DL ~ 10 ppm
Air absorbed on silica gel	n-Heptane	UV
Air absorbed on silica gel	Ethanol	GC DL ppm
Air absorbed on silica gel	Isopropyl alcohol	UV DL: 2-40 ppm
Air absorbed on silica gel	--	Detector tube DL: ~0.005 mg/liter
Air absorbed on charcoal	Solvent	GC DL: 20 ppm GC-MS
Air absorbed on charcoal	Carbon disulfide	GC-FI DL: 20 ppm GC-MS DL: 50 ppb

Source: From Ref. 4.

The most important aspect of any portable instrument is calibration and reproducibility. Calibration charts and graphs are required as shown in Fig. 5.3. As is the case with all instruments, if the calibration is in error, the results are worthless.

Detector tubes are measurement devices commonly used. Their accuracy is limited, but they can indicate the presence of a hazardous pollutant such as benzene. They can be effective when used in conjunction with a full-scale laboratory sampling program [51].

Figure 5.4 pictures a typical detector tube kit. A typical system consists of a detector tube and sampling pump. The tube is inserted into the pump and an air sample is drawn through the tube. A color change is effected upon contact with benzene. Some detector tubes are calibrated for direct reading. Others are supplied with color comparison charts. In either case the results must be determined quickly because colors may fade with time. Detector tubes are quick and economical, but those using them should consider their limitations.

Method for Air Sampling and Analytical Procedures for Determination of Benzene

The following sampling and analytical method for analysis of benzene in air employs adsorption on charcoal, followed by desorption, and gas chromatographic measurement. This recommended method is taken from the NIOSH publication, "Criteria for a recommended standard . . . Occupational Exposure to Benzene." NIOSH derived this method through the modification of work presented in Refs. 54 and 55.

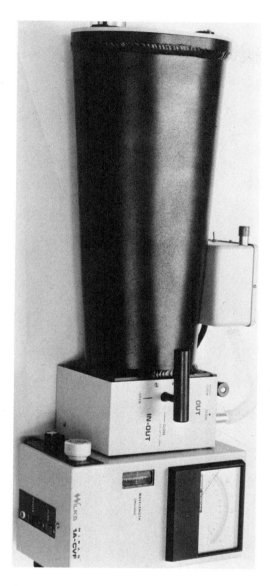

FIG. 5.2 Foxboro/Wilks MIRA-1A Gas Analyzer for benzene. (Courtesy Foxboro/Wilks, South Norwalk, Conn.)

Atmospheric Sampling

Equipment used   The sampling train is composed of a charcoal tube, a vacuum pump, and a flowmeter. A personal sampler pump or a dependable hand pump, e.g., a detector tube pump, may be calibrated to produce the desired volume of air.

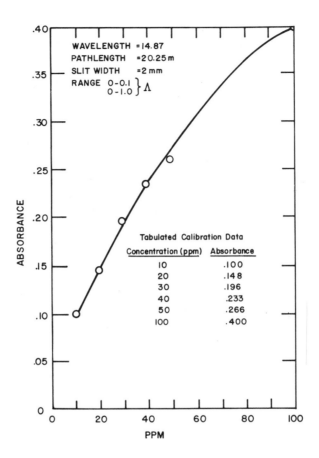

FIG. 5.3 Calibration chart and graph for benzene. (Courtesy of Foxboro/ Wilks, South Norwalk, Conn.)

Calibration of sampling instruments   Air-sampling instruments may be calibrated with a wet test meter or other suitable reference over a normal range of flowrates and pressure drops. The calibration is conducted at least annually and at any time following repairs or modifications to the sampling system. Similarly, wet test meters should be calibrated upon procurement, at least annually, and after each repair. Calibration curves shall be established for each sampling pump and shall be used in adjusting the pumps prior to field use. The volumetric flowrate through the sampling system shall be spot checked and the proper adjustments made before and during each study to assure obtaining accurate airflow data.

Flowmeter calibration test method:

1. With the wet test meter in a level position, check to ascertain that the water level just touches the calibration point on the meter. If the water level is

FIG. 5.4 Matheson-Kitagawa benzene gas detector system with detection limit of 0.5 ppm in air. (Courtesy of Matheson, Lyndhurst, N.J.)

low, add water 1–2° F warmer than room temperature to the fill point and run the meter for 30 min before calibration.

2. Check the voltage of the pump battery with a voltmeter to assure adequate voltage for calibration. Charge the pump battery if needed.

3. Break the tips of a charcoal tube to produce openings of at least 2 mm in diameter.

4. Assemble the calibration train in series, with the test meter, then the charcoal tube, and finally the pump.

5. Turn the pump on, adjusting the rotameter float to a selected reading on the rotameter scale. Wait until the float indicates a steady reading.

6. The pointer on the meter should turn clockwise and indicate a pressure drop of not more than 1.0 in. water. Operate the system 10 min before starting the calibration. If the pressure is greater, recheck the system.

7. Data for the calibration include the serial number; meter reading—start and finish; starting time, finish time, and elapsed time; air temperature; barometric pressure; serial number of the pump and rotameter; the name of the person performing the calibration; and the date.

8. Adjust the rotameter float to at least three other readings and record the pertinent data in step seven at each reading.

9. Correct the readings to standard conditions of pressure and temperature by means of the gas law equation.

10. Use graph paper to plot the actual airflow and the rotameter readings. Determine the rotameter reading which will result in a 1-liter/min flowrate for the pump being calibrated.

<u>Sampling procedure</u> The equipment should be set up in a proper locale. The tips of the charcoal tube are broken off producing openings of at least 2 mm in diameter; the filled end of the tube is inserted toward the pump. The tube should always be in a vertical position during sampling. The pump is started and a 10-liter sample is taken at a flowrate of 1 liter/min. Slower flowrates may be used to lengthen the sampling period but the 1-liter/min rate should not be exceeded. After the sample is taken, each end of the tube should be capped (plastic caps are provided with commercial tubes). The samples will remain stable for at least 2 wk which permits shipment for analysis; however, samples should be analyzed as soon as possible in keeping with good laboratory practices.

Analytical

<u>Principle of the method</u> A known volume of air is drawn through a charcoal tube to trap the organic vapors present. The charcoal in the tube is transferred to a small test tube and desorbed with carbon disulfide and an aliquot of the desorbed sample is injected into a gas chromatograph. The area of the resulting peak is determined and compared with areas obtained from the injection of standards.

<u>Range and sensitivity</u> The lower limit for benzene with instrument attenuation and splitter techniques is 0.01 mg for each sample. This value can be

Monitoring

lowered by reducing the attenuation or by eliminating the splitter. The upper limit value for benzene is 6.0 mg per sample. This value is the number of milligrams of benzene which the front section will collect before a significant amount passes to the backup section. The charcoal tube consists of two sections of activated charcoal separated by a section of urethane foam (see description in Apparatus, item no. 2, p. 90). If a particular atmosphere is suspected of containing a large amount of contaminant, it is recommended that a smaller than normal sampling volume be taken.

Interferences

1. When the amount of water in the air is so great that condensation actually occurs in the tube, organic vapors will not be trapped. Only water present as a mist is a problem, not water vapor.
2. Any compound in the gas chromatograph with the same retention time as benzene at the operating conditions described in this method could be considered an interference. This type of interference can be overcome by changing the operating conditions of the instrument.

Accuracy and precision   The accuracy and precision determined by a representative laboratory test with benzene was found to be:

	Accuracy	Precision
Motor-driven laboratory pump	7.6%	4.2%
Approved coal mine personal sampling pump (calibrated with no in-line resistance)	13.6%	10.1%
Approved coal mine personal sampling pump (calibrated with charcoal tube in line)	8.8%	11.6%

The accuracy includes single-day systematic error by one operator. Precision represents the single-day accuracy on several different tubes and includes tube-to-tube deviation under controlled laboratory conditions.

Advantages and disadvantages of the method   The sampling device is small, portable, and involves no liquids: one basic method is provided for determining many different organic solvents. Interferences are minimal and most can be eliminated by altering chromatographic conditions. In addition, the analysis is accomplished using a rapid instrumental method.

One disadvantage of the method is that the amount of sample which can be obtained is limited by the amount of benzene which the tube will hold before overloading as indicated by benzene recovery at the outlet end of the tube.

Also, the precision is limited by the reproducibility of the pressure drop across the tubes, which affects the flowrate, thus causing the volume to be imprecisely measured.

Apparatus

1. An approved coal mine dust personal sampling pump or any vacuum pump whose flow can accurately be determined at 1 liter/min or less for an area sample.
2. Charcoal tubes: Glass tubes with both ends flame sealed, 7 cm long with a 6-mm OD and a 4-mm ID, containing two sections of 20/40 mesh activated charcoal separated by a 2-mm portion of urethane foam. The absorbing section contains 100 mg of charcoal, the backup section, 50 mg. A 3-mm portion of urethane foam is placed between the outlet end of the tube and the backup section. A plug of glass wool is placed in front of the absorbing section. The pressure drop across the tube must be less than 1 in. of mercury at a flowrate of 1 liter/min. Tubes with the above specifications are commercially available.
3. Gas chromatograph equipped with a flame ionization detector.
4. Column (20 ft × 1/8 in.) with 10% FFAP stationary phase on 80/100 mesh acid washed DMCS Chromosorb W solid support.
5. A mechanical or electronic integrator or a recorder and some method for determining peak area.
6. Small glass-stoppered test tubes or equivalent tubes.
7. 10-μl syringe (and other convenient sizes) for preparation of standards.

Reagents

1. Spectroquality carbon disulfide
2. Benzene, preferably chromatoquality grade
3. Bureau of Mines Grade A helium
4. Prepurified hydrogen
5. Filtered compressed air

Procedure

Cleaning of equipment: All equipment used for the laboratory analysis should be washed in detergent followed by tap and distilled water rinses.

Collection and shipping of samples: Both ends of the charcoal tube are broken to provide openings of at least 2 mm (one-half the ID of the tube). The smaller section of charcoal in the tube is used as a backup section and is, therefore, placed nearest the sampling pump. Tubing may be used to connect the back of the tube to the pump, but no tubing must ever be placed on the front of the charcoal tube. Because of the high resistance of the charcoal tube, the sampling method places a heavy load on the personal sampling pump; therefore, it should not be assumed that the pump will run a full 8 hr without a recharging of the battery.

One or more charcoal tubes serving as blanks are treated in the same manner as the sample tubes (break, seal, ship) except that no air is drawn through them.

If bulk samples are submitted in addition to charcoal tubes, they are to be shipped in a separate container.

Analysis of samples:

1. Preparation   Each charcoal tube is scored with a file and broken open in front of the first section of charcoal. The glass wool is removed and discarded, the charcoal in the first (larger) section is transferred to a small stoppered test tube, the foam separating section is removed and discarded, and the second section is transferred to another test tube. The two charcoal sections are then analyzed separately.

2. Desorption   Prior to analysis, 0.5 ml of carbon disulfide is pipetted into each test tube to desorb the benzene from the charcoal. Desorption is complete in 30 min if the sample is stirred occasionally. EXTREME CAUTION MUST BE EXERCISED AT ALL TIMES WHEN USING CARBON DISULFIDE BECAUSE OF ITS HIGH TOXICITY AND FIRE AND EXPLOSION HAZARDS. IT CAN BE IGNITED BY HOT STEAM PIPES. ALL WORK WITH CARBON DISULFIDE MUST BE PERFORMED UNDER AN EXHAUST HOOD.

3. Gas chromatographic conditions   Typical operating conditions for a gas chromatograph are:

> 85 cc/min (70 psig) helium carrier gas flow, 65 cc/min (24 psig) hydrogen gas flow to detector, 500 cc/min (50 psig) airflow to detector, 200°C injector temperature, 200°C manifold temperature (detector), 90°C oven temperature isothermal, and either dual column differential operation or uncompensated mode.

4. Injection   To eliminate difficulties arising from blowback or distillation within the syringe needle, the solvent flush injection technique is employed to inject the sample into the gas chromatograph. The 10-μl syringe is first flushed with solvent several times to wet the barrel and plunger, then 3 μl of solvent is drawn into the syringe to increase the accuracy and reproducibility of the injected sample volume. Next, the needle is removed from the solvent and the plunger is pulled back about 0.2 μl to separate the solvent flush from the sample with an air pocket to be used as a marker. The needle is then immersed in the sample and a 5-μl aliquot is withdrawn. Prior to injection in the gas chromatograph, the plunger is pulled back a short distance to minimize sample evaporation from the needle tip. Duplicate injections should be made of each sample and the standard. No more than a 3% difference should result in the peak areas that are recorded.

5. Measurement of area   The area of the sample peak is measured by an electronic integrator or some other suitable form of area measurement and preliminary sample results are read from a standard curve prepared as outlined below.

Standards preparation and desorption efficiency

Preparation of standards: It is convenient to prepare standards in terms of mg/0.5 ml of carbon disulfide because this is the quantity used for benzene desorption from the charcoal. To prepare a 0.3 mg/0.5 ml standard, 6.0 mg of benzene (converted to microliters for easy measurement) is injected into exactly 10 ml of carbon disulfide in a glass-stoppered flask. The excess quantity of benzene is used to minimize error due to carbon disulfide volatility. A series of standards is then prepared, varying in concentration over the desired range, and analyzed under the same gas chromatographic conditions and during the same time period as the unknown samples. Curves are established by plotting concentration vs average peak area.

Determination of desorption efficiency: The desorption efficiency, i.e., the percentage of benzene desorbed from the charcoal, is determined only once, provided the same batch of charcoal is always used.

Activated charcoal, equivalent to the amount in the first section of the sampling tube (100 mg), is measured into a 2-in., 4-mm ID glass tube, flame sealed at one end, and capped with a paraffin film or equivalent at the open end. A known volume of benzene, usually equivalent to that present in a 10-liter sample at a concentration equal to the federal standard, is injected directly into the activated charcoal with a microliter syringe and the tube again capped with more paraffin film. A minimum of five tubes are prepared in this manner and allowed to stand for at least 1 day to assure complete adsorption of the benzene onto the charcoal. These tubes are desorbed and analyzed in exactly the same manner as the sampling tubes.

The results of each analysis are compared to the standards to determine the average percentage (desorption efficiency) that is desorbed. The desorption efficiency is then used as a factor in all sample analyses. The desorption efficiency, determined in this manner, has been shown to be essentially the same as that obtained by analysis of a known amount of benzene vapor trapped on the charcoal and the determined value, therefore, is used because of its simplicity. Each laboratory should determine its own desorption efficiency. For comparison purposes, NIOSH determined a value of 96% for benzene on one batch of charcoal.

Calculations

1. Read the weight in milligrams corresponding to each peak area from the standard curve. No correction is necessary for the volume injected, since it is the same for both the sample determination and the standard curve.

2. The weight of benzene on the front section of the blank is subtracted from the weight determined for the front section of each sample; a similar procedure is followed for the backup sections. Amounts present on the front and backup sections of the same tube are then added together to determine the total amount detected in the sample. This total weight is then divided by the desorption efficiency to determine the corrected total number of milligrams in the sample.

Milligrams are converted into ppm by volume in the air sampled by the following equation at 25°C and 760 mmHg:

$$\text{ppm} = \frac{24{,}450 \text{ ml/mol} \times \text{mg/liter}}{\text{mol wt}}$$

For a 10-liter air sample of benzene:

$$\text{ppm} = \frac{24{,}450 \text{ ml/mol} \times \text{mg in sample/10 liters}}{78.11 \text{ g/mol}}$$

$$\text{ppm} = 31.30 \times \text{mg in sample}$$

6
Health Effects

ENTRANCE TO THE BODY

Benzene has been considered a bone marrow poison for more than 100 years and has long been suspected to be a cause of leukemia. Acute exposure to benzene affects the central nervous system. Chronic exposure results in damage to the blood-forming (hematopoietic) systems. These effects are the result of benzene entering the body.

Benzene can enter the body through ingestion, inhalation, and skin absorption. Accidental ingestion of benzene is treated as a poisoning. Swallowing benzene can cause sickness and even death.

Dermal absorption of benzene is relatively slight. When in direct contact with the skin, it can cause congestion in the capillaries effecting a reddish color. This condition is known as erythema. A more progressive condition could result in blisters. Continuous contact with benzene could cause the skin to dry out and scale. Dermatitis is a skin condition which can be derived from skin contact with benzene. These conditions are common precursors to other skin infections. Mucous membranes of the eyes, nose, and lungs are irritated by high atmospheric concentrations of benzene [4,15,56].

Inhalation is the most common form in which benzene is admitted to the body. Benzene vapor levels in excess of 20,000 ppm have been fatal within minutes. Acute exposure to benzene is the inhalation of high concentrations; chronic exposure is the prolonged exposure to low concentrations of benzene vapors (< 5000 ppm). Generally, acute exposure effects are considered to be reversible while chronic exposure effects are considered to be irreversible.

ACUTE EXPOSURE EFFECTS

High concentrations of benzene inhaled affect the central nervous system. At high levels benzene has an anesthetic effect [4]. Fatalities due to benzene are the result of its concentrating in the brain lipid cells effecting circulatory problems, in some instances convulsions and/or paralysis, then loss of consciousness, coma and finally death. Damage to brain tissue has been found to be a primary cause of death. However, damaged tissues have also been found in the lungs, around the heart, urinary tract and mucous tissues.

Nonfatal acute exposures are characterized initially by excitation, nervous energy and exhilaration. This period of euphoria may be followed by headaches, insomnia, tiredness, nausea, depression, and finally vertigo.

Heart attacks and/or unconsciousness are not uncommon. Circulatory problems have effected nervous disorders that have lasted several weeks after exposure. Such effects might include loss of breath, irritability, dizziness, and balance disorders [14,15,56,58,59].

Exposure to benzene concentration on the order of 250-500 ppm has been found to cause nausea, drowsiness, and/or a confused state of mind [14].

These physical symptoms of acute exposure show no preference to gender or age. They are normally short term, with the victim quickly recovering when brought to clean, fresh air (or given oxygen).

CHRONIC EXPOSURE EFFECTS

Long-term exposures to low concentrations of benzene are of primary concern because of the damage done to the blood-making (hematopoietic) system. Chronic effects generally result from benzene levels below those that might produce acute effects making them even more sinister.

Benzene is detrimental to the bone marrow, damaging the blood-making systems. Its toxicity is generally associated with blood abnormalities, including:

Anemia
Thrombocytopenia
Leukopenia
Low red blood cell count
Low platelets
Low white blood cell count

Other blood alterations may include low hemoglobin levels, inhibition of enzyme activity, cellular disruptions [60], and chromosome aberrations [61].

These abnormalities may be severe or mild; they may be intermittent disfunctions or prove fatal. However, the basic toxic mechanism of benzene is unknown.

Conclusive evidence is available concerning benzene exposure and leukemia [14]. Leukemia, a cancer of the blood-making system; and nonfunctioning bone marrow, aplastic anemia, are related to benzene exposure. It has shown a strong correlation between benzene and blood abnormalities which has not been indicated for other substances in the same work environment. Furthermore, hematoxicity cases have increased in industries in which benzene is introduced and decreased in those in which benzene has been removed. Studies have shown animal exposure to benzene to cause marrow disorders [14,15].

Leukemia

Leukemia is listed in the International Lists of Causes of Death as a cancer of the white blood cells. It is the term for a group of diseases characterized by abnormal, immature white blood cells in the circulating blood, leukemic cells in the bone marrow, and anomalies of the liver, spleen, and other tissue.

There is virtually no recovering from leukemia once its presence has been discovered. There are quite a number of variations of leukemic-related diseases. These variations may be acute or chronic, produce abnormal cells, or effect cellular aberrations.

The most common benzene-associated leukemia is myelogenous leukemia, also known as acute myeloid leukemia, acute granulocytic leukemia, and acute myeloblastic leukemia. Studies linking benzene exposure to leukemia are quite prevalent and have produced evidence that is considered conclusive [14, 62-68]. These studies have been performed on groups of workers in particular industries and have indicated a rise in leukemia cases with the usage of benzene. Upon the removal of benzene from the work environment, leukemia cases were reduced.

Shoemakers were found to progress through the effects of benzene after it was introduced in adhesives. They developed aplastic anemia, and then acute myelogenous leukemia over a 10-yr period. The number of leukemia instances decreased after benzene was removed from the adhesive.

Other shoemakers exposed to 200-500 ppm of benzene were found to develop leukemia in abnormal numbers.

The rotogravure industry in Milan, Italy experienced a significant increase in leukemia cases after certain benzene laden inks and solvents were utilized. Ambient benzene levels were about 200-1500 ppm.

The painting industry has also experienced an increase in leukemia cases when benzene is used as a solvent in the paints.

In some studies the associated risk of benzene-exposed workers to develop leukemia is 20 times greater than that of nonexposed workers [69].

Aplastic Anemia and Pancytopenia

Aplastic anemia, while considered by many to refer to benzene toxicity, is a condition of reduced hematopoietic (blood-making) system cells in the bone marrow. Pancytopenia is a reduced level in the blood of red and white blood cells and platelets.

Pancytopenia is the most common chronic benzene exposure effect. A more common form is known as anemia. The effect of anemia is a reduction in the blood's capacity to deliver oxygen to the body. Physical symptoms include weakness and fatigue. However, more extensive anemic conditions result in disruptions of the cardiovascular system.

Another form of pancytopenia is leukopenia. This reduction in white blood cells is also a reduction in the body's defense forces against infections. Infections are of utmost concern for individuals with leukopenia.

Pancytopenia, in the form of a low platelet count, reduces the blood's ability to clot resulting in bleeding and its associated problems.

Health Effects

Numerous studies have been performed which link benzene exposure to pancytopenia and aplastic anemia [14,42,62,70-80].

Rotogravure printing plants utilizing benzene-containing inks and solvents reported benzene concentrations in the workplace to average over 100 ppm. The workers experienced headaches, tiredness, breathlessness, and other acute benzene poisoning effects. Nearly 20% were discovered to have some form of pancytopenia.

Rubber manufacturing and rubber products industry with a high usage of benzene have found a significant number of cases of hematological toxicity.

Other work environments where benzene is used and workers are exposed have been shown to have higher than normal hematological toxicity rate.

Benzene exposure studies on dogs, mice, rats, guinea pigs, and rabbits have all shown these animals to develop blood disorders. Low red blood cell counts, low white blood cell counts, low platelet numbers, and various bone marrow diseases were determined to be effected in these animals by their exposure to benzene.

Generally, pancytopenia cases begin when benzene is introduced into the environment and decline after its removal. An interesting note is the Italian rotogravure industry. It has been reported that upon substitution of benzene with toluene, not a single case of work-related leukemia or aplastic anemia has been determined [69].

Other Benzene Exposure Effects

Although benzene exposure is more commonly associated with pancytopenia, aplastic anemia, and leukemia, other effects have also been cited. Benzene exposure, through epidemiological studies, has been reported to contribute to a number of disorders.

1. Leukocyte function abnormalities. These adversely influence the body's ability to fight infections [14,81,82].
2. Diseases of the lymphatic system. The lymphatic system functions to protect the body from infectious agents and possibly inhibit the development of malignancies. Benzene is believed by some to alter the immune surveillance system such that "enemy" versus "friendly" substances become indistinguishable [14,83,84].
3. Abnormal platelet function. Platelet abnormalities could result in coagulation anomalies and bleeding problems.
4. Destruction of red blood cells. Red blood cell damage results in various blood processes [14].
5. Chromosomal damage. Benzene has been observed to effect chromosomal aberrations. Evidence suggests that benzene is not only a carcinogen but a mutagen also. Chromosomal damage can result in the replication of a cell lesion, thus effecting a mutagen. It can interfere with a cell's reproduction and lead to death. DNA and RNA interferences due to benzene exposure have also been reported [14,70,85].

Apparently there is no exposure dosage or duration that can be conclusively pinpointed as generating benzene-associated problems. It is not known whether the culprit is benzene or any of its metabolic derivatives in causing the damage. Nor is the manner in which its toxicity acts understood. Further, benzene shows no affinity for any particular gender, age, or race.

BENZENE TOXICITY DATA

The following toxicity data have been reported in the NIOSH Registry of Toxic Effects of Chemical Substances [86].

Route of exposure	Species	Description of exposure
Inhalation	Human	LDL_0: 20,000 ppm/5M
Inhalation	Human	TCL_0: 210 ppm TFX:BLD
Inhalation	Man	TDL_0: 2100 mg/m^3 4YI TFX:CAR
Oral	Rat	LD_{50}: 3800 mg/kg
Inhalation	Rat	LC_{50}: 10,000 ppm/7H
Intraperitoneal	Rat	LDL_0: 1150 mg/kg
Oral	Mouse	LD_{50}: 4700 mg/kg
Skin	Mouse	TDL_0: 1200 gm/kg/49WI TFX:NEO
Intraperitoneal	Mouse	LD_{50}: 468 mg/kg
Oral	Dog	LDL_0: 2000 mg/kg
Intraperitoneal	Guinea pig	LDL_0: 530 mg/kg
Subcutaneous	Frog	LDL_0: 1400 mg/kg
Aquatic toxicity rating		$TL_m 96$: 100-10 ppm

Tables 6.1, 6.2, 6.3, 6.4, and 6.5 explain toxicity data abbreviations and give various definitions of toxicology terms.

TDL_0 Toxic Dose Low

The lowest dose of a substance, as published, introduced by any route other than inhalation over any given period of time and reported to produce any toxic effect in humans or to produce carcinogenic, teratogenic, mutagenic, or neoplastigenic effects in humans or animals.

Health Effects

TABLE 6.1   Key to Abbreviations

ALR—allergenic effects
AQTX—aquatic toxicity
BCM—blood clotting mechanism effects
bdw—wild bird species
BLD—blood effects
BPR—blood pressure effects
brd—bird (domestic or lab)
C—continuous
cc—cubic centimeter
CL—ceiling concentration
CAR—carcinogenic effects
cat—cat
chd—child
ckn—chicken
CNS—central nervous system effects
COR—corrosive effects
ctl—cattle
CRIT DOC—criteria document
CUM—cumulative effects
CVS—cardiovascular effects
D—day
dck—duck
DDP—drug dependence effects
DEF—definition
dog—dog
dom—domestic
DOT—Department of Transportation
EPA—Environmental Protection Agency
EYE—eye effects

frg—frog
GIT—gastrointestinal tract effects
GLN—glandular effects
g—gram
gpg—guinea pig
grb—gerbil
H—hour
ham—hamster
hmn—human
I—intermittent
IARC—International Agency for Research on Cancer
iat—intraarterial
ial—intraaural
ice—intracerebral
icv—intracervical
idr—intradermal
idu—intraduodenal
ihl—inhalation
imp—implant
ims—intramuscular
inf—infant
ipc—intraplacental
ipl—intrapleural
ipr—intraperitoneal
irn—intrarenal
IRR—irritant effects
isp—intraspinal
itr—intratracheal
ivg—intravaginal

TABLE 6.1 (continued)

ivn—intravenous
kg—kilogram (one thousand grams)
LC_{50}—lethal concentration 50% kill
LCL_0—lowest published lethal concentration
LD_{50}—lethal dose 50% kill
LDL_0—lowest published lethal dose
mam—mammal (species unspecified)
man—man
M—minute
m^3—cubic meter
mg—milligram (one thousandth of a gram; 10^{-3} g)
mky—monkey
ml—milliliter
MMI—mucous membrane effects
mppcf—million particles per cubic foot
MSK—musculo-skeletal effects
MTH—mouth effects
mul—multiple routes
mus—mouse
MUT—mutagenic effects
NEO—neoplastic effects
ng—nanogram (one billionth of a gram; 10^{-9} g)
ocu—ocular
orl—oral
OSHA—Occupational Safety and Health Administration
par—parenteral

pg—picogram (one trillionth of a gram; 10^{-12} g)
pgn—pigeon
pig—pig
Pk—peak concentration
PNS—peripheral nervous system effects
ppb—parts per billion (v/v)
pph—parts per hundred (v/v) (percent)
ppm—parts per million (v/v)
ppt—parts per trillion (v/v)
preg—pregnancy
PSY—psychotropic effects
PUL—pulmonary system effects
qal—quail
rat—rat
RBC—red blood cell effects
rbt—rabbit
rec—rectal
SCP—Standards Completion Program
scu—subcutaneous
SKN—skin effects
sql—squirrel
sup—super script
SYS—systemic effects
TCL_0—lowest published toxic concentration
TDL_0—lowest published toxic dose
TER—teratogenic effects
TFX—toxic effects
TLV—threshold limit value

Health Effects

TABLE 6.1 (continued)

TOX REV—toxicology review	USOS—U.S. Occupational Health Standard
trk—turkey	
TWA—time-weighted average	W—week
TXDS—qualifying toxic dose	
ug—microgram (one millionth of a gram; 10^{-6} g)	WBC—white blood cell effects
	Y—year
unk—unreported	
UNS—toxic effects unspecified in source	

Source: From Ref. 86.

TCL_0 Toxic Concentration Low

Any concentration of a substance in air to which humans or animals have been exposed for any given period of time, which has been reported to produce any toxic effect in humans or to produce a carcinogenic, teratogenic, mutagenic, or neoplastigenic toxic effect in animals or humans.

LDL_0 Lethal Dose Low

The lowest dose of a substance other than LD_{50} introduced by any route other than inhalation over any given period of time and reported to have caused death in humans or animals introduced in one or more divided portions.

LD_{50} Lethal Dose Fifth

A calculated dose of a chemical substance which is expected to cause the death of 50% of an entire defined experimental animal population, as determined from the exposure to the substance, by any route other than inhalation, of a significant number from that population. Other lethal dose percentages, such as LD_1, LD_{10}, LD_{30}, LD_{99}, and others, may be published in the literature for the specific purposes of the author.

LCL_0 Lethal Concentration Low

The lowest concentration of a substance, other than LC_{50}, in air which has been reported to have caused death in humans or animals. The reported concentrations may be for periods of time less than 24 hr (acute) and greater than 24 hr (subacute and chronic).

TABLE 6.2 Routes of Administration to, or Exposure of, Animal Species to Toxic Substances

Route	Abbreviation	Definition
Intraarterial	iat	Administration into the artery
Intraaural	ial	Administration into the ear
Intracerebral	ice	Administration into the cerebrum
Intracervical	icv	Administration into the cervix
Intradermal	idr	Administration within the dermis by hypodermic needle
Intraduodenal	idu	Administration into the duodenum
Inhalation	ihl	Inhalation in chamber, by cannulation, or through mask
Implant	imp	Placed surgically within the body—location described in reference
Intramuscular	ims	Administration of dose into the muscle by hypodermic needle
Intraplacental	ipc	Administration into the placenta
Intrapleural	ipl	Administration of dose into the pleural cavity by hypodermic needle
Intraperitoneal	ipr	Administration into the peritoneal cavity
Intrarenal	irn	Administration into the kidney
Intraspinal	isp	Administration into the spinal canal
Intratracheal	itr	Administration into the trachea
Intravaginal	ivg	Administration into the vagina
Intravenous	ivn	Administration of dose directly into the vein by hypodermic needle
Ocular	ocu	Administration directly onto the surface of the eye or into the conjunctival sac
Oral	orl	Per os, intragastric, feeding introduction with drinking water
Parenteral	par	Administration into the body through the skin. Reference cited is not specific concerning the route used. Could be ipr, scu, ivn, ipl, ims, irn, or ice
Rectal	rec	Administration of dose by way of rectum to the rectum or colon in form of enema, suppository
Skin	skn	Application to the intact skin, dermal, cutaneous
Subcutaneous	scu	Administration under the skin
Unreported	unk	Dose, but not route, is specified in the reference

Source: From Ref. 86.

TABLE 6.3 Units of Time for Dose Administration

Unit	Abbreviation	Limits of use and definitions
Minute	M	1-99 minutes
Hour	H	1-99 hours
Day	D	2-99 days
Week	W	2-99 weeks
Year	Y	1-99 years
Continuous	C	Indicates that the exposure was continuous over the time administered, such as ad lib. feeding exposures or 24-hr, 7 days/wk inhalation exposures
Intermittent	I	Indicates that the dose was administered during discretely separate periods, such as daily, weekly

Source: From Ref. 86.

TCL_0 Toxic Concentration Low

Any concentration of a substance in air to which humans or animals have been exposed for any given period of time, which has been reported to produce any toxic effect in humans or to produce a carcinogenic, teratogenic, mutagenic, or neoplastigenic toxic effect in animals or humans.

LDL_0 Lethal Dose Low

The lowest dose of a substance other than LD_{50} introduced by any route other than inhalation over any given period of time and reported to have caused death in humans or animals introduced in one or more divided portions.

LD_{50} Lethal Dose Fifty

A calculated dose of a chemical substance which is expected to cause the death of 50% of an entire defined experimental animal population, as determined from the exposure to the substance, by any route other than inhalation, of a significant number from that population. Other lethal dose percentages, such as LD_1, LD_{10}, LD_{30}, LD_{99}, and others, may be published in the literature for the specific purposes of the author.

TABLE 6.4 Notations Descriptive of the Toxicology

Abbreviations	Definitions (not limited to effects listed)
ALR	Allergic systemic reaction such as might be experienced by individuals sensitized to penicillin
BCM	Blood clotting mechanism—any effect which increases or decreases clotting time
BLD	Blood effects—effect on all blood elements, electrolytes, pH, protein, oxygen carrying or releasing capacity
BPR	Blood pressure effects—any effect which changes any aspect of blood pressure away from normal—increased or decreased
CAR	Carcinogenic—producing cancer—a cellular tumor the nature of which is fatal or is associated with the formation of secondary tumors (metastasis)
CNS	Central nervous system—includes effects such as headaches, tremor, drowsiness, convulsions, hypnosis, anesthesia
COR	Corrosive effects—burns, desquamation
CUM	Cumulative effect—where substance is retained by the body in greater quantities than is excreted, or the effect is increased in severity by repeated body insult
CVS	Cardiovascular effects—such as an increase or decrease in the heart activity through effect on ventricle or auricle; fibrillation; constriction or dilation of the arterial or venous system
DDP	Drug dependence—any indication of addiction or dependence
EYE	Eye effects—irritation, diploplia, cataracts, eye ground, blindness by affecting the eye or the optic nerve
GIT	Gastrointestinal tract effects—diarrhea, constipation, ulceration
GLC	Glandular effects—any effect on the endocrine glandular system
IRR	Irritant effect—any irritant effect on the skin, eye, or mucous membrane
MMI	Mucous membrane effects—irritation, hyperplasia, changes in ciliary activity
MSK	Musculoskeletal effects—such as osteoporosis, muscular degeneration
MUT	Mutation or mutagenic—transmissible changes produced in the offspring
NEO	Neoplastic effect—the production of tumors

TABLE 6.4 (continued)

Abbreviation	Definitions (not limited to effects listed)
PNS	Peripheral nervous system effects
PSY	Psychotropic—exerting an effect upon the mind
PUL	Pulmonary system effects—effects on respiration and respiratory pathology
RBC	Red blood cell effects—includes the several anemias
SKN	Skin effects—such as erythema, rash, sensitization of skin, petechial hemorrhage
SYS	Systemic effects—effects on the metabolic and excretory function of the liver or kidneys
TER	Teratogenic effects—nontransmissible changes produced in the offspring
TFX	Toxic effects—used to introduce the principal organ system affected as reported or the pathology
UNS	Unspecified effects—the toxic effects were unspecific in the reference
WBC	White blood cell effects—effects on any of the cellular units other than erythrocytes, including any change in number or form

Source: From Ref. 86.

LC_{50} Lethal Concentration Fifty

A calculated concentration of a substance in air, exposure to which for a specified length of time the death of 50% of an entire defined experimental animal population as determined from the exposure to the substance of a significant number from that population.

Aquatic Toxicity Rating ($TL_m 96$)

The concentration of a substance which will, within a specified period of time (generally 96 hr) kill 50% of the exposed test organisms. The concentration is usually expressed in parts per million (mg/liter). The bioassay may be conducted under static or continuous flow conditions [86]. Toxicity data for benzene is summarized in Table 6.6.

TABLE 6.5 Species—With Assumption for Toxic Dose Calculation from Nonspecific Data[a]

Species	Abbrev.	Age	Weight	Consumption Food g/day	Water ml/day (Approx.)
Bird—any domestic or laboratory bird reported but not otherwise identified	brd		1 kg		
Bird—wild bird species	bwd		100 g		
Cat, adult	cat		2 kg	100	100
Cattle	ctl		500 kg	10,000	
Chicken, adult (male or female)	ckn	8 wk	500 g	200	200
Child	chd	1–10 yr	20 kg		
Dog, adult	dog	52 wk	100 kg	250	500
Domestic animals: goat, sheep, horse	dom		100 kg	2500	
Duck, adult (domestic)	dck	8 wk	2.5 kg	250	500
Frog, adult	frg		33 g		
Gerbil	grb		100 g	5	5
Guinea pig, adult	gpb		500 g	30	85
Hamster	ham	14 wk	125 g	15	85
Human	hmn	13 yr	70 kg		
Infant	inf	4 wk	5 kg		

Health Effects

Mammal—species unspecified in reference					
Man	man	13 yr	70 kg		
Monkey	mky	2.5 yr	5 kg	400	500
Mouse	mus	8 wk	25 g	5	5
Pig	pig				
Pigeon	pgn	8 wk	500 g		
Quail (laboratory)	qal		100 g		
Rabbit, adult	rbt	12 wk	2 kg	100	330
Rat, adult female	rat	14 wk	200 g	10	20
Rat, adult male	rat	14 wk	250 g	15	25
Rat, adult sex unspecified	rat	14 wk	200 g	15	25
Rat, weaning	rat	3 wk	50 g	15	25
Squirrel	sql		500 g		
Toad	tod		100 g		
Turkey	trk		5 kg		
Woman	wmn	13 yr	50 kg		

[a] Values given here are within reasonable limits usually found in the published literature and are selected to facilitate calculations for data from publications in which toxic dose information has not been presented for an individual animal of the study. Data for lifetime exposure are calculated from the assumptions for adult animals for the entire period of exposure. For definitive dose data the reader must review the reference publication.

Source: From Ref. 86.

TABLE 6.6 Benzene Toxic Effects Summary

	Exposure route	Exposure concentration (ppm)	Exposure duration	Effects
Human	Oral	--	Acute	Death
	Oral	--	Acute	Mucous membrane irritation and systemic intoxication
	Dermal	Immersion of	Acute	Erythema, skin defatting, dry scaling, secondary infections
	Inhalation	20,000	5-10 min	Convulsions, paralysis, coma, and death
	Inhalation	High	Acute	Petechial hemorrhage in body tissues, respiratory tract infections, hypoplasia and hyperplasia in sternal bone marrow, kidney congestion, and cerebral edema, death
	Inhalation	Sublethal	Chronic	Insomnia, agitation, headache, dizziness, drowsiness, breathlessness, unsteadiness, irritability, vertigo, nausea, loss of appetite
	Inhalation	0.875 (odor threshold)	Acute	Brain electropotential enhancement
	Inhalation	Sublethal	Chronic	Anemia, thrombocytopenia, thrombocytopathy, leukopenia, low hemoglobin concentration, increased cell size, eosinophil count elevation
	Inhalation	Sublethal	Chronic	Stable and unstable chromosome aberrations

Health Effects

Subject	Route	Dose	Duration	Effect
Human	Inhalation	100	Chronic (work hours)	Leukopenia
Human	Inhalation	25	Chronic (work hours)	Lower hemoglobin levels, minor hematological deviations
Dog	Inhalation	Sublethal	Acute	Hypertension and vasomotor paralysis
Dog	Oral	Sublethal	Acute	Mucous membrane irritation, pulmonary edema and hemorrhage
Rat	Inhalation	20	6 hr/day 6 days/wk 5 1/2 months	Delay in conditioned response time
Rat	Inhalation	100	6 hr/day 5 days	Decreased incidence of spontaneous behavior
Rat	Inhalation	450-500	5 hr/day 10 days	Increase in cytechrome P450 and amino-pyrine demethylase activity
Rat	Inhalation	Threshold	Acute	Disturbed oxidation-reduction and albumin production
Rat (Pregnant)	Inhalation	0.3125-19.84	Chronic	Significant biochemical alterations in both pregnant female and fetus
Rat (Young adult)	Oral	0.71-1.23 g/kg	Acute	LD_{50}
Rat (Young adult)	Oral	3.4 g/kg	Acute	LD_{50}
Rat (Older adult)	Oral	4.9 g/kg	Acute	LD_{50}
Rat (older adult)	Oral	5.6 g/kg	Acute	LD_{50}
Rat (older adult)	Inhalation	44	7 hr/day 5 days/wk	Leukopenia

TABLE 6.6 (continued)

	Exposure route	Exposure concentration (ppm)	Exposure duration	Effects
Rat (older adult) (continued)	Inhalation	200	8 hr/day 5 days/wk 90 days	Leukopenia
	Inhalation	1000	23.5 hr/day, 105 hr	Body weight loss, nose and mouth hemorrhage, stomach distention, engorged blood vessels, reversal of polymorphonuclear: lymphocyte ratio
	Inhalation	88	204 days	Blood, bone marrow, spleen, and testes histopathological alterations
	Inhalation	40,000	20-35 min (5 exposures)	3/8 male Long Evans rats died
	Inhalation	10,000	12.5-30 min (1-17 days)	2/10 male Long Evans rats died
	Inhalation	44	5 hr/day (4 days/wk) (5-7 wk)	Slight leukopenia
Guinea pig	Inhalation	88	269 days	Blood, bone marrow, spleen, and testes histopathological alterations
Rabbit	Inhalation	80	243 days	Leukopenia and degeneration of seminiferous tubules

Inhalation	35,000–45,000	3.7 min	Lightly anesthetized
Inhalation	35,000–45,000	5 min	Excitation and tremors
Inhalation	35,000–45,000	36 min	Death
Inhalation	Sublethal	Chronic	Fatty bone marrow tissue, nucleic acid per unit weight of tissue decreased, synthesis of RNA and DNA increased
Inhalation	3.125	6 hr/day 14 days	Hypertrophy of smooth endoplasmic reticulum ribosome loss, disappearance of segmental distention of ergastoplasm, swelling and myelin degeneration of mitochondria
Inhalation	Sufficient to induce leukopenia		Reduced resistance to pneumonia and tuberculosis
Injection	Sufficient to induce leukopenia		Active acute infections
Inhalation	15.625	Chronic	Decreased phagocytic index and phagocytic number

Source: From Ref. 4.

BENZENE SUBSTITUTES' TOXICITY

The effect of stringent standards regulating the use of benzene is to force industry to seek substitutes. The most probable substitutes for benzene include heptane, hexane, methyl ethyl ketone, methylene chloride, toluene, and xylene [23].

Heptane and Hexane

These substitutes are common glue, varnish, and ink solvents. Inhalation of these compounds can cause central nervous system depression and chronic exposure can lead to degeneration of the nerves. Skin contact with benzene can produce erythema and hyperemia. Tables 6.7 and 6.8 are NIOSH summaries of the effects of hexane and heptane exposures on humans [87].

Methyl Ethyl Ketone (MEK)

This is a common industrial solvent but it apparently does not possess the extreme toxicology problems as benzene. Inhalation of MEK can cause nausea, headaches, and narcosis. It has an extremely pungent odor. Skin contact can be irritating and cause dryness.

Methylene Chloride

This is another common industrial solvent often substituted for benzene in paint removers. Methylene chloride interferes with oxygen transport between tissues and produces abnormalities in the central nervous system. Inhalation of methylene chloride can produce a wide range of body disorders. These effects may include visual impairment, headaches, muscular pains of the extremities, fatigue, breathlessness, chest pains, and heart palpitations [88]. Table 6.9 summarizes the effects of methylene chloride exposure.

Toluene

This is a solvent that is similar to the physical and chemical properties of benzene. However, its toxicological effects are not nearly as extreme. Acute exposure to toluene may result in stronger effects than benzene. Chronic toluene exposure effects are not as extreme as benzene [23].

Xylene

This is often substituted for benzene as a solvent and in rubber cement and paint removers. The acute toxicity of xylene is considered to be greater than benzene. Yet its chronic effects are considered to be less severe than benzene. However, research toxicology data are not complete on xylene and toluene, making relative toxicology ratings difficult [23].

TABLE 6.7 Summary of Hexane Exposure Effects on Humans

Routes of exposure	Subjects	Exposure concentration and duration	Effects
Respiratory	3-6 men and women	5000 ppm 10 min	Marked vertigo
	6 men and women	2500-1000 ppm 10-12 hr/day	Drowsiness in 0.5 hr, fatigue, loss of appetite in some, paresthesia in distal extremities
	93 men and women	2500-500 ppm	Sensory impairment in distal portion of extremities, muscle weakness, cold sensation of extremities in some, blurred vision, headache, easy fatigability, anorexia, weight loss by onset of polyneuropathy
	3-6 men and women	2000 ppm 10 min	No symptoms
	11 men and women	1000-500 ppm 3-6 months	Fatigue, loss of appetite in some, paresthesia in distal extremities
Dermal	5 men and women	Undiluted 5 hr	Blister formation, no anesthesia
	5 men and women	Undiluted 1 hr	Irritation, itching, erythema, pigmentation, swelling, painful burning sensation, reduced pain 90 min after removal
Respiratory, dermal, and oral	1 woman, 27 yr	1300-650 ppm 2 months	Frequent headaches, abdominal cramps, burning sensation of face, numbness of distal extremities, decreased left ulnar nerve conduction rate
	1 woman, 47 yr	1300-650 ppm 2 months	Abdominal cramps, numbness of distal extremities, paresthesia, bilateral foot and wrist drop, sensory impairment of extremities, decreased left ulnar nerve conduction rate
	1 woman, 46 yr	1300-650 ppm 4 months	Weakness in extremities, moderate weakness and sensory impairment of distal extremities, decreased left ulnar nerve conduction time

Source: From Ref. 87.

TABLE 6.8 Summary of Heptane Exposure Effects on Humans

Routes of exposure	Subjects	Exposure concentration and duration	Effects
Respiratory	3–6 men and women	5000 ppm 15 min	Marked vertigo, incoordination, hilarity for 30 min
	3–6 men and women	5000 ppm 7 min	Marked vertigo, incoordination of space, hilarity in some
	3–6 men and women	5000 ppm 4 min	Marked vertigo, inability to walk straight, hilarity
	3–6 men and women	3500 ppm 4 min	Moderate vertigo
	3–6 men and women	2000 ppm 4 min	Slight vertigo
	3–6 men and women	1000 ppm 6 min	Slight vertigo
Dermal	5 men and women	Undiluted 5 hr	Blister formation, no anesthesia
	5 men and women	Undiluted 1 hr	Irritation, itching, erythema, pigmentation, swelling, painful burning sensation in skin, reduced pain 120 min after removal

Source: From Ref. 87.

TABLE 6.9 Summary of Methylene Chloride Inhalation Exposure Effects

Concentration (ppm)	Exposure variables	Effects
Humans (Experimental)		
50-1000	1-7.5 hr/day 5 days/wk	COHb percentages proportional to exposure concentration and time
50-500	7.5 hr/day 5 days/wk	Increased affinity of Hgb for oxygen in proportion to exposure concentration
100 and 500	7.5 hr/day 5 days/wk	Slight blood lactic acid increase from exercise at 500 ppm, not 100 ppm
317 and 751	4 hr	Depressed CFF, decreased auditory vigilance performance
317, 470, 751	3-5 hr	Decreased performance of CFF, auditory vigilance, psychomotor tasks
Humans (Occupational)		
Unknown	1 subject, 13 yr (intermittent)	Irregular, severe leg and arm pains, hot flashes, vertigo stupor, poor night vision, anorexia, precordial pain, rapid pulse, shortness of breath, fatigue, 4,910,000 RBC, 6200 WBC; punctuate basophilia of 3500/million improved 6 wk after removal from work
Unknown	1 worker, 20 yr (intermittent)	Drowsiness, pains in head, tingling in hands and feet
Unknown	4 workers, acute exposure, probably 1-3 hr	3 workers hospitalized with eye, lung, and respiratory tract irritation, reduced Hgb and RBC counts; 1 worker died with veins of pia-arachnoid conspicuously engorged
Unknown	4 hr	Oppressive odor, irritation of eyes, excessive fatigue, weakness, sleepiness, light headedness, chilly sensations, nausea, shortness of breath, substernal pain, weakness, dry rales in chest, pulmonary edema

TABLE 6.9 (continued)

Concentration (ppm)	Exposure variables	Effects
Humans (Occupational)		
28-4896	33 workers, average of 2 yr exposure	Headache, fatigue, irritation of upper respiratory tract, conjunctiva, neurasthenic disorders, mild acute poisoning in 3, with unconsciousness in 1, sweet taste, heart palpitations
660-3600	1 worker, several hr/day for 5 yr	After 3 yr: burning pain around heart, restlessness, feeling of pressure, palpitations, forgetfulness, insomnia, feeling of drunkeness. After 5 yr: auditory and visual hallucinations, slight erythema of hands and underarms, diagnosed as having toxic encephalosis.
159-219 (average 183)	4 workers, 3 investigators	Increased alveolar CO at end of work day.

Source: From Ref. 88.

7

Benzene in the Environment

BENZENE FATE

The low boiling point and high vapor pressure of benzene cause its rapid evaporation at standard pressures and temperatures. Because of its ring structure benzene is basically stable in the atmosphere. It is relatively photochemically inactive. Thus, its contributions to photochemistry-induced pollutants are minor.

Benzene lends itself to precipitation scavenging and atmospheric rain wash out of benzene to the oceans is a most probable fate. However, it rapidly separates from water due to its high vapor pressure and evaporates. The half-life of benzene in water was found to be 37 min. This cycle would occur indefinitely were it not for its absorption or adsorption onto available surfaces [4,24,89].

Vegetation plays an important role in the absorption of atmospheric benzene. Some species of plant life can readily break down the benzene molecule into other compounds [25].

SOURCES EMITTING BENZENE INTO THE ENVIRONMENT

A summary of benzene annual emissions from various sources is given in Table 7.1. These sources of emissions range from the production of benzene, to its processing into its derivatives, to its storage and handling, and to its usage. In all activities associated with benzene, significant emissions are derived.

A significant source of benzene spill into the environment is oil spills. The benzene content of crude oil is approximately 0.2%. Tanker washings, accidents, exploration, production, transportation, handling, and wastes account for nearly 6 million tons of crude oil being discharged into the oceans each year. Of this, nearly 120,000 tons is benzene [4,90].

Another substantial source of benzene leakage is gasoline. The benzene content of gasoline varies between less than 1% to over 2%. Benzene is highly volatile and evaporates relatively fast as compared to other gasoline components.

TABLE 7.1 Summary of Benzene Annual Emissions During 1976

No.	Source	U.S. benzene emissions (tons/yr)
1	Petroleum refineries gasoline production (1976)	2,050
2	Crude oil operations	1,100
	Chemical products derived from benzene (1975)	
3	Nitrobenzene	3,821
4	Aniline	?
5	Styrene	3,358
6	Cumene	250
7	Maleic anhydride	17,400
8	Dichlorobenzene (o & p)	2,650
9	Monochlorobenzene	
10	Phenol	845
11	Solvent operations	80,500
	Storage tanks	
12	Gasoline	850
13	Benzene	300
14	Gasoline distribution (loading and unloading of storage tanks and transfer to service stations	1,903
15	Benzene distribution	1,000
16	Service stations (loading, car refueling and spillage)	7,260
17	Car exhausts (27% uncontrolled, 73% controlled)	186,000
18	Car evaporative losses	34,286
	Total	343,573

Source: From Ref. 25.

It is estimated that nearly 27,000 tons of benzene are annually lost to the environment due to evaporation from gasoline. In 1971, over 470,000 tons of benzene were emitted to the atmosphere due to motor vehicle exhausts [4].

The production of coke in 1974 effected benzene emissions of about 61,000 tons. Benzene emissions due to usage in commerce during 1971 were approximately 379,000 tons. Table 7.2 gives the quantities of benzene emissions to the environment during 1971 during benzene consumption practices. These totals were calculated through the use of the material balance method.

Table 7.3 gives the sources and amounts of benzene discharged to the environment during 1971. The total amount of benzene emitted should be reduced as a result of the OSHA benzene worker regulation. The low level of the benzene exposure standard (1 ppm TWA) and the strict monitoring requirements have directed many benzene users to utilize substitutes or develop alternate processing methods.

Benzene usage as a solvent has been widespread. It has been utilized in the rubber and tire, organic chemical, pharmaceutical, paint, varnish and lacquer, and various other industries.

A typical example of the problems associated with benzene is the paint utilized on the New Jersey Turnpike. Approximately 26% of the traffic paint type 111A is benzene. Each year about 27,000 gal of this paint is used, effecting 27 tons of benzene emissions [25].

BENZENE EXPOSURES

Benzene concentrations in the atmosphere due to gasoline bulk loading operations ranged from 0.3 to 3.2 ppm for gasoline with a benzene content of from 3.1 to 5.8% [57,94]. Another study at a bulk loading facility for gasoline

TABLE 7.2 Benzene Environmental Emissions During 1971

Source	Quantity (tons)	Environmental entry point
Commercial benzene production, storage, and transport	39,500	Water, air
Commercial benzene usage	329,000	Water, air
Oil spills	12,000	Water
Motor vehicle emissions	500,000	Air
Coke ovens	61,000	Air

Source: From Refs. 4 and 92.

TABLE 7.3   Mass Balance Accounting for Benzene Losses from By-product Manufacturing Facilities

By-product	Benzene unaccounted for (tons)
Ethylbenzene	55,500
Phenol	145,000
Cyclohexane	0
Maleic anhydride	70,000
Detergent alkylate	32,500
Aniline	10,500
Dichlorobenzene	7,000
DDT	8,500
Other nonfuel uses	--
	329,000

Source: From Refs. 4 and 93.

containing up to 33% benzene showed workers to be exposed to between 1.4 and 9.4 ppm. Conclusions were drawn from these studies that the benzene exposure hazard at service stations is minimal [6]. Table 7.4 gives the results of a NIOSH report studying the extent of exposure to benzene via benzene in gasoline at service stations. The benzene level in gasoline is not to be regulated under the Consumer Product Safety Commission's proposed standard, limiting the benzene content in consumer products to be below 0.1 ppm.

Benzene exposures in various industries and activities are given in the following tables. Table 7.5, storage and transfer operations; Table 7.6, benzene coating plant; Table 7.7, benzene plant; Table 7.8, coal chemical plant; Table 7.9, petrochemical plants; Table 7.10, chemical production plants identify benzene exposure levels.

Interest in controlling benzene through regulation has increased steadily. A historical perspective of benzene control through regulation is summarized in Table 7.11.

TABLE 7.4 NIOSH Study of the Extent of Exposure to Benzene via Benzene in Gasoline at Service Stations

Benzene	Toluene	Xylene	Bulk sample concentration vol. % (weighted average) Benzene	Toluene	Xylene	Average wind speed (mph)	Wind direction	Average temperature (°F)	Average humidity (%)	Average barometric pressure	Amount of gas pumped (gal)
0.123	0.313	0.544	1.6	7.2	9.8	2.4	N-E	35	81	29.20	153
0.123	0.313	0.724	1.6	7.2	9.8	2.4	N-E	35	81	29.20	153
BLD[a]	0.120	0.519	1.6	7.2	9.8	2.8	N-E	36	84	29.20	153
0.119	0.101	BLD	1.5	7.9	8.9	3.5	N	27	75	29.54	1054
BLD	BLD	BLD	1.5	7.9	8.9	4.3	N	27	75	29.52	1054
0.121	0.102	BLD	1.5	7.9	8.9	5.0	N	27	76	29.55	1054
BLD	BLD	BLD	1.7	4.5	7.3	6.9	S	45	45	29.31	411
BLD	BLD	BLD	1.7	4.5	7.3	11.0	S-W	47	41	29.25	137
0.054	0.046	BLD	1.1	5.7	7.1	6.3	Variable	55	55	29.40	720
BLD	BLD	BLD	1.7	4.5	7.3	6.9	S	49	58	29.29	411
BLD	BLD	BLD	1.7	4.5	7.3	6.9	S	49	58	29.29	411
BLD	BLD	BLD	1.7	4.5	7.3	6.9	S	49	58	29.29	411
0.033	0.028	BLD	1.1	5.7	7.1	5.2	Variable	56	54	29.40	955
BLD	0.028	BLD	1.1	5.7	7.1	5.2	Variable	56	54	29.40	955
0.033	0.093	BLD	1.1	5.7	7.1	5.2	Variable	56	54	29.40	955
BLD	0.028	BLD	1.1	5.7	7.1	5.2	Variable	56	54	29.40	955
0.039	0.033	BLD	1.0	6.8	8.1	7.4	Variable	62	59	29.19	1831
0.039	0.035	0.057	1.0	6.8	8.1	7.4	Variable	62	59	29.19	1831

[a]BLD = below detection limit.
Source: From Ref. 95.

TABLE 7.5 Benzene Levels at Facilities During Storage and Transfer Operations

Mode of transportation	Location	Exposure level (ppm)
Tank truck		
1	Loading point (upwind)	ND[a]
2	Control house (during hookup)	< 1.0
3	3-5 ft from connection	0-0.9
4	Loading tank truck	30
5	Sampling shipment	30
6	Directly above hatch	400
7	Disconnect lines	75
Tank car		
1	25 ft downwind from loading point	60
2	50 ft downwind from loading point	10
3	Hatch opening	104,000
4	Downwind of tank care	100
	Downwind of tank car	260
5	Sample shipment	30
Barge/tank ship	10-20 ft from vent stack	7-20
	Barge personnel	1
	Barge personnel downwind	30
	Dock samples	1
	Dock house	< 1
	Hatchway	100
	8-hr TWA (estimated)	1-5
	Dockman's station	trace
	Dock manifold before docking	Nil
	Dock manifold during hookup	Nil
	Walkway downwind from vent	11
	Ullage at breathing zone of gauger	130
	Making connections	10
	Loading tank	12
	At vent line	1400
	Ullage port	600
	20 ft downwind from dock	7
	Connecting load line	3
	Disconnecting load line	10
Storage facilities	Top of tank	375
	Downwind of tank vent	< 10
	25 ft downwind from floating roof tank	0
	Gauger	30

[a]ND = not detectable.
Source: From Ref. 24.

TABLE 7.6 Benzene Exposure Levels at a Benzene Coating Plant

	Benzene vapor (ppm)			
Location	December 1938	July 1946	August 1946	Average
Coating room—machine No. 1	60	70	50	60
Coating room—average	45	40	40	40
Coating room—maximum	60	70	55	60
Mixing room—average	80	80		80

Source: From Ref. 95.

TABLE 7.7 Benzene Exposure Concentrations at a Benzene Plant

	Benzene in air (ppm)	
Occupation	8-hr TWA	Range
Agitator operator	6.0	0.5-20
Benzol loader and loader helper	4.0	0.5-15
Benzol still operator	4.0	1-15
Light oil still operator	2.5	1-15
Naphthalene operator	10.0	2-30
Analyst	10.0	2-30
Chemical observer	10.0	4-50
Foreman	1.5	1-10

Source: From Ref. 95.

TABLE 7.8 Benzene Exposure Levels at Coal Chemical Plants

Position	Concentrations (ppm) Benzene	
Report by one plant		
Naphthalene operator	5.9	
Benzol agitator operator	16.0	
Benzol loader	2.8	
Benzol still operator	8.0	
Light oil still operator	1.2	
Repairman	2.0	
Oiler	3.6	
Benzol chemist	4.7	
Chemical observer	8.6	
Report by another plant		
Benzol still operator	2.3	
Exhauster engineer	2.2	
Exhauster engineer's helper	1.8	
Saturator operator	1.9	
Tar still operator	0.75	
Tar loader	1.6	
Benzol loader	29.0	
Swingman	3.1	
Laborers	2.6	
Benzene light oil plant		
Agitator operator	0.5–20 ppm	Range
Benzene loader and helper	0.5–15 ppm	
Benzene still operator	1–15 ppm	
Light oil still operator	1–15 ppm	
Naphthalene operator	2–30 ppm	
Analyst	2–30 ppm	
Chemical observer	4–50 ppm	
Foreman	1–10 ppm	
Operator of benzol plant		
Stillman	4.59 ppm	10-min TWA
Agitator operator	9.0 ppm	
Loader—laborer	11.91 ppm	

Source: From Ref. 24.

TABLE 7.9 Benzene Exposure Levels in the Petrochemical Industry (ppm)

Plant No.	1965-1970			1970-1975		
	1	1-5	5-10	1	1-5	5-10
1	25	0	0	25	0	0
2	20	15	0	17	10	0
3	34	24	0			
4				50	0	0
5	30	0	0	30	0	0
6	111	0	0	155	0	0
7	13	32	7	13	32	7
8				32	4	0
9	30	0	0	30	0	0
10	21	0	4	21	4	0
11				37	3	0
12	26	0	0	26	0	0
Total	310	71	11	436	53	7

Source: From Ref. 24.

TABLE 7.10 Benzene Exposure Levels at Various Chemical Production Plants

Operation	Individual samples (ppm)
Ethylbenzene	
Class 1 operator	
1	0.6
2	1.0
3	0.5
Class II operator	1.0
	0.5
	2.0
	2.0
Benzene recovery	
Operator	0.6
	1.0
	2.0
	<0.5
Olefin plant	
Operator	<0.5
Ethanol plant	<0.5
	15
	0.8
Aromatic unit	
Operator I	<0.5
	<0.5
	<0.7
Operator II	0.8
	2.0
Olefin plant	
2 samples	4.0
1 sample	1.0
5 samples	0
Cumene plant	
2 samples	5.0
1 sample	2.0
6 samples	0

TABLE 7.10 (continued)

Operation	Individual samples (ppm)
Caustic addition station	
1 sample	5.0
1 sample	4.0
2 samples	3.0
3 samples	1.0
1 sample	6.0
Aniline production	
Benzene filter area	0.2
Benzene feed pump area	0.06
Spin acid wash pump area	0.06
Benzene unit	
Feed pumps	5
Reflux pumps	Nil
Collecting product sample	Nil
Collecting crude sample	90
Feed pump	Nil
Sewer basin	75
Chlorobenzene production	
Mechanics and still operators	4.6-6.2
Other production operations	0.1-4.6
Alkyl benzene production	
Operators (alkylating)	0.3-13.0 ⎫
Mechanics	1.3-14.7 ⎬ Range
Other production operations	0.6-5.3 ⎭
Ethyl cellulose resins	
Manufacturing area	3.8-4.0
Cumene plant	
Outside control room	1,0,0,3,4,2,4 (7 samples)
Benzene pumps	6
Benzene reboiler	5,1,4,6,1,1,3 (7 samples)
Condensate pump	2
Between udex pumps	2
Near pump	1,2,2,2,5 (5 samples)

Source: From Ref. 24.

TABLE 7.11 Benzene Control Through Regulation

Date	Action taken
1927	Exposure limit first proposed at 100 ppm.
1939	State of Massachusetts proposed maximum allowable concentration of 75 ppm.
1946	American Conference of Governmental Industrial Hygienists (ACGIH) recommends 100 TLV (threshold limit value).
1947	State of Massachusetts lowers maximum allowable concentration to 35 ppm.
1947	ACGIH recommends 50 ppm level of exposure.
1948	ACGIH recommends 35 ppm level of exposure
1957	ACGIH recommends 25 ppm level of exposure.
1966	Los Angeles County Rule 66 controls the emission of volatile substances into the atmosphere. Certain compounds in hydrochemical solvents and other compounds used as solvents were found to contribute to photochemical smog.
1971	OSHA adopts American National Standards Institute (ANSI) standard of 8-hr TWA of 10 ppm and ceiling concentration of 25 ppm over a 10-min period.
1976	ACGIH lowers TWA to 10 ppm.
1976	ACGIH recommends 8-hr TWA of 10 ppm, a ceiling concentration of 25 ppm, and an acceptable peak above the ceiling of up to 50 ppm for a duration not to exceed 10 min in any 8-hr work period.
April 1976	United Rubber, Cork, Linoleum, and Plastic Workers of America urges Emergency Temporary Standard (ETS) regulating occupational exposure to benzene be issued.
June 1976	National Academy of Sciences concludes benzene must be considered as suspect leukemogen.
August 1976	NIOSH concludes that benzene is leukemogenic, and that since no safe level for benzene exposure can be established, no worker be exposed to benzene in excess of 1 ppm in the air.

8

Controlling Benzene Emissions

YEN-HSIUNG KIANG

Trane Thermal Company
Conshohocken, Pennsylvania

Methods for benzene emission control can be readily classified into the following two categories:

1. Nondestructive type—in this group, the control methods remove benzene and other contaminants in their original form. The recovery of benzene is possible.
2. Destructive type—benzene and other organic contaminants are converted into inorganic forms. Heat recovery is possible.

The basic benzene removal methods are carbon adsorption, oil absorption, thermal oxidation, and catalytic oxidation. The first two methods are nondestructive; the latter two methods are destructive.

CHARACTERISTICS OF BENZENE EMISSIONS

Sources of air pollution from benzene include continuous or intermittent process vents, equipment decommissioning vents, and storage tank and loading operation vents. The compositions and flow rates of the vent gases usually change with time. The vent gases can be classified in two categories:

Category 1—Benzene-Contaminated Air

This group covers waste gases which are air contaminated by benzene and other hydrocarbons. Since the flammable limits of benzene in air are between 1.4 and 8% by volume, one must be careful when handling this group of waste gases.

Category 2—Benzene-Contaminated Inerts

In this group, benzene and other hydrocarbons are carried by inert gases such as nitrogen or carbon dioxide.

There is a third group, benzene mixed with other hydrocarbons. This group will not be included in the following discussion.

CARBON ADSORPTION

A flow diagram for a typical carbon adsorption benzene recovery plant is illustrated in Fig. 8.1 [96,97,98]. Two adsorbers are normally used. This arrangement will enable the continuous operation of the system, with one unit adsorbing while the other is being regenerated. Steam can be used to regenerate the carbon. The heat carried by the steam is used to raise the carbon bed temperature, thus reducing its adsorption capacity, and therefore providing heat of vaporization for benzene. The latent heat of benzene is shown in Fig. 8.2 [99]. The mass flow of the steam reduces the partial pressure of benzene in the vapor phase. The steam and benzene mixture is passed into a condenser. The condensate is collected, separated, and the benzene recovered. If the contaminants are benzene plus other organics, distillation is required to recover benzene. Distillation systems would be designed for specific applications.

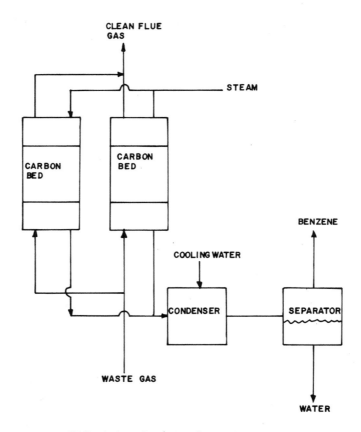

FIG. 8.1 Carbon adsorption system.

Controlling Benzene Emissions

The regenerated carbon bed contains steam condensate. This must be removed to prevent water buildup in the bed. Usually, this is accomplished by passing dry air through the unit.

Capacity of Activated Carbon

The capacity of a carbon bed is dependent upon the bed material selected. Basic capacity data, such as the adsorption isotherm, are available from the activated carbon manufacturer. A typical adsorption isotherm for a commercially available carbon, the BPL granular-activated carbon by Calgon Corp., is shown in Fig. 8.3. In this section, the analytical method to determine the capacity and bed properties is presented. This method can be readily applied to different carbon materials.

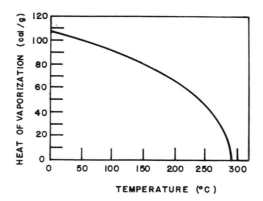

FIG. 8.2 Heat of vaporization of benzene.

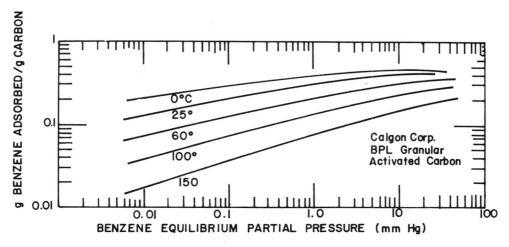

FIG. 8.3 Benzene adsorption isotherm (Calgon Corp./BPL Granular-Activated Carbon).

Adsorption Isotherm

The quantity of benzene that can be adsorbed is dependent upon the carbon benzene equilibrium. A typical equilibrium relationship, adsorption isotherm for carbon and benzene, is shown in Fig. 8.4 [100]. The adsorption capacity is a function of the bed materials used. For the purpose of illustration, these special carbon benzene equilibrium data will be used throughout this section.

A typical adsorption/desorption cycle can be explained by using the isotherm (Fig. 8.5). The carbon bed, initially at 60°C, contains C_1 gram of benzene per gram of carbon, and adsorbs benzene following the 60°C isothermal line. Due to heat of adsorption, the adsorption equilibrium line deviates from the isotherm. The actual adsorption line follows the dashed line in Fig. 8.5. The final concentration of the carbon bed is C_2 g benzene/g carbon. This point is selected based on the partial pressure of benzene in the waste gas stream. The desorption cycle starts to heat the bed to 300°C, the equilibrium carbon capacity follows the solid line. The temperature is gradually decreasing during the desorption process. The actual desorption line is the dashed line as indicated in Fig. 8.5. A final cooling is required to reduce the bed temperature to 60°C. The actual adsorption capacity is the difference between C_2 and C_1, i.e., (C_2-C_1) g benzene/g carbon. In practice, only 50% of this value is considered to be the net bed capacity. Other factors which may effect the bed capacity are fluid velocity, bed depth, and cycle timing. The maximum gas velocity for Calgon BPL carbon is shown in Fig. 8.6.

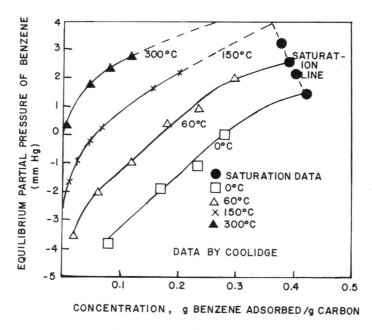

FIG. 8.4 Benzene adsorption isotherm.

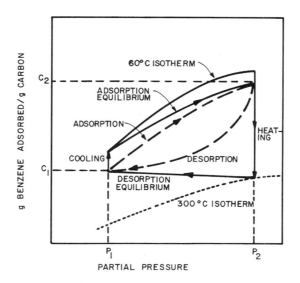

FIG. 8.5   Adsorption/desorption operation.

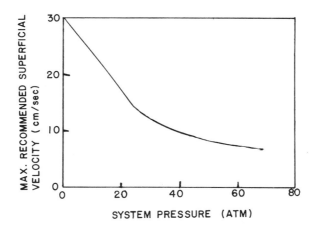

FIG. 8.6   Maximum fluid velocity (Calgon Corp./BPL Granular-Activated Carbon).

Adsorption Isosteres

The adsorption isotherms usually do not form a straight line. A different curve, defined as isosteres, usually gives a straight line with few exceptions. The isosteres are a plot of equilibrium pressure against vapor pressure. Figure 8.7 illustrates the isosteres for benzene/carbon system described in Fig. 8.4. The vapor pressure data of benzene is shown in Fig. 8.8 [13].

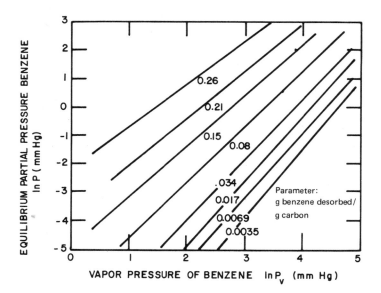

FIG. 8.7 Benzene isosteres.

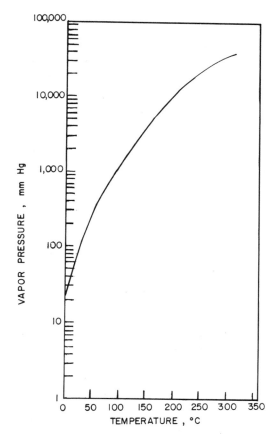

FIG. 8.8 Vapor pressure of benzene.

By introducing the "adsorption potential," the equilibrium relationship can be reduced to a single line, as shown in Fig. 8.9. The adsorption potential is defined as

$$RT \log \frac{p_v}{p} \quad (1)$$

where R is the ideal gas constant, T is the temperature in °K, P_v is the vapor pressure in mmHg; and p is the equilibrium pressure in mmHg. The adsorption potential is the free energy of compression of 1 mol of gas from the equilibrium pressure to the vapor pressure.

Heat of Adsorption

When a vapor is adsorbed, heat is usually released. The heat of adsorption can be determined by experiment, or through the use of isosteres and heat of

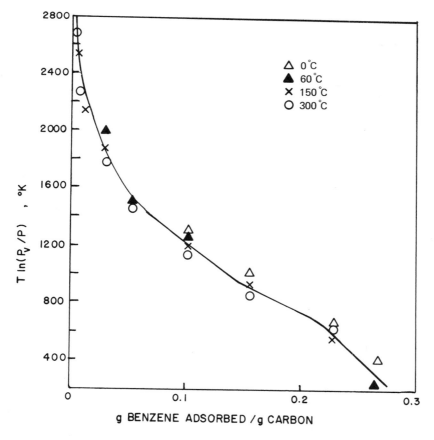

FIG. 8.9 Adsorption potential of benzene.

vaporization. The differential heat of adsorption is defined as the heat released by 1 g of vapor adsorbed in an infinite quantity of adsorbent. The integral heat of adsorption is defined as the heat evolved when the concentration of a finite amount of adsorbent changes between two given concentrations.

The differential heat of adsorption can be calculated by

$$Q = mL \tag{2}$$

where m is the slope of isostere and L is the heat of vaporization. The heat of adsorption of benzene is plotted in Fig. 8.10, for benzene/carbon system described in Fig. 8.4.

Rate of Adsorption

The time required to reach equilibrium is extremely variable. For benzene, the rate of adsorption is fast at the beginning and slows down gradually. The data determined by Tryhorn and Wyatt are presented in Figs. 8.11 and 8.12 [102].

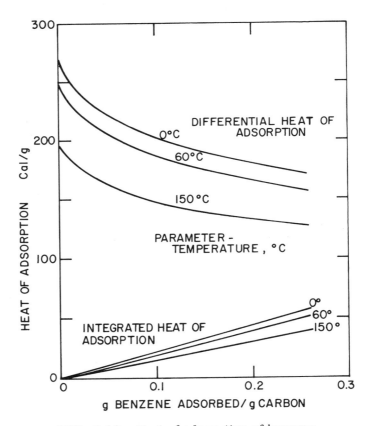

FIG. 8.10 Heat of adsorption of benzene.

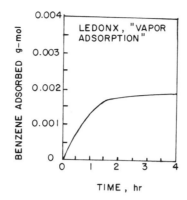

FIG. 8.11   Total benzene adsorption.

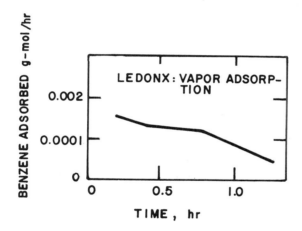

FIG. 8.12   Benzene adsorption rate.

Process Design

Selection of Carbon Material

As mentioned before, the bed material controls the capacity of the system. To select the best material, the adsorption isotherm has to be compared as well as the operating condition. In Fig. 8.13 there are two carbon materials available, carbon "B" has a higher equilibrium capacity. However, the net capacity of carbon "A" is greater because of the shape of the isotherms. Thus, a detailed analysis of isotherms is necessary to ensure optimal bed material selection.

Heat Effect

As carbon is adsorbed, the heat of adsorption is carried by the gas throughout the bed. Thus, the temperature of the bed is raised even before the bed is

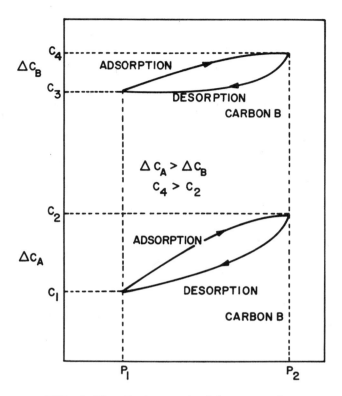

FIG. 8.13 Carbon materials comparison.

actively adsorbed. The higher temperature will eventually reduce the capacity of the bed. Therefore, temperature is also a factor in determining the adsorption/desorption cycle time.

Desorption Process

There are different methods to regenerate the carbon bed, as follows: steam or gas stripping, vacuum, heating, and electrodesorption. The most commonly used method is steam stripping, because of its simplicity and its easy to separate steam from most hydrocarbons. Electrodesorption has been tested successfully by Fabass and Dubris [103].

Multicomponent Contaminants

The existence of other hydrocarbons will effect the desorption capacity of benzene. The isotherms shown in Fig. 8.14 can be used to illustrate the effect. If the mixture contains methane and benzene, the benzene adsorption is substantially unaffected by the presence of the poorly adsorbed methane. The adsorption isotherm for benzene is applicable, provided that the partial pressure is used as the equilibrium pressure.

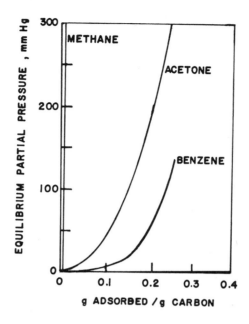

FIG. 8.14   Multicomponent adsorption.

If the two isotherms are compatible, as in the case of acetone and benzene, a more elaborated analysis is required to determine the capacity.

Operating Problems

The major operating problems encountered with carbon bed adsorption are contamination, corrosion, and oxidation. Contamination occurs when the gas stream carries particulates and polymerizable materials. They must be removed to ensure a reasonable service life of the carbon bed. Corrosive problems occur mostly in the desorption cycle. Hydrolysis is the main reason for the formation of corrosive materials. For example, if vinyl chloride is one of the contaminants other than benzene, in the regeneration cycle, the combination of high temperature and catalytic action of carbon will hydrolyze vinyl chloride to hydrochloric acid, which is a rather corrosive chemical. The presence of an oxidizing agent may oxidize carbon at a higher temperature. Thus, a favorable oxidation condition should be avoided.

WASH OIL ABSORPTION

Since benzene is not water-soluble, the conventional water scrubber cannot be used to remove benzene from gas streams. In order to scrub the benzene vapor, wash oil must be used. Table 8.1 shows the properties of some wash oils as well as the solubility of benzene in them [96].

TABLE 8.1 Properties of Some Wash Oils

Wash oil	Distillation range (°C)	Specific gravity (20/20°C)	Viscosity (20°C, cp)	Solubility of benzene (v.p. 6.5 mmHg, wt%)
Creosote oils				
Fraction 1	205-265	0.9535	25	4.55
Fraction 2	200-300	1.0135	35	3.95
Fraction 3	200-350	1.031	51	3.6
Petroleum Gas				
Fraction 1	214-285	0.8295	36	3.2
Fraction 2	210-400	0.8638	70	2.9
Fraction 3	260-365	0.849	92	2.8

Source: From Ref. 96.

Absorption Equipment

Absorber types which can be used to remove benzene are conventional spray towers, packed towers, and tray columns. The design methods for the towers are available elsewhere [13,104].

The typical benzene absorption and recovery system is illustrated in Fig. 8.15. In the absorber, benzene is absorbed in the wash oil. The benzene rich oil is then passed into a distillation-stripper. In the distillation-stripper, steam is used to evaporate benzene. A condenser and a separator are then used to recover benzene. The distillation-stripper can be either a vacuum or a pressure column. Bubble cap tray towers are common distillation-stripper equipment.

Absorber Design

Simmons and his co-workers [105,106] have performed a series of tests on the absorption of benzene by a wash oil. The specifications of the wash oil are

Olefins	18%
Flash point (open)	25°C
Viscosity	146
Specific heat	0.5
Specific gravity	0.875
Molecular weight	260

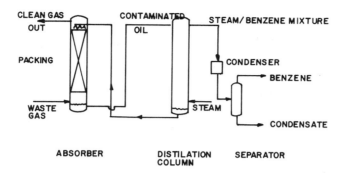

FIG. 8.15 Flow scheme of benzene absorption plant.

The data in Simmons and colleagues' work were recalculated to determine $K_g a$. The mass transfer coefficients are defined by

$$K_g a = \frac{N}{V \Delta p_{1m}} \tag{3}$$

where

$K_g a$ = gas film mass transfer coefficient, g(mol)/cm^3(hr)(atm)

N = moles benzene transferred, g-mol/hr

V = volume of packing, cm^3

Δp_{1m} = logarithm mean of benzene partial pressures, atm

The calculated $K_g a$ data are shown in Fig. 8.16. Since the gas rate variations in the test are minimum, it is not possible to determine the effect of the gas rates on mass transfer coefficients. The packings used in the test are of small size. However, the $K_g a$ number is smaller than in the conventional water/air system. This is because of the relatively viscous wash oil blocking some of the packing surface, by accumulating at the point of contact of the packing materials, and thus reducing the area of contact between phases.

Oil absorption for benzene has been long applied to recovering light oil from coal gas. Benzene is the most abundant compound in the light oil. A number of absorber designs have been developed throughout the years.

Wood-grid packing is frequently used because of its low pressure drop. For high-pressure systems, bubble cap or sieve trays have been used. For wood-grid packing, the optimal liquid-to-gas ratio is determined to be 1.7 m^3/1000 m^3 for benzene removal. The mass transfer coefficient for a 2.75 m diameter wood-grid packed tower is 38 kg(mol)/(hr)(m^3)(atm) [96,107]. The operating temperature of this tower is 23°C, the oil gas ratio is 1.9 m^3 oil per 1000 m^3 gas, and the liquid rate is approximately 1190 kg/(hr)(m^2).

Hixon and Scott [108] used straw oil as the absorption liquid in a spray for benzene removal. The straw oil has a gravity of 350 API, viscosity 40 saybolt at 38°C, and a molecular weight of 226. The mass transfer coefficient was determined to be

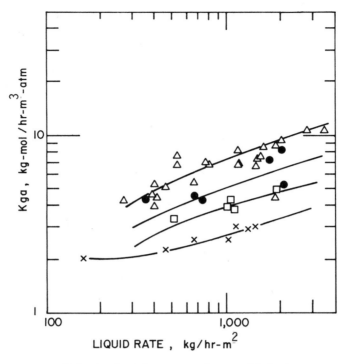

FIG. 8.16 Mass transfer coefficients.

$$K_g a = 1.32 \times 10^{-7} \frac{G^{0.8} L^{0.9}}{H^{0.5}} \quad (4)$$

where

G = gas rate, g/(hr)(cm^2)

L = liquid rate, g/(hr)(cm^2)

H = tower height, cm

Gross and Simmons determined the mass transfer coefficient for kerosene absorption of benzene. The packing material used is 1 in. ceramic Berl saddle. They used the relation [109]

$$K_g a = a G^b \quad (5)$$

to correlate the coefficients. The constants for three kerosene rates are presented in Table 8.2. There is no correlation between $K_g a$'s and the liquid rate.

Other packings have also been applied to benzene absorption. Efficiency data may be available through the packing manufacturers.

TABLE 8.2 Constants for Equation (5)

Kerosene rate ($g/hr-cm^2$)	a	b
510	0.000272	0.937
900	0.0002757	0.912
1465	0.0002364	0.962

Source: From Ref. 100.

THERMAL OXIDATION

Thermal oxidation is the most widely employed method to control hydrocarbon emissions. Contaminants are heated in the presence of oxygen to a temperature sufficient to allow complete oxidation in the residence time available. Thermal oxidation equipment consists of a refractory lined furnace and a fuel-fired burner to supply heat and stabilize the oxidation reaction. The major drawbacks of thermal oxidation methods are the relatively high operating expense caused by the requirement of supplementary fuel and the significant initial cost. However, thermal oxidation is chosen for its trouble free operation and, if heat recovery is employed, its lower total energy costs.

The typical adiabatic oxidation temperature of benzene vent gases are shown in Fig. 8.17. If benzene is the only contaminant in the vent gases, about 700°C is required to ensure benzene destruction at required residence time. However, the existence of other hydrocarbons may increase the oxidation temperature requirement. Note that these curves are based on both air and waste gases at ambient condition (assuming 20% excess air with no auxiliary fuel used) and no combustion product decomposition considered. Depending upon the design requirement, the excess air rate may be higher.

Thermal Oxidation Process

The basic process for benzene emission control includes a thermal oxidation section, a heat recovery section, and, if required, a gas treatment section. Because of the wide variation of both flow rates and composition of the waste gases, a two-stage combusion process is required to ensure stable oxidation reaction [110]. The combustion process can be either a standard off-gas oxidation system as in Fig. 8.18A, or a modified thermal vortex burner as in Fig. 8.18B.

In the standard off-gas oxidation process [111], the waste gases bypass the burner and are introduced into the secondary chamber. Auxiliary fuel is used in the burner. For category 1, waste gases containing only low concentration of organics, they can be used as combustion air in the primary chamber. However,

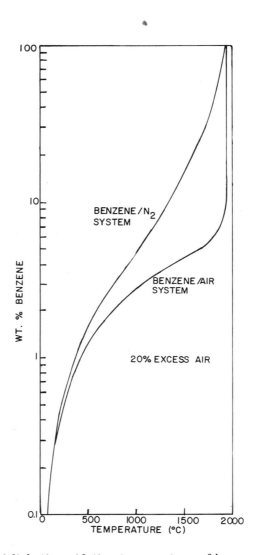

FIG. 8.17 Adiabatic oxidation temperature of benzene waste gases.

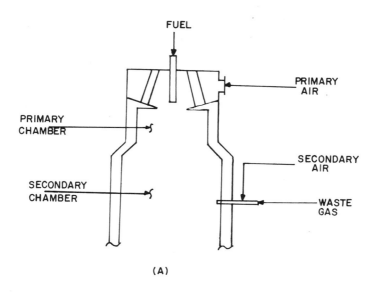

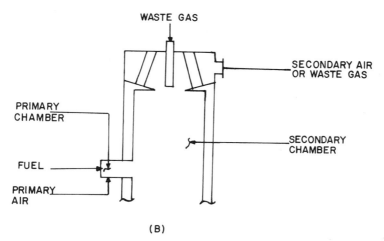

FIG. 8.18 Two-stage combustion process. (A) standard off-gas oxidation process. (B) modified thermal vortex burner.

the majority of the gases are introduced into the secondary chamber to minimize fuel consumption. The secondary chambers are usually at a lower temperature than the primary chambers.

The modified thermal vortex burner [112], developed by Trane Thermal Company, has a side-mounted burner as the primary chamber. Waste gases, as well as required combustion air, will be introduced into the burner either through the tangential inlet or the central nozzle. Auxiliary fuel and air, or possibly some category 1 gases, are used in the primary chamber.

Because of the variation in both composition and flow rate of the waste gases, a well-designed control system is required to ensure trouble-free operation. The control system will ensure not only that the oxidation temperature be kept constant, but also that enough excess air be supplied. A different approach is to base load the system with auxiliary fuel. Temperature and stack oxygen content will then be used to fine tune the system. This approach will increase operating cost for a "disposal only" system. However, if heat recovery is used, this method will give a more stable system.

Process Design

The important design parameters for an oxidation system are: temperature, excess air, mixing, and residence time. These four parameters are interrelated.

As described before, the minimum oxidation temperature requirement for benzene is 700°C. The actual oxidation temperature must be determined by a study of all the contaminants as well as be balanced between the operating (fuel requirement) and the capital (residence time) costs.

Excess air requirement is dependent upon the mixing and concentration of organics. As a general rule, a better mixed system or a higher organic content waste requires less excess air. Usually, the excess air requirement varies with the design of the waste gas injection system. It is advisable to study the mixing characteristics of waste gas injection systems to determine the excess air requirement.

The residence time is another important design parameter. In general, the higher the temperature or the better the mixing, the lower the residence time requirement. Note that the residence time defined here is the "effective residence time." It is defined as the time the combustion gases traveled at the designated temperature after complete mixing. This is different from conventionally defined residence time, which is a function of the oxidation chamber volume only, and does not take into consideration the mixing and flame length of the burner. Figure 8.19 illustrates the effect of flame length on the effective residence time. Assuming identical conventionally defined residence time, the short flame burner application will give a longer effective residence time than that of a long flame burner application.

Heat Recovery System

There are two economic approaches to recovering waste heat, the recovery of energy for process application or the reduction of waste disposal cost. The

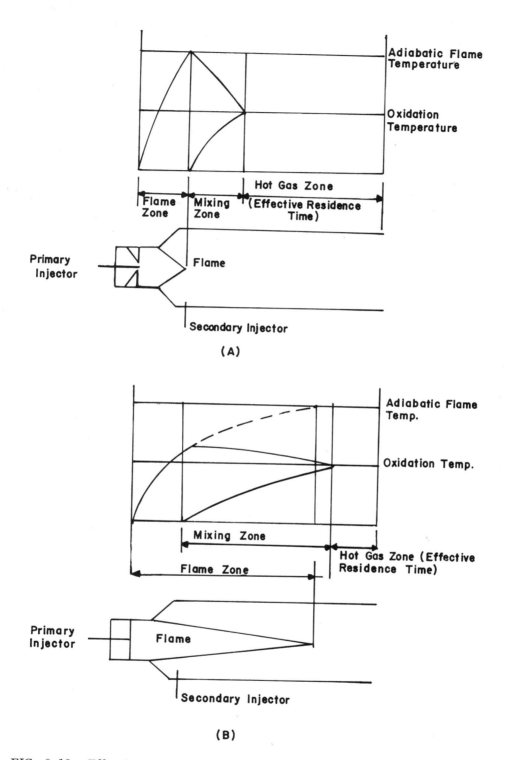

FIG. 8.19 Effective residence time. (A) Short flame burner. (B) Long flame burner.

system illustrated in Fig. 8.20A is one example of energy recovery. A boiler is used to generate steam for process use. Besides steam generation, the energy could be recovered as hot process fluid; both liquid and gas.

The auxiliary fuel requirement can be reduced by preheating either air or waste gases, Fig. 8.20B and 8.20C, using combustion hot gases. The three systems illustrated in Fig. 8.20 are basic processes. A combination of them can be used to fit the requirement of any individual plant [113].

Contaminants Other Than Hydrocarbons

Sometimes the waste gases will contain contaminants other than pure hydrocarbons, such as chlorinated and sulfonated organics. Final flue gas treatment is required to ensure a clean stack. The stack gas treatment equipment can be wet scrubbers and the like [114,115].

The existence of these contaminants will also affect the heat recovery system selection [113,114]. For example, for chlorinated hydrocarbons, gas to gas heat exchangers are not recommended due to the corrosion of hydrogen chloride gas [116]. In boiler application, special attention must be given to hydrochloric acid condensation corrosion. The oxidation chamber design is

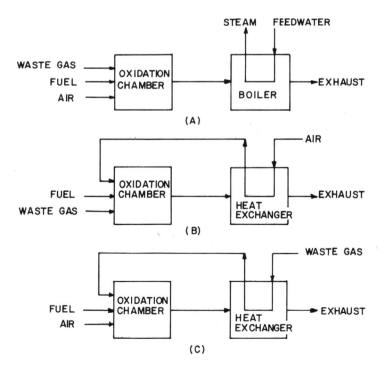

FIG. 8.20 Heat recovery system. (A) steam generation. (B) air preheating. (C) waste gas preheating.

Controlling Benzene Emissions

also affected by the existence of these nonhydrocarbon contaminants [117]. Thus, it is necessary to determine all the possible contaminants before a final emission control process can be selected.

Reaction Kinetics

The oxidation reaction is a very complicated process which involves a series of decomposition and free radical reactions. Lee et al. [118] have performed experimental study for the destruction of category 1 benzene waste gases. Since category 1 gases are contaminated air, the oxygen concentration in this study was assumed to be constant. The oxidation reaction is assumed to be

$$B + O_2 \longrightarrow \text{end products} + O_2 \tag{5}$$

where B represents benzene. The rate equation becomes

$$\frac{dC_B}{dt} = -kC_B \tag{6}$$

where

C_B = the concentration of benzene

k = the kinetic rate constant

t = time

The rate constant is defined as

$$k = A\exp\left(-\frac{E}{RT}\right) \tag{7}$$

where

E = the activation energy, g cal/g mol

R = the gas constant, 1.987 g cal/g mol, °K

T = the reaction temperature, °K

The rate constant for benzene destruction is determined to be

$$k = 2.51 \times 10^{22} \exp\left(-\frac{99650}{RT}\right) \tag{9}$$

Note that this rate constant can only be used to predict benzene destruction. Other hydrocarbon intermediates and carbon monoxide destruction cannot be predicted.

The reaction rates of a plug flow reactor are illustrated in Fig. 8.21. In practice, some backmixing exists in the combustion chamber. The reaction rates of a completely mixed reactor are shown in Fig. 8.22. These two sets

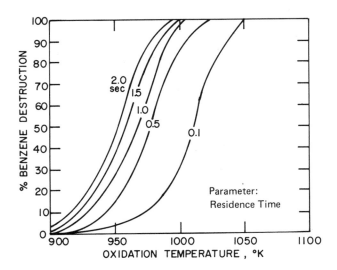

FIG. 8.21 Benzene oxidation rate (plug flow reactor).

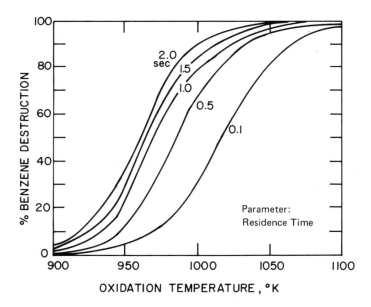

FIG. 8.22 Benzene oxidation rate (completely mixed reactor).

of curves represent two extremes in the combustion chamber design. The actual oxidation rate is between these two extremes. Figure 8.23 shows the oxidation rates for plug flow and completely mixed reactors at 1000°K oxidation temperature.

The rate data presented above adhere to the assumption that the combustibles and air are completely mixed. In practice, this is not possible. Thus, the rate

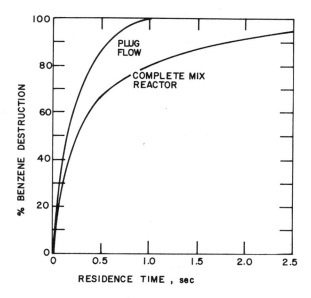

FIG. 8.23   Benzene oxidation rate comparison.

equation can be used to define the minimum requirement of a benzene oxidation system. The actual residence time requirement is different with each oxidation system designed.

CATALYTIC OXIDATION

Catalytic oxidation is an alternative to thermal oxidation as a means to oxidize gaseous hydrocarbons to carbon dioxide and water. Catalytic oxidation always requires a lower oxidation temperature when compared with thermal oxidation. The catalyst is used to accelerate the oxidation reaction without itself being changed in any way. The oxidation reaction produces the same combustion products and liberates the same heat of combustion as does thermal oxidation.

Catalytic Oxidation Process

The basic process of catalytic oxidation is illustrated in Fig. 8.24. Auxiliary fuel and waste gases are introduced into a preheating/combustion/mixing chamber. The purpose of this chamber is to achieve a uniformly distributed mixture of combustion products and waste stream. The mixed gases are then passed through the catalyst bed. The catalyst bed is typically a metal mesh-mat, ceramic honeycomb, or other ceramic matrix structure with a surface deposit or coating of catalysts.

Recovery of heat from the cleaned effluent stream may be included in the system using conventional techniques. Because the fuel requirement and

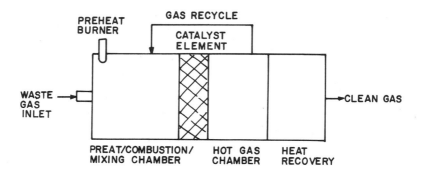

FIG. 8.24 Basic catalytic oxidation process.

temperature are low, there is often less economic incentive to include heat recovery in the catalytic oxidation systems.

In order to operate a catalytic oxidation system successfully, it is necessary to study the contaminants carefully to ensure that catalyst poisoning and fouling are not possible. During the operation, careful monitoring of the unit is required to ensure that the catalyst has not lost its activity. Also, the temperature in the catalyst bed must be kept constant to avoid overheating the catalyst.

Process Design

The oxidation performance of a catalytic oxidation system depends on the catalyst used, operating temperature, hydrocarbon type and concentration, and flow velocity. The reduction of benzene and other hydrocarbons depends strongly on the amount and type of catalyst used. For benzene destruction, a platinum material is the most widely used catalyst.

The catalytic ignition temperature is based on the commercial catalysts used. For Al_2O_3-based catalysts, the ignition temperature of benzene is 180°C and for Nichrome Ribbon-based catalyst, the ignition temperature is 230°C. The minimum preheat temperature for 90% benzene destruction is 240-300°C for Al_2O_3-based catalyst [111]. The reaction rate and catalyst requirement is also a function of catalyst selected.

The basic differences between the various commercially practical catalytic oxidation processes are the type of catalyst used and the form of the catalytic element. There are various process schemes in the catalytic oxidation process. Recirculation or heat exchange can be used to reduce fuel requirement. Waste heat boilers have also been used in connection with catalytic oxidation process. Several different schemes are illustrated in Fig. 8.25.

Catalyst Deactivation

Since the quality of waste gases may not be a controllable property, the activity of the catalyst for promoting oxidation will reduce with time and exposure. The

Controlling Benzene Emissions

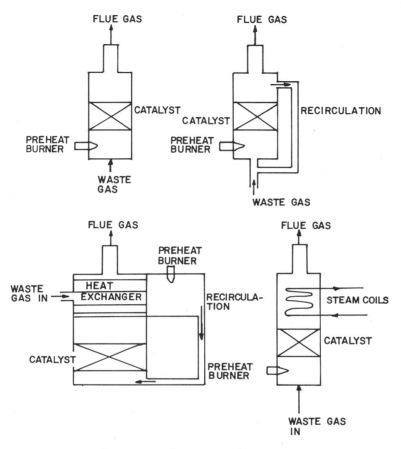

FIG. 8.25 Schematics of some catalytic oxidation processes.

oxidation efficiency usually declines along with catalyst activity. To compensate for the decreasing catalyst activity, the following methods can be used:

1. Overdesign the catalyst requirement.
2. Increase the preheat temperature.
3. Reactivate the catalyst.
4. Replace the catalyst.

The appropriate compensation action is usually dependent upon the cause and the nature of the deactivation process.

Catalyst Poisoning

By far, the most serious restriction in the application of catalytic oxidation for air pollution control is the poisoning of the catalyst. To minimize the poisoning effect, a provision must be made to remove the poisonous materials before they come in contact with the catalyst.

COMBINATION SYSTEMS

The emission control systems described in previous sections can be combined to give a more efficient system. In Fig. 8.26, the system consists of carbon adsorption and thermal oxidation. The carbon adsorption system is used to concentrate the dilute benzene waste gases. The concentrated benzene waste gases are then combusted in the thermal oxidation system. The system can be designed so that no auxiliary fuel is required.

If kerosene is used as the wash oil in benzene absorption system. The contaminated oil can then be used as fuel for boilers or other process equipment. This approach eliminates the need of distillation-stripper. These are the two possible combination benzene emission control systems. By carefully studying the plant requirement, an optimal system can be derived for every application.

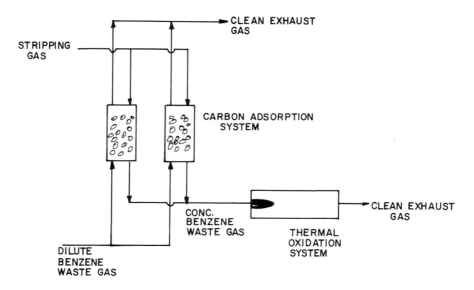

FIG. 8.26 Schematics of carbon adsorption/thermal oxidation system.

References

1. Werner Herz, <u>The Shape of Carbon Compounds</u>, Benjamin, New York, 1963.
2. R. O. C. Norman and R. Taylor, <u>Electrophilic Substitution in Benzeneoid Compounds</u>, Elsevier, New York, 1965.
3. H. W. Gerade, <u>Toxicology and Biochemistry of Aromatic Hydrocarbons</u>, Elsevier, New York, 1960.
4. P. Walker, <u>Air Pollution Assessment of Benzene</u>, U.S. Environmental Protection Agency, PB-256, Washington, D.C., April 1976.
5. <u>Guidelines for Control of Occupational Exposure to Benzene</u>, Occupational Safety and Health Administration, Washington, D.C., January 1970.
6. <u>Criteria for a Recommended Standard: Occupational Exposure to Benzene</u>, U.S. Department of Health, Education, and Welfare, Public Health Service, NIOSH, 1974.
7. <u>Chemical Economics Handbook</u>, Stanford Research Institute, Menlo Park, California, 1972.
8. <u>Handbook of Chemistry and Physics</u>, 37th ed., Chemical Rubber Publishing Company, 1955-1956.
9. J. Smith, <u>Trans. Am. Soc. Mech. Engrs. 58</u>:719, 1936.
10. G. Moser, Dissertation, Berlin, 1913.
11. Karel Verschueren, Handbook of Environmental Data on Organic Chemicals, Van Nostrand Reinhold, New York, 1977.
12. Chemical Safety Data Sheet SDZ-Benzene, Revision 3, Manufacturers Chemists Association, Washington, D.C., 1960.
13. R. H. Perry, C. H. Chilton, and S. D. Kirkpatrick (Eds.), <u>Chemical Engineer's Handbook</u>, 4th ed., McGraw-Hill, New York, 1963.
14. Consumer Product Safety Commission: Benzene. <u>Fed. Reg. 43</u> (Pt. IV, no. 98), May 19, 1978.
15. Occupational Safety and Health Administration, Occupational exposure to benzene. <u>Fed. Reg. 43</u> (Pt. II, no. 29), February 10, 1978.
16. <u>Economic Impact Statement: Benzene (Vol. II)</u>, U.S. Department of Labor, Occupational Safety, and Health Administration, Washington, D.C., May 1977.

17. D. E. Schoeffel and B. Dmuchovsky, The outlook for benzene, 1977-85. Chem. Eng. Prog. 73:13-16, August 1977.
18. Key chemicals—benzene. Chem. Eng. News 55:16, November 1977.
19. R. F. Goldstein and A. L. Waddens, The Petroleum Chemicals Industry, E. & F. N. Spon, London, 1967.
20. T. C. Ponder, Benzene outlook through 1980. Hydrocarbon Proc. 11:217-218, 1976.
21. Air Pollution Assessment of Benzene, U.S. Environmental Protection Agency, PB-256 734, April 1976.
22. Occupational Safety and Health Administration, Occupational exposure to benzene. Fed. Reg. 43 (Pt. II, no. 29), February 10, 1978.
23. Economic and Environmental Issues Related to a Proposed Ban on Consumer Products Containing Benzene—Final Report, Consumer Product Safety Commission, Washington, D.C., February 1978.
24. Economic Impact Statement: Benzene (Vol. I), U.S. Department of Labor, Occupational Safety, and Health Administration, Washington, D.C., May 1977.
25. M. Graber, Environmental assessment of benzene in New Jersey. "Toxic Air Contaminants: Their Measurement, Evaluation and Control," Proceedings in Mid-Atlantic States Section Semi-Annual Technical Conference, Newark, New Jersey, October 21, 1977, pp. 14-48.
26. J. C. Verchin, The outlook for benzene derivatives in Europe. Chem. Eng. Prog. 73:35-39, November 1977.
27. Fire Protection Guide on Hazardous Materials, 6th ed., National Fire Protection Association, Boston, Massachusetts, 1975.
28. Guidelines for reducing exposure to benzene. Job Safety and Health, July 1977, pp. 4-5.
29. Environmental Protection Agency, Water programs—hazardous substances. Fed. Reg. 43, March 13, 1978.
30. D. J. Behrendsen, Guidelines to the Handling of Hazardous Materials, Source of Safety, Denver, Colorado, 1976.
31. Department of Transportation, Code of Federal Regulations, Title 49, Pts. 100-199, revised December 31, 1976.
32. Hazardous Materials Placarding Guide, published material, UNZ & Company, Jersey City, New Jersey, 1976.
33. R. B. Conn, "Normal Laboratory Values of Clinical Importance—Normal Hematologic Values," in Cecil-Loeb Textbook of Medicine (P. B. Reeson and W. McDermott, eds.), 13th ed., Saunders, Philadelphia, 1971.
34. R. Truhaut, Determination of a tolerable limit of benzene an work environment. Arch. Mal. Prof. (French) 29:5-22, 1968.
35. D. W. Parke and R. T. Williams, Studies in detoxication—the metabolism of benzene—(a) the determination of benzene; (b) the elimination of unchanged benzene in rabbits. Biochem. J. 46:236-242, 1953.
36. H. Schrenk, et al., Absorption, distribution and elimination by body tissues and fluids of dogs exposed to benzene. J. Ind. Hyg. Toxicol. 23:20-34, 1941.
37. L. Greenburg, Benzol poisoning as an industrial hazard—VII. Public Health Reports 41:1526-1539, 1926.

38. Benzene: uses, toxic effects, substitutes. Meeting of Experts on the Safe Use of Benzene and Solvents Containing Benzene, May 16-22, 1967, International Labor Office, Geneva, 1968.
39. R. D. Sewart, et al., Chronic overexposure to benzene vapor. Toxicol. Pharmacol. 10:381, 1967.
40. C. G. Hunter, Solvents with reference to studies on the pharmacodynamics of benzene. Proc. Roy. Soc. Med. 61:913-915, 1968.
41. R. J. Sherwood, One man's elimination of benzene. Proceedings of the 3rd Annual Conference on Environmental Toxicology, AMRL, TR-72-130, Dayton, Ohio, 1972.
42. L. Greenburg, et al., Benzene poisoning in the rotogravure printing industry in New York City. J. Ind. Hyg. Toxicol. 21:395-420, 1939.
43. J. E. Walkley, et al., The measurement of phenol in urine as an index of benzene exposure. Am. Ind. Hyg. Assoc. J. 22:362-367, 1961.
44. R. J. Sherwood and F. Carter, The measurement of occupational exposure to benzene vapor. Ann. Occup. Hyg. 13:125-146, 1970.
45. H. Buchwald, The expression of urine analysis results. Ann. Occup. Hyg. 7:125-136, 1964.
46. R. C. Theis and S. R. Benedict, The determination of phenols in the blood. J. Biol. Chem. 61:67-71, 1924.
47. R. Buchwald, The colorimetric determination of phenol in air and urine with a stabilized diazonium salt. Ann. Occup. Hyg. 9:7-14, 1966.
48. H. D. Gibbs, Phenol tests III—the indophenol test. J. Biol. Chem. 72:649-664, 1927.
49. S. G. Rainsford and T. A. Lloyd Davies, Urinary excretion of phenol by men exposed to vapor of benzene—a screening test. Br. J. Ind. Med. 22:21-26, 1965.
50. A. B. Van Haaften and S. T. Sie, The measurement of phenol in urine by gas chromatography as a check on benzene exposure. Am. Ind. Hyg. Assoc. J. 26:52-58, 1965.
51. P. N. Cheremisinoff and A. C. Morresi, Air Pollution Sampling and Analysis Deskbook, Ann Arbor Science, Ann Arbor, Michigan, 1978.
52. H. B. Elkins, et al., The ultraviolet spectrophotometric determination of benzene in air samples adsorbed on silica gel. Anal. Chem. 34:1797-1801, 1962.
53. P. A. Maffett, et al., A direct method for the collection and determination of micro amounts of benzene or toluene in air. Am. Ind. Hyg. Assoc. Quart. 17:186-188, 1956.
54. L. D. White, et al., A convenient optimized method for the analysis of selected solvent vapors in the industrial atmosphere. Am. Ind. Hyg. Assoc. J. 31:225-232, 1970.
55. R. E. Kupel and L. D. White, Report on a modified charcoal tube. Ann. Ind. Hyg. Assoc. J. 32:456, 1971.
56. H. W. Gerarde, "The Aromatic Hydrocarbons," in Industrial Hygiene and Toxicology, Interscience, New York, 1963, pp. 1220-1225.
57. Health Effects of Benzene—A Review, National Research Council for Environmental Protection Agency, PB-254388, Washington, D.C., June 1976.

58. C. Averill, Benzole poisoning. Br. Med. J. 1:709, 1889.
59. S. Imamiya, The effects of hydrocarbons in human bodies in work environments. Kankojo Sozo 3:62-69, September 1973.
60. L. Zorina and T. M. Sukharevskaya, The mechanism of the development of the Hemorrhagi syndrome in chronic benzene intoxications (Russian). Gigiena Gruda I Prof. Zabolevaniya 10(8):26-31, August 1966.
61. A. M. Forni, et al., Chromosome changes and their evolution in subjects with past exposure to benzene. Arch. Environ. Health 23(5):385-391, November 1971.
62. R. Girard, et al., Malignant hemopathies and benzene poisoning. Med. Lavoro 62:71-76, 1971.
63. M. Aksoy, et al., Acute leukemia due to chronic exposure to benzene. Am. J. Med. 52:160-166, 1972.
64. M. Aksoy, et al., Types of leukemia in chronic benzene poisoning, a study in thirty-four patients. Acta Haematol. 55:65-72, 1976.
65. M. Aksoy, et al., Leukemia in shoe workers exposed chronically to benzene. Blood 44:837-841, 1974.
66. M. Aksoy, et al., Acute leukemia in two generations following chronic exposure to benzene. Human Heredity 24:70-74, 1974.
67. E. C. Vigliani, Leukemia associated with benzene exposure. Ann. N.Y. Acad. Sci. (U.S.A.) 271:143-151, 1976.
68. E. C. Vigliani and A. Forni, Benzene and leukemia. Environ. Res. 11: 122-127, 1976.
69. Cancer and the Worker, New York Academy of Sciences, New York, 1977.
70. M. Kissling and B. Speck, Further studies in experimental benzene-induced aplastic anemia. Blut 25:97-103, 1972.
71. S. Moeschlin and B. Speck, Experimental studies on the mechanism of action of benzene in the bone marrow. Acta Haematol. 38:104-111, 1967.
72. F. T. Hunter, Chronic exposure to benzene. J. Ind. Hyg. Toxicol. 21: 331-354, 1939.
73. L. J. Goldwater, Disturbances in the blood following exposure to benzene. J. Lab. Clin. Med. 26:957-973, 1941.
74. J. L. Hamilton-Peterson and E. Browning, Toxic effects in women exposed to industrial rubber solutions. Br. Med. J. 1:349-352, 1944.
75. L. D. Pagnotto, et al., Industrial benzene exposure from petroleum napth-I in the rubber coating industry. Am. Ind. Hyg. Assoc. 22:417-421, 1961.
76. T. A. Doskin, Effect of age on the reaction to a combination of hydrocarbons. Hygiene and Sanitation 36:379-384, 1971.
77. S. Hernberg, Prognostic aspects of benzene poisoning. Br. J. Ind. Med. 23:204-209, 1966.
78. T. W. Li and S. Freeman, The effect of protein and fat content of the diet upon the toxicity of benzene for rats. Am. J. Physiol. 145:158-165, 1945.
79. H. W. Gerarde and D. B. Ahlstrom, Toxicologic studies on hydrocarbons. Toxicol. Appl. Pharmacol. 9:185-190, 1966.
80. E. W. Lee, et al., Acute effects of benzene on Fe incorporation into circulating erythrocytes. Toxicol. Appl. Pharmacol. 27:431-433, 1974.

References

81. T. A. Kaslova and A. P. Volkova, Blood picture and phatocytic activity of leukocytes in workers having contact with benzene. Gig. Sanit. 25: 29-34, 1960.
82. R. Girard, et al., Leukocyte alkaline phosphatase and benzene exposure. Med. Lavora 61:502-508, 1970.
83. W. A. Kagan, Aplastic anemia: presence in human bone marrow of cells that suppress myelopoiesis. Proc. Natl. Acad. Sci. (U.S.A.) 73:2890-2894, 1976.
84. R. Hoffman, et al., Suppression of erythroid colony formation by lymphocytes from patients with aplastic anemia. New Eng. J. Med. 296:10-13, 1977.
85. Y. Dobashi, Influence of benzene and its metabolites on mitosis of cultured human cells. Jap. J. Ind. Health 16:453-461, 1974.
86. NIOSH Registry of Toxic Effects of Chemical Substances, U.S. Department of Health, Education, and Welfare, Public Health Service, 76-191, NIOSH, Rockville, Maryland, June 1976.
87. NIOSH Criteria for a Recommended Standard: Occupational Exposure to Alkanes, U.S. Department of Health, Education, and Welfare, Public Health Service, NIOSH, March 1977.
88. NIOSH Criteria for a Recommended Standard: Occupational Exposure to Methylene Chloride, U.S. Department of Health, Education, and Welfare, Public Health Service, NIOSH, March 1976.
89. D. MacKay and A. W. Wolkoff, Rate of evaporation of low-solubility contaminants from water bodies to atmosphere. Environ. Sci. Tech. 7(7): 611-614, 1973.
90. A. D. Green and C. E. Morrell, "Petroleum Chemicals," in Kirk-Othmer Encyclopedia of Chemical Technology, Vol. 10, 1st ed., Interscience, New York, 1973, pp. 177-210.
91. T. M. Barnes, Evaluation of Process Alternatives to Improve Control of Air Pollution from the Production of Coke, MITRE Corp., Washington, D.C., 1975.
92. P. H. Howard and P. R. Dustin, Benzene-Environmental Sources of Contamination, Ambient Levels, and Fate, Syracuse University Research Corporation, Syracuse, New York, December 1974.
93. Benzene Environmental Sources of Contamination, Ambient Levels, and Fate, Life Sciences Division, Syracuse University Research Corporation, Syracuse, New York, 1974.
94. G. S. Parkinson, Benzene in motor gasoline—an investigation into possible health hazard in and around filling stations and in normal transport operations. Ann. Occupl. Hyg. 14:145-153, 1971.
95. Occupational Exposure to Benzene at Service Stations, NIOSH Progress Report, June 9, 1976.
96. A. L. Kohl and F. C. Riesenfeld, Gas Purification, McGraw-Hill, New York, 1960.
97. G. Nonhebel, Gas Purification Processes, George Newnes Ltd., London, 1964.
98. Y. H. Kiang, Controlling benzene emissions. Paper No. 78-58.1, APCA 71st Annual Meeting, Houston, June 1978.

99. C. L. Yaws, Physical Properties, McGraw-Hill, New York, 1977.
100. A. S. Coolidge, The adsorption of vapors by charcoal. J. Am. Chem. Soc. 46:609, 1924.
101. D. F. Othmer and F. G. Sawyer, Correlating adsorption data. Ind. Eng. Chem. 35(12):1269, 1943.
102. E. Ledonx, Vapor Adsorption, Chemical Rubber Publishing Company, New York, 1945.
103. B. M. Fabass and W. H. Dubris, Carbon adsorption-electrodesorption process. Paper No. 70-68, APCA 63rd Annual Meeting, St. Louis, Missouri, 1970.
104. M. Leva, Tower Packings and Packed Tower Design, U.S. Stoneware Company, Akron, Ohio, 1953.
105. C. W. Simmons and J. D. Long, Tower-absorption coefficients. III—absorption of benzene by mineral oil. Ind. Eng. Chem. 22(7):718, 1930.
106. C. W. Simmons and H. B. Osborn, Jr., Tower-absorption coefficients. V—determination and effect of free volume. Ind. Eng. Chem. 26(5):529, 1934.
107. L. Silver and G. V. Hopton, J. Soc. Chem. Ind. (London) 61:37, 1942.
108. A. W. Hixon and C. E. Scott, Absorption of gases in spray towers. Ind. Eng. Chem. 27(3):307, 1935.
109. W. F. Gross and C. W. Simmons, Countercurrent absorption in non-aqueous systems. Trans. AIChE 40:121, 1944.
110. Y. H. Kiang, Total hazardous waste disposal through combustion. Ind. Heating, December 1977.
111. R. W. Rolke, R. D. Hawthorne, C. R. Garbett, E. R. Slate, T. T. Phillips, and G. D. Tower, Afterburner system study. EPA Report No. EPA-R2-72-062, 1972.
112. Y. H. Kiang, Liquid waste disposal system. Chem. Eng. Prog. 72(1):71, 1976,
113. Y. H. Kiang, Technology for the utilization of waste energy. IEC 23rd Annual Meeting, Los Angeles, April 1977.
114. Y. H. Kiang, Controlling vinyl chloride emissions. Chem. Eng. Prog. 72(12):37, 1976.
115. J. J. Santoleri, Chlorinated hydrocarbon waste disposal and recovery system. Chem. Eng. Prog. 69(1):68, 1973.
116. W. Hung, Results of a firetube test boiler in flue gas with hydrogen chloride and fly ash. ASME Winter Annual Meeting, Houston, November 1975.
117. Y. H. Kiang, Prevent shell corrosion for chlorinated hydrocarbon incineration. Seminar on Corrosion Problems in Air Pollution Control Equipment, sponsored by APCA, IGCI, and NACE, Atlanta, January 1978.
118. K. Lee, H. J. Jahnes, and D. C. McCauley, Thermal oxidation kinetics of selected organic compounds. Paper No. 78-58.6, APCA 71st Annual Meeting, Houston, June 1978.

Appendix A

EXPOSURE TO BENZENE; LIQUID MIXTURES: OCCUPATIONAL SAFETY
AND HEALTH STANDARDS

(Department of Labor—Occupational Safety and Health Administration; Reprinted from the Federal Register, Tuesday, June 27, 1978, Part III)

[4510-26]

Title 29—Labor

CHAPTER XVII—OCCUPATIONAL SAFETY AND HEALTH ADMINISTRATION, DEPARTMENT OF LABOR

PART 1910—OCCUPATIONAL SAFETY AND HEALTH STANDARDS

Occupational Exposure to Benzene; Liquid Mixtures

AGENCY: The Occupational Safety and Health Administration, Department of Labor.

ACTION: Final rule.

SUMMARY: This document amends the recently issued occupational safety and health standard for exposure to benzene by: (1) Exempting from all the provisions of the permanent standard for benzene (29 CFR 1910.1028), for the first three years following the effective date of this amendment, liquid mixtures containing 0.5 percent or less benzene, and thereafter liquid mixtures containing 0.1 percent or less benzene; and (2) exempting from the labeling requirements liquid mixtures containing benzene which are already packaged and which contain 5.0 percent or less benzene. The amendments are in response to several petitions concerning the applicability of the benzene standard to liquid mixtures.

EFFECTIVE DATE: June 27, 1978.

FOR FURTHER INFORMATION CONTACT:

Mr. Gail Brinkerhoff, Office of Compliance Programs, OSHA, Third Street and Constitution Avenue NW., Room N3112, Washington, D.C. 20210, telephone 202-523-8034.

SUPPLEMENTARY INFORMATION: These amendments are issued pursuant to sections 4(b) and 6(b) of the Occupational Safety and Health Act of 1970 (the Act) (84 Stat. 1592, 1593; 29 U.S.C. 653, 655), the Secretary of Labor's Order No. 8-76 (41 FR 25059) and 29 CFR Part 1911. These amendments appear at 29 CFR 1910.1028 (a)(2)(iii) and at 29 CFR 1910.1028 (k)(2)(iii).

I. BACKGROUND

Benzene is a naturally occurring constituent of crude oil and national gas produced from underground reservoirs and surfacing through wells. There are approximately 630,000 wells in some 10,000 oil and gas fields in the U.S. (exhibit 29D1). One or more reservoirs underlies each of these oil and gas fields (exhibit 29f, p. 2). Since crude oil and natural gas vary in composition and physical properties from reservoir to reservoir (exhibit 29f, p. 2), the benzene content of the fluid at a particular oil or gas field may vary from well to well.

From the wellhead, the reservoir fluid is delivered by flow lines or gathering lines to separation facilities (exhibit 29D1) or field treatment plants (exhibit 29f, p. 4) where it undergoes a number of production treatment steps necessary to produce marketable crude oil, condensates and natural gas streams, as well as a variety of hydrocarbons (exhibit 29f, p. 4).

The benzene content of crude oil ranges from below detectable limits to greater than 1 percent (exhibit 29f, p. 3). Condensates produced from natural gas liquids present in both crude oil and natural gas (exhibit 29f, pp. 2-3) have a higher percentage of benzene than crude, with that percentage ranging from approximately 0.2 to 1.0 percent by volume (exhibit 29D1). The benzene in natural gas varies from 0 to about 4 percent (exhibit 29D1).

From the separation facilities or treatment plants, the crude oil, condensates and liquid and gaseous hydrocarbons are delivered to refineries either by waterway or by pipelines (exhibit 29D1; Comeaux). It is common practice for liquids from a number of fields to be combined into one pipeline stream for transportation do refineries (exhibit 29D1; Comeaux, p. 4).

At the refineries, the crude oil and field condensate liquids are stored in holding tanks prior to processing (exhibit 29f, p. 5). Typical petroleum refining processes are many. Not all refineries have all of these processes since there is specialization, such as fuels, lubes or petro-chemical operations, within refineries (exhibit 29f, p. 5). At all refineries, however, additional benzene is generated during refining by catalytic cracker, reformer and coker operations (exhibit 29f, p. 2; tr. 284).

Petroleum refined products are many and their benzene content varies according to the content of the crude taken into the refinery, the nature or the efficiency of the refining process, and the balance of product demands on the refinery. Thus, the benzene content of product streams within a single company may vary from refinery to refinery or within a single refinery from year to year. The same product from different refineries or at different times from the same refinery, consequently, may have a different benzene content (tr. 284-5). Motor gasoline ranges from 1 to 3 percent benzene by volume (Bailey). Aviation gasoline, specialty naptha solvents and naptha-based, (type B) jet fuels may exceed 1 percent benzene concentrations. Heavier jet fuel (type A), light fuel oils and cutback asphalts may occasionally exceed 0.01 percent benzene. Petrochemical feedstock napthas and certain aromatics, such as toluene may contain up to 1 percent benzene (exhibit 29f, pp. 7-8). This petrochemical feedstock is used to produce ethylbenzene, styrene monomer, cumene, phenol, cyclohexane, and nitrobenzene.

Processed refined petroleum products are distributed through marketing channels, including bulk terminals and bulk plants, to consumers. Initial distribution from the refinery storage facilities may be by pipeline, barge, marine tankers, tank truck or tank car (exhibit 29f, p. 11). These refined products have numerous uses, such as fuels, extractants, processing aids, and solvents in paints, surface coatings, adhesives and pesticides, inks, etc. The benzene content of these products varies from less than one-tenth to a few percent (tr. 284, comment 53, 58, 21, 35).

There are some substitutes for benzene; however, most of the solvent substitutes themselves contain benzene. As indicated below, the percentage of benzene in some substitutes may range up to 4 percent (18-45; tr. 304).

Benzene is also derived from coal. Recovery of coal-derived benzene, primarily as a by-product of the coking process in steel mills, accounts, however, for only 6 percent of the total U.S. production (43 FR 5918). The light oil, which is condensed from coke gases, contains up to 70 percent benzene. This light oil is distilled to produce benzene. Most light oil plants do not produce benzene but sell their light oil to petroleum refineries for further processing. Only 10 light oil plants do produce benzene and these sell it to other users.

II. HISTORY OF REGULATION

On February 10, 1978, a permanent occupational safety and health standard regulating occupational exposure to benzene was published in the FEDERAL REGISTER (43 FR 5918) as 29 CFR 1910.1028. (A correction document was published on March 31, 1978, at 43 FR 13561). This standard required employers to take prescribed measures to control employee exposure to benzene. The standard applied to each place of employment where benzene in any quantity was produced, reacted, released, packaged, repackaged, stored, transported, handled or used. The standard contained no percentage exclusion and applied to work operations involving any amount of benzene.

The permanent benzene standard was developed, pursuant to sections 6(b) and 6(c) of the act, after exhaustive rulemaking which commenced with publication on May 3, 1977 of an Emergency Temporary Standard for Occupational Exposure to Benzene (42 FR 22516). The emergency temporary standard exempted liquid mixtures containing 1 percent or less of benzene volume, or the vapors released from

Exposure to Benzene; Liquid Mixtures

these liquids. On May 27, 1977, OSHA published a proposed permanent standard to control occupational exposure to benzene (42 FR 27452). This proposed standard would also have exempted work operations where the only exposure to benzene was from liquid mixtures containing 1.0 percent (0.1 percent after 1 year) or less of benzene by volume, or the vapors released from these liquids. On the basis of the record developed in the rulemaking on the permanent benzene standard (Docket H-059), OSHA concluded that there was no consistent predictable relationship shown between the percentage of benzene in a liquid mixture and the resultant airborne exposure to benzene and that, consequently, the percentage proposed exclusion could not be supported (43 FR 5942).

After promulgation of the final standard which contained no percentage exclusion, OSHA received requests from several employers and employer groups for an administrative stay of the standard or other relief from the provisions of the standard as it applied to work operations where exposure to benzene resulted from liquid mixtures containing small or "trace" amounts of benzene. For reasons set forth in its notice (43 FR 12890, March 28, 1978), OSHA proposed to amend the benzene standard to exclude from its coverage work operations where exposure to benzene is from liquid mixtures containing 0.1 percent or less benzene or the vapors from such liquids. Skin contact with such mixtures would also be excluded from the standard. Pending final action on this proposed amendment, OSHA stayed the application of the provisions of the benzene standard to such work operations (43 FR 12891). Since this stay was immediately effective, OSHA commenced an expedited rulemaking to resolve the percentage exclusion question so that employers could know which, if any, of their operations were excluded from the permanent benzene standard. (A correction document was published on April 4, 1978, at 43 FR 14071). On April 28, 1978, OSHA published a notice of hearing on the proposed amendment (43 FR 18215). The public hearing on this proposal was held on May 23 and May 24 at Washington, D.C. Approximately 40 individuals participated at this hearing. Furthermore, more than 100 comments, arguments, and views were received from interested parties. The verbatim transcript of this hearing, as well as the numerous comments, exhibits, and briefs submitted to OSHA before, during and after the hearing, are part of this rulemaking record, along with portions of the record in the earlier benzene proceeding which were relevant to the issues herein. The rulemaking record was originally scheduled to close on June 8, 1978, but, at the request of participants, was kept open until June 12, 1978.

These amendments are based on a careful consideration of the entire record of the informal rulemaking hearing, including the transcript, exhibits, and prehearing and post-hearing written comments. Copies of the official list of hearing exhibits, comments, and notices of intent to appear at the hearing can be obtained from the Docket Office, Docket H-059A, Room S6212, U.S. Department of Labor, 3rd Street and Constitution Avenue NW., Washington, D.C. 20210.

III. Regulatory Analysis and Environmental Impact

Since an economic impact statement for the permanent standard (43 FR 5918) was prepared by OSHA pursuant to Executive Orders 11821 and 11949, OSHA has not performed a separate regulatory analysis pursuant to Executive Order 12044 (43 FR 12661, March 24, 1978). The economic analysis for the permanent standard considered the economic impact of compliance on employers with operations utilizing liquid mixtures containing any amount of benzene (43 FR 5934-5941). This amendment exempts from the permanent standard operations utilizing 0.5 percent or less benzene (0.1 percent or less after 3 years) and, therefore, reduces the cost of compliance with the standard.

In view of the fact that the amended standard does not require that employers reduce the amount of benzene in their products, costs involved in reducing the percentage of benzene in liquid mixtures in order to avoid coverage by this standard have not been attributed to the amended standard.

The final environmental impact statement published January 1978, prior to the issuance of the permanent benzene standard concluded that the standard would have a beneficial impact on the workplace environment and also that beneficial effects on the environment external to the workplace may also be anticipated. This amendment to the permanent benzene standard may result in employers switching to other solvents (containing lower benzene contamination levels). However, the atmospheric impact of such action is not expected to be significant. Where reformulation is accomplished to reduce organic solvent usage, the atmosphere will realize a reduction in total hydrocarbon load. As these types of actions were addressed in the final environmental impact statement, this amendment is not expected to alter the basic conclusions contained in that document. Accordingly, it was concluded that no new environmental impact statement was necessary for this amendment.

IV. Principal Issues Involved

The following is a discussion of the major issues involved in the rulemaking on the proposed percentage exclusion and an analysis of the evidence submitted into the record. The exhibit numbers refer to the certified exhibit list of docket H-059A. The first number designates the particular exhibit on that list. Where the exhibit contains more than one item, the second number references the particular item of the exhibit. The designation "tr." refers to the transcript of the hearing on percentage exclusion and indicates the pages of that transcript which are referenced. Furthermore, the designation "PC" refers to post hearing comments submitted by interested parties. All references are intended to provide examples of record support for the information stated.

(1) WHETHER WORK OPERATIONS SHOULD BE EXEMPT FROM § 1910.1028 WHERE THE SOLE OCCUPATIONAL EXPOSURE TO BENZENE IS FROM LIQUID MIXTURES CONTAINING LESS THAN A SPECIFIED PERCENTAGE OF BENZENE

The majority of those responding recommended that the benzene standard be amended to exclude from its coverage liquid mixtures containing less than a specified percentage of benzene. Reasons in support of this form of exemption varied. Some participants argued that exposure to low levels of benzene does not result in any health hazard, particularly leukemia (tr. 225, 391, 313, and exhibits 18-34, 63, 77, 84, 91). Other participants pointed out that they do not add benzene to their products but that benzene is an unavoidable contaminant in all petroleum based products (tr. 68, 78, 88, 106, 107, 120, 129, 132, 163, 168, 187, 215, and exhibits 18-41, 93, PC-5-8). Oil and gas producers stressed that benzene is a natural contaminant of crude oil and gas and that it is technologically infeasible to remove benzene at the wells (exhibits 18-77, 85, 1, 19). Refiners contended that it is economically prohibitive to remove benzene from their process streams, particularly within a short timeframe (exhibit 18-48). Most industry participants presented data indicating that the low levels of benzene in the liquid mixtures present in their workplaces produced airborne concentrations below the standard's permissible exposure limit of 1 ppm. (This data is discussed below under issue 6.)

Objections to any percentage exclusion for airborne exposure were raised by the Oil, Chemical & Atomic Workers (tr. 400). OCAW argued that, even where low percentages of benzene were present in liquid streams, refinery process units were subject to frequent leaks resulting in employee exposures above 1 ppm. OCAW contended that only where process units are well maintained, can exposures be reduced to the permissible exposure limit, but that such high quality maintenance cannot be guaranteed (tr. 402).

The United Rubber, Cork, Linoleum & Plastic Workers of America (URW) contended by way of comment that while reduction of the amount of benzene in a liquid can result in reduction of airborne concentrations to conform with a PEL of 1 ppm, reduced levels cannot be attained in certain processes unless ventilation control is provided. URW also objected to permitting dermal contact with benzene mixtures where there has been no documentation of a "no-risk" health factor to benzene on skin contact, and recommended that temporary variances be sought by industry in situations where the use of protective clothing is a problem (exhibit 18-33).

The Public Citizen Health Research Group opposes the proposed amendment on the grounds that there is no safe level for exposure to a carcinogen and that liquid mixtures containing 0.1 percent benzene have been shown to give rise to exposure levels in excess of the PEL of 1 ppm (exhibit 18-94).

For different reasons, the American Petroleum Institute also urged that a percentage exclusion not be adopted, other than possibly for dermal exposure. API's objections to a percentage exclusion were based on the view that the medium or matrix in which benzene is found, the environmental conditions of temperature, humidity, air movement, and physical volume of space, the physical nature of the system in which benzene is found—namely whether in a closed or open system, and work practices are all critical to reduction of benzene exposure. In lieu of a percentage exclusion, API proposed that OSHA limit the scope of the benzene standard by adopting appropriate triggering mechanisms for activation of individual provisions of the standard. Thus, initial monitoring of airborne exposures should occur, in API's view, only when professionals, acting for the employer, have reason to believe that employee exposure is above the level which triggers the monitoring requirement (tr. 353).

Dr. Hervey B. Elkins testified on behalf of AISI that a percentage exemption would be appropriate inasmuch as naturally occurring benzene is found as a contaminant in many liquids employed in industry, formulation of many products will involve ingredients which often contain benzene in small quantities, and many products are made by chemical reactions employing benzene as a raw material. Dr. Elkins indicated that to completely eliminate the last traces of benzene from these materials may be very difficult (PC 82).

(2) WHETHER 0.1 PERCENT BENZENE IS THE APPROPRIATE PERCENTAGE FOR EXEMPTION, OR WHETHER THE PERCENTAGE FIGURE SHOULD BE HIGHER OR LOWER

Some participants supporting a percentage exemption indicated that 0.1 percent would be of no significant benefit to them since the benzene contaminated mixtures they used contained benzene in excess of that figure. They also argued that a higher percentage exemption would be appropriate since the airborne concentrations arising from the higher benzene content mixtures did not generally exceed 1 ppm.

Further, industry participants provided evidence that, at present, many suppliers cannot furnish the various benzene-contaminated materials used in their processes with a benzene content as low as 0.1 percent. (This evidence is discussed in detail in issue 5.) For example, tire manufacturers testified that the rubber solvents used in tire building could not be obtained with any confidence that the benzene content would not be up to at least 0.7 percent. They maintained, therefore, that a percentage exemption of at least 0.7 percent would be appropriate, especially to provide relief from the requirement for protective clothing which is infeasible in these operations (tr. 68, 87, 108, 170, 188, 212, 215).

The Adhesives & Sealant Council (tr. 224), Wilhold Glues (tr. 224), St. Clair Rubber Co. (tr. 260), Miracle Adhesive Corp. (tr. 268), DuPont (tr. 313), the American Iron & Steel Institute (tr. 396), and others (exhibits 18-16, 21, 61, 53, 55, 56, 45, 108) suggested that, because of the unavailability of low benzene solvents and the lack of potential for exposure above 1 ppm, that a 1.0 percent exemption would be appropriate. The B. F. Goodrich Co. indicated that while they would be willing to comply with a 0.1 percent exemption as it would apply to generation of airborne concentrations, a special action level of 0.5 percent for activation of dermal protection requirements would be necessary for them to be able to comply with the standard. Other participants argued that 0.1 percent is inappropriate since exposures below 1 ppm occur in such operations as oil and gas production with crude oil and gas liquids containing up to 4 percent benzene (exhibit 18-71, PC 81).

Industry participants endorsing a 0.1 percent exclusion did so primarily because the benzene containing materials used in their operations generally contained less than 0.1 percent benzene and they would thus be exempt from the standard (exhibits 18-29, 37, 46, 70, 89, 106, 107).

Dr. Hervey Elkins testified that, while it may not be possible to establish liquid percentage limits that will guarantee, with complete certainty in every possible situation, airborne concentrations that will not exceed a specified level, it would be logical and appropriate to accept a particular percentage exemption which would most likely, based on calculations of available data, be consistent with a permissible exposure limit of 1 ppm. The calculations provided by Dr. Elkins were based on theoretical considerations, and on extrapolation from data in papers written by him. The mean of all values of benzene percentage limits in liquid mixtures consistent with a permissible exposure level of 1 ppm was calculated by Dr. Elkins to be 0.5 percent (PC 82).

(3) WHAT IS THE CURRENT PERCENTAGE (OR RANGE OF PERCENTAGES) OF BENZENE IN LIQUID PRODUCTS, SUCH AS PRINTING INKS, PETROLEUM SOLVENTS SUITABLE FOR TIRE BUILDING, ADHESIVES, SEALANTS, PAINTS, COATINGS, DETERGENTS, INSECTICIDES, DISINFECTANTS, WAXES, FLOOR FINISHES, CRUDE OIL, PETROLEUM SOLVENTS OF VARYING GRADES, GAS LIQUIDS, LUBRICATING OILS, PETROCHEMICALS, GASOLINE IN REFINERY STREAMS, AND SIMILAR PRODUCTS

Participants who furnished information on this issue generally indicated that, since benzene is primarily a contaminant rather than an intended final ingredient, the benzene content of liquid products would vary not only from supplier to supplier but also from the same supplier. The American Petroleum Institute has indicated that production factors, which cause varying benzene content in their products, include differences in crude runs, the nature and efficiency of individual refinery process and the specific balance of final products produced at any point in time (PC 104).

Tire manufacturers reported rubber solvents used in tire building as containing up to 1.0 percent benzene, with only one supplier claiming capability of supplying solvents with 0.1 percent benzene or less. However, industry analysis of that company's shipments indicated benzene content from 0.097 to 0.137 percent (tr. 88, 95, exhibits 18-45). Petroleum based solvents used in the formulation of adhesives and glues was reported to vary from 0.3 percent to 3 percent (exhibits 18-26), while the finished products and other products such as paints, waxes, floor coatings, and printing inks, normally contain benzene well below 0.1 percent. The benzene content of gasoline was reported as covering a range from 0.5 to 3.5 percent with most samples containing less than 2.0 percent benzene (tr. 344). The American Petroleum Institute submitted results of analyses for benzene on 1,007 samples of crude oil, liquid condensate, natural gas liquids, and gas plant streams. A summary of this data reveals 49 percent of the samples containing less than 0.1 percent benzene, 34 percent containing between 0.1 and 0.5 percent benzene, 10 percent containing between 0.5 and 1.0 percent benzene, 5 percent containing between 1.0 and 2.0 percent benzene, and 2 percent of the samples containing greater than 2.0 percent benzene (PC 81). Rohm & Haas Co. reported that none of their 2,768 products, primarily synthesized organic chemicals, contained greater than 0.1 percent benzene and that benzene was an unintended contaminant in all but 2 of those products (exhibits 18-31).

(4) WHAT IS THE CURRENT PERCENTAGE (OR RANGE OF PERCENTAGES) OF BENZENE IN OTHER SOLVENTS COMMONLY USED AS SUBSTITUTES FOR BENZENE SUCH AS TOLUENE, XYLENE, HEXANE, AND SOLVENT NAPTHAS

Data provided on this issue came primarily from companies not using benzene as a raw material but using materials such as toluene and xylene and other solvents contaminated with benzene. It was indicated, however, by the rubber manufacturers and adhesives and sealant manufacturers, that reformulation or substitution of materials containing no benzene would be infeasible and could possibly diminish the integrity of the final product (tr. 101, 213, 254, 315).

The benzene levels reported in other solvents were somewhat varied, again due to different suppliers, process conditions and equipment used in their manufacture. As a result, it is difficult to identify specific percentages of benzene content in substitute substances with any degree of accuracy. For example, the Adhesives & Sealant Council, Inc., reported that members whose hexane suppliers are located on the west coast show benzene levels of 0.3 percent or higher, while members in other parts of the country purchase hexane with benzene content typically below 0.1 percent. Ranges of benzene impurity reported by the Council include hexane 0.001-2.3 percent; toluene 0.005-0.24 percent; rubber solvent up to 1.0 percent; naptha 0.1-1.0 percent; and mineral spirits up to 1.0 percent (exhibits 18-45). The National Association of Printing Ink Manufacturers, Inc., indicated the following range of benzene content: heptane 0.1 percent-0.75 percent; heptane ("aromatic free") 0.01-0.02 percent; toluene 0.02-0.1 percent; lactol spirits 0.02-0.1 percent; xylene 0.15 percent; and aliphatic hydrocarbon blend 0.01 percent to 0.02 percent (exhibits 18-83). Other participants reported benzene at or below the 0.1 percent level in toluene, xylene, hexane and solvent napthas (exhibits 18-15, 28, 35, 38, 39, 46, 47, 58, 69, 88) while others reported levels in excess of 0.1 percent (exhibits 18-35, 39, 61, 70, 84, 90) with one report of the benzene content of toluene to be 15 percent (exhibits 18-61).

(5) TO WHAT LEVELS CAN THE PERCENTAGE OF BENZENE BE FEASIBILY REDUCED

Those participants, such as the rubber manufacturers and the adhesives and sealant manufacturers, who obtain their solvents from other sources, could not provide any detailed feasibility assessment since they did not have knowledge of the technological potential of their suppliers to reduce the percentages further than present levels. Further complications involved the refusal of solvent suppliers to guarantee a specific benzene level and the fluctuating benzene content found in a given solvent from supplier to supplier. Some suppliers appear reluctant to certify particular solvents since they have not designed their processes with benzene content specification as a consideration. The Manufacturing & Chemists Association testified that a feasibility assessment can be made only after their determination of the benzene content of streams and products is complete (tr. 286). Solvent purchasers generally indicated that, if the solvent industry is capable of reducing the percentage of benzene over some period of time, they would support a graduated exemption setting lower acceptable levels in the future to coincide with the technological ability of the solvent industry to deliver those percentages in dependable volume shipments sufficient to maintain production and employment (exhibit 18-41, 80).

The only projection relative to benzene content reduction was in terms of costs. Texaco estimated that benzene reduction in their refinery streams from present levels to 1.0 percent would cost $85 million, a reduction to 0.5 percent would cost $508 million, and a reduction to 0.2 percent would cost $1,381 million. Detailed analysis of how these figures were arrived at however, was not made available (exhibit 18-48). Another study, which addressed the economic impact of reducing benzene content, is the Arthur D. Little study on gasoline, prepared for EPA in February of 1978 (exhibit 12b).

(6) WHAT AIRBORNE CONCENTRATIONS OF BENZENE RESULT FROM THE PERCENTAGE OF BENZENE IN EACH PRODUCT

Comments and testimony presented on this issue generally revealed that in the industrial setting the benzene content of the liquid mixture plays only a part in the resultant airborne concentration to which employees may be exposed. The evidence regarding the other factors affecting airborne concentration are discussed under issue 7.

Monitoring data submitted to the record has provided a reasonable description of expected exposure levels in the various industrial segments affected by the proposed amendment. As previously discussed, Dr. Hervey Elkins submitted calculations indicating that a 0.5 percent benzene content in liquid mixtures would appropriately be consistent with a 1 ppm permissible exposure limit. The monitoring data submitted tends, for most all cases, to support Dr. Elkins' calculations. For example, General Tire & Rubber Co. submitted data for tire building operations which indicated resultant benzene exposures from rubber solvents containing greater than 0.5 percent benzene did exceed 1 ppm, while solvents with benzene content below 0.5 percent did not (exhibit 18-41). Exposures reported by B. F. Goodrich resulting from up to 0.3 percent benzene content indicate that the companion air levels of benzene were below 0.5 ppm in 95 percent of the cases and only one out of 255 samples was as high as 1.1 ppm (exhibit 18-75). While the Goodyear data tended to support the proposition that exposure levels, in practice, will generally not exceed 1 ppm with low benzene content liquid mixtures, Goodyear's data from one tire manufacturing plant using from 0.263 to 0.280 percent benzene content rubber solvents indicated exposures between 1.11 and 1.98 ppm. Goodyear attributed these exposure levels to inadequate ventilation since other monitoring data they submitted from other tire building plants using up to 0.4 percent benzene rubber solvents showed exposure levels not in excess of 0.4 ppm. The adhesive and sealant manufacturers reported that, with liquid mixtures containing up to 1.0 percent benzene, exposure levels never exceeded 1 ppm (exhibit 18-45). Printing ink manufacturers submitted data indicating that, while solvents used contained in some cases in excess of 0.1 percent benzene, exposure levels never exceeded 0.3 ppm (exhibit 18-83). DuPont provided results of tests from paint spraying operations showing that the use of liquid mixtures of 1 percent benzene will not result in airborne concentrations of benzene greater than 5 ppm, in most cases less than 0.5 ppm. Tests were performed with DuPont topcoat paints with 0.05 percent, 0.1 percent, and 1.0 percent benzene added. Benzene was intentionally added to the paints in an attempt to correlate benzene levels in the liquids with airborne benzene concentrations. The highest benzene level obtained was 4.3 ppm with a 1.0 percent benzene spike. Averaged over an 8-hour day, DuPont estimates that normal spraying operations with nonspiked paint will result in employee exposures below 1 ppm. Other DuPont data involving handling of 0.2 to 4.3 percent benzene in gasoline and 0.19 to 0.45 percent benzene in a p-xylene process stream resulted in time-weighted averages up to 0.38 ppm and 0.15 ppm respectively (exhibit 18-39).

Edison Electric Institute reported that exposures resulting from handling fuel oils containing 0.003 to 0.005 percent benzene resulted in exposures less than 0.2 ppm (exhibit 18-56). Husky Oil Co. reported refinery streams with 0.71 percent benzene and finished gasoline with 0.35 percent benzene as producing exposures of less than 0.11 ppm and 0.042 ppm respectively (exhibit 18-64). Earlier data submitted by NIOSH indicated that fuel oil containing less than 0.1 percent benzene gave rise to benzene concentrations of 60 ppm under conditions of elevated temperature, confined space and possibly inadequate ventilation. NIOSH indicated, however, that the sampling was done with detector tubes which are relatively inaccurate and subject to numerous interferences. (tr. 753–755, July 25, 1977, hearing)

While other participants (exhibits 18-24, 35, 49, 84, 87) supported the proposition that benzene levels in liquid mixtures greater than 0.1 percent would still not generally produce exposures to benzene above 1 ppm, some participants provided data to the contrary. The United Rubber, Cork Linoleum & Plastic Workers of America (URW) (exhibit 18-33) referenced testimony presented by Dr. Robert T. Harris, of the University of North Carolina at the public hearing on the proposed benzene standard (docket H-059), in which Dr. Harris stated that bulk solvent percentages of benzene from 0.5 to 1.07 percent can produce exposure levels below and above 1 ppm, some as high as 12 ppm. The reliability of the results of this study were questioned by Dr. Curtis Smith of the Manufacturing Chemists Association who indicated that the findings as to the presence of benzene in the ambient air could have been artificially high because of interference of ketones which are present in the ambient air around rubber plants (tr. 299). URW submitted other exposure data from an Armstrong tire manufacturing facility (the "Harvard study") also indicating exposure levels above 1 ppm with solvents containing low levels of benzene. A number of participants, including Armstrong, questioned the reliability of the Harvard study (PC 89, 90, 91, 92, 105). Armstrong commented on the study as follows:

The report itself makes it clear that no analyses were made by the Harvard study during the week of January 9, 1978, with respect to the bulk samples. Reference is made to the analysis by the Armstrong laboratory in June 1977 that benzene in Texol at that time was less than 0.1 volume percent. More than 6 months elapsed between such sampling of the solvent and the air sampling. It is to be regretted that in the only monitoring done to date with respect to an Armstrong facility, and a very limited sampling at that, the monitoring was not accompanied by an analysis of the solvent in use at the time of such monitoring.

In addition, the sample period did not cover the full 8-hour period for computation of the time weighted average (TWA). The TWA, in fact, was estimated from a sampling time of approximately 2 to 3 hours. NIOSH in its Occupational Exposure Sampling Strategy Manual, DHEW (NIOSH) Publication No. 77-173 (January 1977) is critical of partial period sampling. The Manual states, at page 40, that the "sampled portion of the period should cover at least 70 percent to 80 percent of the full period." Indeed, in discussing the validity of a 6-hour TWA exposure average as compared with an 8-hour TWA standard, the Manual states, at page 41, that "[t]his type of measurement should be avoided if possible." It therefore follows that a significantly shorter period is even less valid (PC-91).

Other participants, sharing reservations relative to a particular percent exemption, did so on the grounds that exposure measurements from liquid mixtures containing 0.1 percent or less benzene did or could theoretically produce airborne concentrations of benzene in excess of 1 ppm or that variables such as ventilation, liquid temperature, work practices, etc. affected resultant benzene airborne concentrations to too great an extent to confidently rely on only benzene content for limitation of employee exposures (exhibits 18-32, 61, 72, 79, 92, 94, PC-S-8).

(7) TO WHAT EXTENT DO VARIOUS FACTORS, SUCH AS TEMPERATURE, DILUTION WITH AMBIENT AIR, WORK FACTORS, NATURE OF OPERATIONS, ETC., AFFECT THE RELATIONSHIP BETWEEN THE PERCENTAGES OF BENZENE IN THE VARIOUS PRODUCTS AND RESULTING AIRBORNE CONCENTRATIONS, AND SHOULD ANY EXEMPTION BE LIMITED TO THE USE OF THESE PRODUCTS UNDER PARTICULAR CIRCUMSTANCES

As previously indicated, most participants agreed that factors other than benzene content can play a significant role in resultant airborne concentrations of benzene. Goodyear stated that ventilation, climate, work habits, etc., would definitely affect the airborne concentration of benzene, and submitted monitoring data from an older tire manufacturing facility and two newer plants for comparison. The two newer facilities using rubber solvent containing 0.4 percent and 0.2 to 0.3 percent benzene experienced airborne concentrations of benzene at 0.45 and 0.25 ppm respectively. The older tire plant using rubber solvent with 0.263–0.280 percent benzene experienced airborne concentrations of 1.11 to 1.98 ppm. Goodyear attributed this discrepancy to inadequate ventilation in the older tire manufacturing facility. Goodyear further indicated that, while they could not determine an exact correlation from their data, variables other than benzene content could cause up to an 80 percent variance in resulting airborne concentrations. (PCS-8). Other participants provided either theoretical or actual data indicating exposures greater than 1 ppm from low benzene content mixtures used under conditions of confined space, inadequate ventilation, elevated temperatures etc. (exhibit 18-32, 92). URW submitted previous testimony of Dr. Robert T. Harris presented at the public hearing on the proposed benzene standard (docket H-059) in which Dr. Harris stated that "* * * the lower the benzene content of the solvent, the lower the potential for exposure, but a low benzene content of solvent alone does not assure that a particular air concentration will not be exceeded" (exhibit 18-33).

As mentioned previously, Dr. Hervey Elkins submitted calculations estimating that a 0.5 percent benzene content would not normally be expected to produce airborne concentrations of benzene greater than 1 ppm. Dr. Elkins addressed the affect of other factors on his calculations:

It is self-evident that the factors mentioned (temperature, ventilation, work practices, nature of operations) will affect the relationship between the percentage of benzene in a liquid product and the resulting concentration of benzene vapor in the air. Important factors not mentioned include the quantity of substance consumed or processed, the area of the liquid surface which is exposed to the air, and the vicosity of the liquid.

The exemption percentages proposed (i.e., 0.5 percent to meet 1 ppm) were based on rather severe conditions: complete or free evaporation of the benzene and other volatile ingredients, limited only by the provision that the permissible limits of the vapors of the other components of the liquid must not be exceeded.

With some high boiling liquids, such as heavy oils and tars, there is little evaporation of the base material, even when heated to temperatures well above the boiling point of benzene. In theory, processes employing such materials at elevated temperatures and with large surface areas exposed (as might occur in certain coating or impregnating processes) could produce concentrations of benzene vapor in the surrounding air in excess of the postulated permissible exposure levels, even when the concentration of benzene in the liquid is less than the limits recommended above.

On the other hand, handling and processing such liquids in enclosed or partially enclosed spaces, with limited exposure of liquid surfaces to air, and at temperature such that the material is not highly fluid, would result in benzene concentrations well below the postulated permissible exposure levels, even when the percentage of benzene in the liquid is in excess of the limits recommended above.

For the sake of simplicity, it is recommended that * * * the 0.5-percent exemption percentage be applied if the airborne permissible exposure limit is 1 ppm—with the proviso that if operations involving heating of the liquids and exposure of large surface areas are carried out, at least a one time monitoring of the area of benzene in air be done, if the benzene content of the liquid exceeds 0.1 percent. (PC-82)

API argued in its post-hearing brief that the record evidence does not identify a specific percentage of benzene which will guarantee that a given PEL will not be exceeded under all work environments and that, therefore, any percentage exemption must also take account of the roles of many other controlling factors. API further stated that, regardless of the benzene content in their streams, the nature of their oil and gas operations is the most important factor in considering potential exposures and need for regulation of their segment of the industry. API witness Dr. William G. Domask testified that:

It is evident from the data and information presented here that well-maintained, closed-system operations in general represent a low risk for exposure to benzene.
Specifically, petroleum production, pipeline, and marketing personnel air monitoring data reflect a very low risk of exposure to benzene vapor at all concentration levels of benzene in the liquids handled by these segments of the industry.
Similarly, data for the closed-system portions of refining operations indicate a very low level of exposure to benzene (PC-81).

While many other participants addressed this issue and acknowledged the significance of factors other than benzene content relative to resultant airborne concentrations of benzene, there were no recommendations relative to a practical translation of requirements for a homogeneous regulatory conclusion which would give full weight to those factors.

(8) TO WHAT EXTENT, AND FOR WHICH LIQUID MIXTURES, ARE BENZENE-FREE SUBSTITUTE AVAILABLE

Most industry participants argued that benzene-free substitutes are not commercially available, are not suitable for their process due to incompatibility with other materials and existing air pollution control systems and methods, are economically prohibitive, or would reduce product performance (exhibit 18-15, 21, 39, 41, 54, 80, 90, 109). The Public Citizen Health Research Group argued that benzene substitutes are available (exhibit 18-94).

However, the record indicates that "benzene-free" substitutes are not generally available for the majority of industrial uses.

(9) IF LIQUID MIXTURES OF SPECIFIED PERCENTAGES ARE EXEMPT FROM THE OTHER PROVISIONS OF THE BENZENE STANDARD, SHOULD THEY NEVERTHELESS BE SUBJECT TO THE LABELLING REQUIREMENTS OF THE STANDARD. CONVERSELY, IF SUCH MIXTURES ARE NOT EXEMPT FROM THIS STANDARD, SHOULD THEY BE EXEMPT FROM THE LABELLING REQUIREMENTS

Industry participants indicated that some relief from the labeling requirements was necessary. They argued that, without some labeling exemption, liquids which contain trace amounts of benzene would require cancer hazard labels. They objected to such labeling on the following grounds: lack of associated health hazard requiring warning; economic burdens; uncontrolled use of warning labels dilutes effectiveness of warning messages; and, the ubiquitous nature of benzene in petroleum distillates. The recommendation of industry participants was that any liquid mixture exemption of a specified percentage should include an exemption from the labeling requirements ((exhibit 18-21, 29, 31, 37, 41, 42, 69, 71, 72, 79, 83).

DuPont testified that, without a labeling exemption to allow additional time to clear existing inventories carrying levels of 0.1 to 1.0 percent benzene, the cost of locating already packaged containers and labeling them would run to about $31 million (tr. 317). DeSoto, a manufacturer of paints, industrial coatings, detergents, furniture and fireplace accessories, estimated their cost of labeling present inventory at $700,000 (exhibit 18-43). Another paint and coatings manufacturer, Pratt & Lambert, Inc., estimated a current inventory of 1.4 million containers, which without an exemption, would cost $700,000 to label (18-44).

(10) OTHER RELATED ISSUES

(a) *Monitoring feasibility.* A number of industry participants provided data indicating potential difficulties in obtaining industrial hygiene and analytical services. The cost burden and reported lack of monitoring service capabilities in some locations which would result if some exemption were not provided to part of the industry, would render compliance with parts of benzene standard impossible (exhibit 18-1, 3, 9, 12, 20, 37, 39, 65, 68, 82, 87).

The Manufacturing Chemists Association stated that analysis for benzene requires equipment which is expensive and which must be operated by specialists. They maintained that, due to interference from other substances, multiple gas chromatography analyses on a single product or use of mass spectrometric equipment together with gas chromatography, available only in large laboratories, are required to produce good analytical results (PC-103). NIOSH addressed the problem of interferences in a post-hearing submission (PC-88) in which they report that work by Levadie and MacAskill, "Analytical Chemistry," 48, 76, 1976, and by Esposito and Jacobs, "American Industrial Hygiene Journal," 38, No. 8, 401, 1977, describes the modifications to the NIOSH sampling and analytical methods for benzene, S311, necessary to solve the problem of interferences arising from ketone materials in the analysis for benzene. NIOSH also noted that the price of mass spectrometer systems has declined steadily over the last five years. It is NIOSH's position, therefore, that benzene can be identified with confidence using the NIOSH recommended method, that interferences can be handled by simple extraction techniques prior to gas chromatography, that detectors or columns are readily available and within the normal financial limitations of most commercial laboratories, and that recourse to mass spectrometry is unnecessary.

(b) *Miscellaneous comments.* Due to the unique nature of the motor carrier industry whose employee exposure to benzene is occasional and then only minimal, the National Tank Truck Carriers, Inc. recommended that if OSHA has any concern over motor carrier's employee exposure to benzene, OSHA should petition the Department of Transportation to develop regulations, thus avoiding regulatory imposition by another agency over an already regulated sector (tr. 414).

The National Agricultural Chemicals Association argued that OSHA has no authority to require labeling of pesticide products and no need to duplicate or interfere with the labeling controls imposed by the Environmental Protection Agency (exhibit 18-53).

The National Retail Merchants Association urged an exemption from the standard for retail stores since employee exposure in those stores is from consumer goods which are in closed containers. They also argued that, while paint cans are opened for mixing and coloring, the brief and intermittent nature of the operation would not result in sufficient benzene exposures to merit regulation. NRMA further stated that the standard should not apply to existing inventories since retailers would then be unable to sell products they already owned (exhibit 18-37).

V. ANALYSIS AND FEASIBILITY

OSHA has concluded that it is appropriate to amend the permanent benzene standard to provide for a percentage exclusion. Specifically the agency has exempted from all the provisions of the standard for the first 3 years following the effective date of this amendment, liquid mixtures containing 0.5 percent or less benzene, and thereafter liquid mixtures containing 0.1 percent or less benzene. Further, OSHA has exempted from the labelling requirements liquid mixtures containing less than 5.0 percent benzene which are already packaged.

OSHA recognizes that the scope of the permanent benzene standard (29 CFR 1910.1028), unamended, is so broad as to encompass work place operations utilizing liquid mixtures with any amount of benzene however small (tr. 30). OSHA's view that there is no level of benzene exposure that is without some attendant health risk remains unchanged. However, because of the ubiquitous nature of benzene, i.e., its presence in a myriad number and type of worksites (benzene is a contaminant in most, if not all petroleum-based liquid mixtures), OSHA believes that it is proper to focus industrial hygiene and medical resources on those operations with higher exposures and which present the greatest potential risk to worker health. This decision is in accord with the evidence developed during the recent rulemaking which revealed the need to and appropriateness of limiting the scope of the standard.

The mechanism chosen to effectuate this relief must in OSHA's view be consistent with the intent of the permanent standard, which is to minimize the risk to worker health to the greatest extent feasible. The framework within which the agency has examined this issue was articulated by the Director of OSHA's health standards programs, Grover C. Wrenn, at the outset of the informal hearing:

Since the standard itself establishes a permissible exposure limit for benzene, we would certainly endeavor to set an exemption which assured that employees exposed to materials exempt from coverage under the regulation are subject to no greater exposure than employees who are subject to the provisions and protection of the standard (tr. 16).

And one of the questions that was raised in the earlier rulemaking and one of the questions that is raised here is the question of the likelihood that exempting any particular category of materials from regulation, under the benzene standard—the likelihood that that exemption would provide a basis for being confident that workers involved with those exempt materials would not be exposed in a manner that was intended to be avoided by the benzene regulation (tr. 16).

The agency has considered and reviewed several approaches for limiting the scope of the permanent benzene standard suggested by participants to the rulemaking. These options included a single percentage exclusion applicable to all provisions of the standard, as was set forth in OSHA's proposed amendment in the FEDERAL REGISTER notices of March 28, 1978 (43 FR 12890) and April 28, 1978 (43 FR 18215). Such an approach was recommended by many participants as cited under the discussion of issue 1. Another option called for a general percentage exclusion, but with different levels for certain provisions of the permanent standard. This type of amendment, which was suggested by the Rubber Manufacturers Association and member companies (Exh. 92), specifically called for a higher percentage exclusion for the dermal and labelling provisions of the standard relative to other sections. A third option was that relief from the standard by based not on a percentage exclusion (with perhaps the exception of the dermal provision) but rather be based upon actual employee exposue levels. This recommendation was made by API, NPRA and member companies and also supported by OCAW (Ech. 104, Tr. 400). A final option would include a general percentage exclusion with the additional proviso that, in the case of severe or unusual work situations, the permissible airborne limits should not be exceeded as determined by monitoring. (Exh 27-C.1, PC-82)

OSHA has carefully evaluated these and other possible approaches and, based upon a review of the evidence and views contained in the rulemaking proceedings, has concluded that a percentage exclusion amendment applicable to all provisions of the standard most adequately satisfies, for regulatory purposes, the dual intention of appropriately limiting the scope of the permanent benzene standard while not exposing exempted employees to greater exposures than employees covered by the standard.

The record evidence of percentage exclusion rulemaking establishes, as shown in the preceding discussion of issues, that a variety of liquid mixtures with small or "trace" amounts of benzene generally result in exposure levels below the permissible exposure limit of the permanent benzene standard in a wide variety of industries. OSHA has determined, by examining the relationship of the percentages of benzene to resultant exposure levels, that generally an appropriate percentage exclusion can assure that employees who would be exempt from the coverage of the benzene standard are not exposed above the level set in the standard (tr. 16). Furthermore, it is OSHA's view that a percentage exclusion will encourage employers to act to reduce the amount of benzene present in liquid mixtures utilized in their workplaces or present in their products and, therefore, reduce the potential health hazard to employees. The record clearly establishes that, since publication of the permanent benzene standard, paint manufacturers and other solvent users have already examined the need for benzene in their liquid mixtures, and have made extensive efforts to obtain solvents with lower percentages of benzene. OSHA believes that an appropriate percentage exclusion will be an incentive to many other employers to reduce benzene levels and thus minimize the leukemia and other health risks to their employees.

In arriving at the conclusion that an "across-the-board" percentage exclusion is the appropriate means to afford an opportunity for relief from the standard without subjecting workers to undue risk, OSHA recognizes that factors other than the benzene content of liquid mixtures can act to significantly modify the resulting levels of airborne exposure. Variables such as the nature of work operations and work practices, and quantifiable parameters such as temperature, size of evaporative area, and especially ventilation all have been shown to play an important role. This data thus supports Dr. Harris's earlier conclusions that there is no necessary correlation between the amount of benzene in liquid mixtures and resulting airborne levels. However, defining the parameters and determining the exact combination of factors which significantly increase exposure, a necessary accomplishment for regulation on this basis, is not possible and no witness could suggest means of doing so. Moreover, the preponderance of evidence submitted to this rulemaking (most of which was not available at the earlier rulemaking) manifestly establishes, based on objective sampling data derived from current industrial settings, that for the vast majority of worksites, small amounts of benzene in liquid mixtures do not result in worker exposures above 1 ppm.

An additional reason for not adopting alternative strategies to amend the permanent standard is that the record did not provide evidence for the need to provide different exemption levels for different provisions. Manufacturers of rubber goods argued that an "action-level" concept be applied to the dermal section of the standard because of the infeasibility of performing certain manual operations in their industry without some skin contact with benzene-containing solvents. However, as was explained in the preamble to the permanent standard, from the point of view of choosing a "safe" level, the permissible exposure limit should be zero (exhibit 3A). In the case of airborne exposure, clearly this was not attainable and their airborne permissible exposure limits established were not "no-effect" levels, rather were based on feasibility considerations. However, with respect to dermal contact, avoidance of skin exposure is feasible for workers in most industrial sectors simply by the use of suitable protective clothing, such as impermeable gloves. OSHA recognizes that in tire-building, the record evidence shows that at present there are no suitable methods available to prevent skin contact with solvents containing a small amount of benzene. However, the record evidence establishes that rubber solvents with 0.5 percent benzene, which are suitable for use in tire operations, are already available in sufficient quantities. Accordingly, the exemption of 0.5 percent liquid mixtures from all the provisions of the benzene standard, will substantially relieve the feasibility problems in tire building operations. Furthermore, the 3 year stepdown provision of the amendment from 0.5 to 0.1 percent exclusion levels will allow time for increased production of solvents containing lower amounts of benzene and for development and evaluation of alternative methods of compliance with the standard's dermal provision.

Some participants suggested that a higher exemption level apply to labeling (Tr. 68, 188). OSHA does not agree. The labeling requirement of the permanent standard serves to inform the worker of the hazard associated with working with benzene containing liquid mixtures. Use of a higher percentage exemption for this provision could result in the employee not being apprised of the danger in situations where exposure might be excessive.

API has argued that an "across the board" percentage exclusion is not an effective method by which to amend the standard, and that, in general, an exemption predicated on exposure levels is a better way of dealing with the problem presented by liquid mixtures containing small quantities of benzene (PC 104). While acknowledging that there are other factors in addition to benzene concentration which may significantly affect airborne exposure levels, OSHA believes that there is sufficient record evidence which demonstrates that with low levels of benzene contamination airborne exposures are generally below the PEL and frequently below the action level. Many participants in the rulemaking also supported this conclusion. Furthermore, API's recommendation that various provisions of the permanent standard be triggered by workers' exposure levels is, to a great degree, already incorporated into the standard by OSHA's use of the action level concept. The objective of the benzene standard is to provide necessary protection to employees from the hazards of benzene exposure, and to this end, the standard imposes upon employers different compliance requirements depending on the level of employee exposure, with minimal requirements imposed where employee exposure is below the action level.

The primary difference between API's suggested use of an "exposure determination" and OSHA's decision reflected in the standard is that API would not require initial monitoring in all cases but rather would rely upon professional judgment to determine whether various provisions of the standard apply. However, it should be noted that API's judgment as to which operations in the petroleum industry have low exposures is based upon objective sampling data submitted to this record; to the extent that such data exists the initial monitoring requirements of the permanent standard may be satisfied.

OSHA recognizes that conditions, such as elevated temperatures, inadequate ventilation, confined space, quantity of material used etc., could in some cases act to produce exposures above the PEL even if the same benzene percentages would in other operations result in exposures less than the PEL. However, as already stated, the record does not provide a sufficient basis upon which to identify and define these variables for regulatory purposes.

The results of monitoring of employee exposure submitted by numerous participants demonstrates that, by and large, the use or presence of liquid mixtures containing benzene with maximum concentrations of 0.5 percent benzene results in airborne concentrations of less than 1 ppm and frequently less than the action level of 0.5 ppm. However, the low airborne concentrations reported are not necessarily due to the inherent nature of the liquid mixtures utilized (exhibit 18-32), but rather are in part the result of the maintenance and effectiveness of engineering controls to limit exposures. Therefore, maintenance of these low exposure levels can be dependent upon the continued use of engineering controls. This is illustrated by the testimony of Goodyear which shows that, in contrast to the majority of their operations, which indicated that use of solvents containing up to 0.4 percent did not result in exposure levels greater than 0.4 ppm, in one plant with inadequate ventilation, exposures greater than 1 ppm were observed when the solvent utilized contained only 0.26 to 0.28 percent benzene. (PC-S-8). In the case of refineries, although the benzene content of liquid streams may range up to 3.5 percent by volume or higher, the great majority of worker exposures are less than 0.5 ppm (tr. 339-345.) It is evident that such low exposures in the petroleum industry are due to the outdoor setting and, importantly, to the use of closed systems. Although not always explicitly stated, the low exposures measured in some situations, such as the use of benzene-containing solvents by rubber manufacturers, appears to be the result of engineering controls as well as effective work practices.

Close examination of Dr. Elkins' calculations also indicates that a percentage exclusion level of 0.5 percent may not be sufficiently conservative. His calculated average of all values of benzene percentage limits, which was consistent with exposures of 10 ppm (the time-weighted average PEL of the previous standard) was approximately 1 percent when utilizing a total upper limit concentration of 500 ppm for petroleum distillates (PC-82). If the same underlying assumption is used (rather than utilizing NIOSH's recommended 350 milligrams per cubic meter which is not presently in effect) to calculate the percentage of benzene in a liquid mixture consistent with the 1 ppm PEL of the permanent standard, the mean value would decline approximately 10-fold or to a level of about 0.1 percent benzene in liquid mixtures (tr. 393-94). Elkins' data takes into account abnormal worst case work situations. Therefore, a 0.1 percent exclusion level should maintain workplace levels at below the permissible exposure limit even in the abnormal work situations cited by Dr. Elkins.

In adopting an across the board exclusion level, two problems arise: (1) In some cases, where the benzene stream content is in excess of the prescribed percentage and the standard there applies, exposures may well be below the PEL or action level; and (2) in certain cases, the exemption of work operations where the benzene content of liquids are below the prescribed exclusion level may produce exposures in excess of the PEL.

In the first case, although exposures may be low, the potential exists for excessive exposures, such as in the case of leaks, spills and process upsets from enclosed systems. The agency believes, that because of this potential, the requirements of the permanent standard are necessary for the protection of employees working in such areas. In the second situation, this agency recognizes that adoption of a 0.1 percent exclusion level does not, in all cases, assure that resulting airborne exposures will necessarily not exceed the action level of 0.5 ppm or even the permissible exposure limit of 1 ppm. OSHA further recognize that 0.5 percent factors other than the percentage of benzene may become more significant in determining exposure levels. However, the record does not indicate that exposures greater than 1 ppm have been demonstrated to be commonly found in existing industrial situations where benzene levels in liquid mixtures are less then 0.1 percent.

Ideally, the percentage exclusion level chosen should be so low as to assure that in all instances, the PEL will not be exceeded. Participants representing users of benzene-containing solvent mixtures who recommended exemption levels of 0.5 to 1 percent, conceded that they would readily utilize solvents containing even less benzene if they were available. While some participants from industry testified that they were able to obtain solvents with less than 0.1 percent on a regular basis, the record evidence demonstrates that for most industrial processes liquid mixtures containing 0.1 percent or less benzene are not commercially available at this time.

Thus, the percentage exemption level chosen by OSHA for amendment purposes must take into account the feasibility of supplying large volumes of liquids containing less than a specific amount of benzene to a multitude of industrial users.

Testimony from the producers of liquid mixtures which contain varying amounts of benzene indicate that the benzene which is present is there only as a contaminant and generally is not intentionally added to produce a particular property in the formulations or products. Furthermore, while it appears to be impossible to exclude very small amounts of benzene in many of these products, MCA witnesses observed that the technology currently exists to produce liquid mixtures containing 0.1 percent or less of benzene (tr. 278). However, their testimony also demonstrated that reduction of benzene content to the 0.1 percent level is not a simple undertaking. Because benzene is a widespread contaminant in most petroleum-based liquid products, some time will be required to complete testing to determine the current levels of benzene in their products. Once this process is complete, industry will require an additional period of time to implement appropriate process changes which would assure low level benzene content. This latter phase is in some instances complex, as the percent of benzene is dependent upon existing production factors, such as differences in crude runs, the nature and efficiency of individual refinery processes and the specific balance of final products produced at a given point in time. In addition, it appears that market conditions may also be a significant factor determining the degree to which benzene has been extracted from petroleum-based streams. The record also indicates that only in the recent past have producers of benzene-containing liquids given significant consideration to the benzene content of their products (tr. 286).

Two studies addressed the economic feasibility of reducing benzene content in certain liquid mixtures (exhibit 18-48; exhibit 12B). However, it is OSHA's view that the cost of reducing benzene content are not attributable to this percentage exclusion amendment. Reducing the benzene content in liquid mixtures in order to avoid coverage by the benzene standard is an option which the employer may exercise; it is not a requirement of the benzene standard that he do so. Employers may prefer to comply with the permanent standard, particularly where their employees are exposed below the action level, in which case the employer would need only to conduct initial measurement, record that measurement and train his employees. Since the standard does not compel reduction of benzene content, it is OSHA's view that costs involved in reducing the percentage of benzene in a liquid are not a consideration in this amendment.

Other than indicating that considerable time would be required to complete analysis of benzene content of their products, (exhibit 18-87) industry did not provide estimates nor specific recommendations as to the time frames required to effect production changes in order to produce liquid mixtures containing 0.1 percent of less of benzene in sufficient quantity to meet the anticipated needs of downstream users. Because of the above considerations, industry participants felt tht OSHA should not at this time adopt a percentage exclusion at the level of 0.1 percent as proposed by the agency. However, it should be noted that the proposed permanent regulation, which did indicate that the agency was considering a 0.1 percent exclusion level, was published over a year prior to the most recent rulemaking. Review of the evidence submitted indicates that while not commercially available on a sufficiently large basis, mixtures with 0.1 percent or less benzene content have been produced and are available on a limited basis and are compatible with most processes requiring such mixtures.

The record evidence further demonstrates that mixtures with a benzene content of up to 0.5 percent are presently available in sufficient quantity to satisfy the needs of affected industrial segments and that suppliers of these liquid mixtures are attempting to reduce the benzene content still further. OSHA has, therefore, concluded that 0.1 percent benzene content can be feasibly attained on a commercial basis at some point in the future.

It is the judgment of the Agency that a period of 3 years is a sufficient and reasonable allowance for development and implementation of means and methods necessary for production of adequate supplies of 0.1 percent benzene content mixtures. Along with the consideration of feasibility, the record evidence indicating a relative lack of suitable "benzene-free" substitutes, dictates the need for providing a period for implementation which would meet the anticipated demand for solvents containing low percentages of benzene. In addition, the 3-year period before the stepdown to the 0.1 percent level, will allow those users of benzene-containing liquid mixtures to test for product integrity before commencing reformulation on a large-scale basis.

By adoption of the 0.5 percent exclusion level initially, many employers engaged in crude oil and gas production activities will be provided relief from all provisions of the permanent standard. The agency is aware that, by decreasing the exclusion level to 0.1 percent after 3 years, many of the facilities and employees in the petroleum production sector will then be covered by the standard. It is also recognized that since benzene is a naturally-occurring constituent of crude oil and natural gas, its level is not under the control of the employer. However, since compliance requirements of the permanent standard are directly related to the exposure level of the employees, and since it has been demonstrated that exposure levels of personnel in oil and gas production are generally below the action level of 0.5 ppm, the requirements of the permanent standard are minimal beyond the taking of an initial representative exposure measurement. The 3-year delay prior to the step-down to the 0.1 percent level also provides additional time for employers to obtain the required sampling information.

This amendment also exempts from the labelling requirement, liquid mixtures containing 5.0 percent or less benzene if the liquid mixture is already packaged on the effective date of this amendment. Record evidence indicates that there may be a large number of containers already in the channels of commerce, particularly consumer products, that would be subject to the labelling requirement if such an exemption were not provided. While the labelling of existing containers may be possible, it is also clear that it would require a substantial effort. Thus, imposition of the labelling requirement for liquid mixtures already packaged might well result in an excessive disruption of the commercial framework. For liquid mixtures packaged after the effective date of these amendments, the 0.5 percent exemption applies. Those employers who utilize or manufacture liquids containing 0.5 percent or less benzene will have 3 years to meet the labelling requirements, and may be exempt entirely if in that period of time they can reduce benzene concentrations below 0.1 percent.

OSHA has chosen a level of 5 percent or less benzene (by volume) contamination as the boundary for exempting liquid mixtures already packaged. Evidence developed during the rulemaking indicated that there may be some products already in the channels of commerce which may contain benzene well in excess of 1 percent and which may not have been analyzed and which would be difficult to track down. In addition, the 5 percent level chosen is similar to the requirement of the Consumer Product Safety Commission which, under the Federal Hazardous Substance Act Regulations (16 CFR 1500.14(a)(3)), mandates that products with 5 percent or more benzene receive a special label. In products with an average molecular weight similar to that of benzene, 5 percent by volume is similar to 5.0 percent by weight.

It is the agency's judgment that much of the National Tank Truck Carrier's Association's concerns regarding the imposition of the provisions of the permanent standard on their members will be substantially relieved as users of various liquid mixtures demand products containing less than the prescribed percentage levels of benzene. In the absence of the exercise of authority by the Department of Transportation in this matter, § 1910.1028, as amended, applies to this industry (sec. 4(b)(1)).

EFFECTIVE DATE

This amendment is effective immediately on June 27, 1978. Since this amendment is a rule granting an exemption, the Administrative Procedure Act (5 U.S.C. 553) does not require a 30-day period before the amendment becomes effective. Section 533(d)(1) of the APA exempts rules which grant an exemption or relieve a restriction from the requirement that publication of a substantive rule be made not less than 30 days before its effective date. This amendment exempts from the permanent benzene standard operations utilizing liquid mixtures of 0.5 percent or less (0.1 percent or less after 3 years). Without this amendment, all employers with such operations would be required to implement the various protective requirements of the permanent standard, such as initial monitoring, training and recordkeeping. The amendment relieves them of this burden.

In addition the amendment exempts from the labelling requirements liquid mixtures which are already packaged in containers and which contain 5 percent or less benzene. Such containers would otherwise be subject to the labelling requirements of the standard. The amendment, therefore, relieves employers of the requirements to label these containers.

Accordingly, these amendments are effective June 27, 1978. As with the other operations exempted from § 1910.1028, the benzene standard contained in Table Z-2 of § 1910.1000 will continue to apply to the operations exempted by the amendments.

Upon the publication of this amendment, the limited administrative stay adopted by OSHA in conjunction with this rulemaking (43 FR 12891) is no longer in effect.

VII. AUTHORITY

This document was prepared under the direction of Eula Bingham, Assistant Secretary of Labor for Occupational Safety and Health, U.S. Department of Labor, 200 Constitution Avenue NW., Washington, D.C. 20210.

Accordingly, pursuant to section 4(b)(2) and 6(b) of the Occupational Safety and Health Act of 1970 (84 Stat. 1592, 1593, 29 U.S.C. 653, 655), the specific statutes referred to in section 4(b)(2), Secretary of Labor's Order No. 8-76 (41 FR 25059), and 29 CFR Part 1911, Part 1910 of Title 29, Code of Federal Regulations, is hereby amended by adding new paragraphs (a)(2)(iii) and (k)(2)(iii) to 29 CFR 1910.1028.

Signed at Washington, D.C., this 21st day of June 1978.

EULA BINGHAM,
Assistant Secretary of Labor.

Part 1910 of Title 29 of the Code of Federal Regulations is hereby amended by adding a new paragraph (a)(2)(iii) and a new paragraph (k)(2)(iii) to § 1910.1028 to read as follows:

§ 1910.1028 **Benzene.**

(a) *Scope and application.* * * *

(2) This section does not apply to:

* * * * *

(iii) Work operations where the only exposure to benzene is from liquid mixtures containing 0.5 percent (0.1 percent after June 27, 1981) or less of benzene by volume, or the vapors released from such liquids.

* * * * *

(k) *Signs and labels.* * * *

(2) The employer shall assure that caution labels are affixed to all containers of benzene and of products containing any amount of benzene, except:

* * * * *

(iii) Liquid mixtures containing 5.0 percent or less benzene by volume which were packaged before June 27, 1978.

(Secs. 4, 6, 84 Stat. 1593 (29 U.S.C. 653, 655); Secretary of Labor's Order 8-76 (41 FR 25059); 29 CFR Part 1911.)

[FR Doc. 78-17633 Filed 6-21-78; 3:16 pm.]

Appendix B

OCCUPATIONAL EXPOSURE TO BENZENE: OCCUPATIONAL SAFETY AND HEALTH STANDARDS

(Department of Labor—Occupational Safety and Health Administration; Reprinted from the Federal Register, Friday, February 10, 1978, Part II)

Appendix B

[4510-26]

Title 29—Labor

CHAPTER XVII—OCCUPATIONAL SAFETY AND HEALTH ADMINISTRATION, DEPARTMENT OF LABOR

PART 1910—OCCUPATIONAL SAFETY AND HEALTH STANDARDS

Occupational Exposure to Benzene

AGENCY: The Occupational Safety and Health Administration, Department of Labor.

ACTION: Permanent standard for the regulation of benzene.

SUMMARY: This standard is based on a determination by the Occupational Safety and Health Administration (OSHA) that the available scientific evidence establishes that employee exposure to benzene presents a cancer hazard—specifically, the hazard of developing leukemia. Therefore, in accordance with OSHA's regulatory approach to the control of employee exposure to carcinogens, this standard limits employee exposure to benzene to the lowest feasible level, in this case 1 part benzene per million parts of air (1 ppm) as an 8 hour time—weighted average concentration, with a ceiling level of 5 ppm for any 15 minute period during the 8 hour day. The standard also prescribes limits on eye and skin contact with benzene.

The standard provides for the measurement of employee exposure, engineering controls, work practices, personal protective clothing and equipment, signs and labels, employee training, medical surveillance and recordkeeping.

EFFECTIVE DATE: March 13, 1978.

FOR FURTHER INFORMATION CONTACT:

Mr. Gail Brinkerhoff, Office of Compliance Programs, OSHA, Third Street and Constitution Avenue NW., Room N3112, Washington, D.C. 20210, telephone 202-523-8034.

SUPPLEMENTARY INFORMATION: This permanent Occupational Safety and Health standard is issued pursuant to sections 6(b), 6(c) and 8(c) of the Occupational Safety and Health Act of 1970 (the Act) (84 Stat. 1593, 1596, 1599; 29 U.S.C. 655, 657), the Secretary of Labor's Order No. 8-76 (41 FR 25059) and 29 CFR Part 1911. The new standard on occupational exposure to benzene which appears at 29 CFR 1910.1028, applies to all employment in all industries covered by the Act. For reasons set out below, the standard does not apply to the distribution or use of gasoline and other fuels, used as fuels, subsequent to discharge from bulk terminals. Moreover, the standard applies labelling and training requirements only to sealed, intact containers of benzene.

This document also amends Table Z-2 of 29 CFR 1910.1000 by adding a footnote which provides that benzene exposures not covered by the new § 1910.1028 are still covered by the exposure level and other requirements of § 1910.1000. Pursuant to section 4(b)(2) of the Act, OSHA has determined that this standard is more effective than corresponding standards now applicable to the maritime and construction industries and currently contained in Subpart B of Part 1910, and Parts 1915, 1916, 1917, 1918 and 1926 of Title 29, Code of Federal Regulations. Therefore, those corresponding standards are superseded by the new standards in § 1910.1028. A new paragraph (c) is added to § 1910.19 to clarify the applicability of this new benzene standard to the construction and maritime industries.

I. BACKGROUND

Benzene (C_6H_6) is a clear, colorless, non-corrosive, highly flammable liquid with a strong, rather pleasant odor. Benzene's low boiling point and high vapor pressure cause it to evaporate rapidly under ordinary atmospheric conditions, giving off vapors nearly three times heavier than air.

Benzene is produced primarily by the petrochemical and petroleum refining industries by a process called catalytic reformation, which converts certain lower octane hydrocarbons into higher octane aromatics. These two industries are responsible for 94 percent of the total U.S. production of benzene. Recovery through catalytic reformation, including the benzene formed from the hydroalkylation of toluene, accounts for almost 80 percent of the total quantity produced. Recovery of coal-derived benzene, primarily as a by-product of the coking process in steel mills, was once the major source of benzene. Today, however, it accounts for only 6 percent of the total U.S. production.

The production of benzene is rapidly expanding with approximately 11 billion pounds produced in 1976. Only eleven other chemicals and only one other hydrocarbon (ethylene) are produced in greater tonnage in the U.S. Approximately 86 percent of this benzene is used chiefly as an intermediate in the production of other organic chemicals, including styrene, phenol, and cyclohexane. The remaining amount is used primarily in the manufacture of detergents, pesticides, solvents and paint removers. Benzene is also present as a component of motor fuels, averaging less than 2 percent in gasoline.

The first major industrial use of benzene, however, was as a solvent in the rubber industry just preceding World War I. During World War I, benzene production was stimulated greatly by the demand for and resulting production of toluene in the manufacture of explosives. The large quantities of benzene which were produced, resulted in its more widespread use as a starting point for the manufacture of various organic compounds. This situation led to greatly increased uses of benzene as a solvent in the artificial leather, rubber goods, and rotogravure industries.

Industries and processes currently using benzene include the chemical, printing, lithograph, rubber cements, rubber fabricating, paint, varnish, stain removers, adhesives, and petroleum industries. Benzene is also used extensively in chemical laboratories as a solvent and as a reactant in numerous chemical applications. Where benzene is produced and used in large amounts, it is generally used in enclosed systems, although exposures can occur during liquid transfer operations, from equipment leakage and carryover losses, and in maintenance operations.

II. HISTORY OF REGULATION

Benzene has been recognized as a toxic substance capable of causing acute or chronic effects since 1900. In 1927, on the basis of extensive examination of exposed workmen and animal inhalation data, Winslow recommended an exposure limit of 100 ppm for benzene (Ex. 156-3, Annex D).[1] In 1934, partially as a result of the fact that benzene toxicity in the shoe leather industries was a serious problem in that state, the Massachusetts Department of Labor and Industries established a Division of Occupational Hygiene (Ex. 156-3, p. 2). Relying on reports by Bowditch, Hunter, Mallory and Elkins of cases of benzene poisoning occurring at concentrations below 100 ppm, the Massachusetts Division of Occupational Hygiene reduced the maximum acceptable limit (MAC) to 75 ppm (Ex. 156-3, p. 6-7). In the 1940's, as a result of blood abnormalities and one death among leather workers exposed to benzene concentrations ranging from 40 to 80 ppm, Massachusetts lowered the permissable limit of benzene exposure to 35 ppm (Elkins Ex. p. 7).

The American Conference of Governmental Industrial Hygienists

---

[1] The exhibit numbers used in this document refer to the certified exhibit list of the benzene rulemaking proceeding. The first number designates the particular exhibit on that list. Where the exhibit contains more than one item, the second number references the particular item of the exhibit. The designation "PC" refers to posthearing comments in Exhibit 217. The designation "Tr" refers to the transcript of the benzene hearing and indicates the pages of that transcript which are referenced.

All references in this document are intended to provide examples of record support for the information stated.

(ACGIH) recommended in 1946 a threshold limit value (TLV) for benzene exposure of 100 ppm. This TLV was reduced in 1947 to 50 ppm. In 1948, following Massachusetts' lead, ACGIH adopted a TLV of 35 ppm. In 1963, a TLV of 25 ppm was proposed by the ACGIH. The effects of benzene noted by ACGIH at this time were blood changes, aplastic anemia and other blood dyscrasias. No mention was made of any association of leukemia with benzene exposure (Ex. 191). It was not until 1974 that the ACGIH adopted the TLV of 10 ppm which had sometime earlier been recommended by the American National Standards Institute (Ex. 156-3, p. 7).

The present OSHA standard for benzene (29 CFR Part 1910.1000, Table Z-2) was adopted in 1971 from the Z 37.4—1969 consensus standard of the American National Standards Institute (ANSI). The OSHA standard was adopted without rulemaking under the authority of section 6(a) of the Act. It prescribes, as the ANSI standard, an 8-hour TWA of 10 ppm with an acceptable ceiling concentration of 25 ppm and, in addition, allows excursions above the ceiling to a maximum peak concentration not to exceed 50 ppm for more than 10 minutes in any 8-hour work period. Neither the ANSI standard nor the resultant OSHA standard was based on the possible leukemogenic effects of exposure to benzene.

In 1974, pursuant to section 22(d) of the Act, the Director of NIOSH submitted to the Secretary of Labor a criteria document concerning occupational exposure to benzene which stated that "the possibility that benzene can induce leukemia cannot be dismissed." (Ex. 32A, p. 1). However, NIOSH recommended retention of the existing permissible exposure limit of 10 ppm and ceiling concentration of 25 ppm as measured over a 10 minute period. This recommendation was not based on benzene's potential leukemia hazard.

In a letter to the Secretary of Labor, dated April 23, 1976, the United Rubber, Cork, Linoleum, and Plastic Workers of America urged that an emergency temporary standard regulating occupational exposure to benzene be issued (Ex. 2-42). This request was denied on May 18, 1976 by then Secretary of Labor, William J. Usery (Ex. 2-43).

Also in 1976, the National Academy of Sciences, under contract with the United States Environmental Protection Agency, reviewed the literature concerning health effects of benzene exposure (Ex. 2-4). The Academy concluded that benzene must be considered a suspect leukemogen.

In August 1976, NIOSH submitted to OSHA an updated criteria document which revised its earlier assessment of 1974 (Ex. 2-6). On the basis of a review of old studies and new data, NIOSH concluded in that document that benzene was a leukemogen. This report further pointed out that "it is apparent from the literature that benzene leukemia continues to be reported." NIOSH, therefore, recommended that since no safe level for benzene exposure could be established that, "no worker be exposed to benzene in excess of 1 ppm in air." Following publication of the updated criteria document, the Director of NIOSH recommended to the Assistant Secretary of Labor, by letter dated October 27, 1976, that OSHA publish an emergency temporary standard for benzene establishing the exposure level at 1 ppm (Ex. 2-6).

Based on the information supplied by NIOSH, OSHA issued on January 14, 1977, voluntary "Guidelines for Control of Occupational Exposure to Benzene," recommending that exposure to benzene in air not exceed an 8-hour time-weighted average to 1 ppm in any 8-hour shift of a 40 hour week (Ex. 2-44).

In January 1977, NIOSH informed OSHA that workplace environments had been found in St. Mary's and Akron, Ohio where a sufficient number of employees had been exposed to benzene for a number of years to facilitate an epidemiological study of health risks (Ex. 2-45). The worksite was a manufacturing plant owned by Goodyear Tire and Rubber Company which utilized benzene at various stages in the production of pliofilm. The preliminary conclusions of the epidemiological study, which NIOSH conducted of the pliofilm workers, were transmitted to OSHA on April 15, 1977. In his letter of April 15, 1977, transmitting this report, the Director of NIOSH again urged that an emergency standard be issued (Ex. 2-7).

On May 3, 1977, the Assistant Secretary for OSHA issued an Emergecy Temporary Standard for Occupational Exposure to Benzene (42 FR 22516), pursuant to sections 6(c) and 8(c) of the Act, Secretary of Labor's Order No. 8-76, and 29 CFR Part 1911. A correction document was published on May 10, 1977 (41 FR 23601), and an amendment to the emergency temporary standard was published on May 24, 1977 (42 CFR 26429). The evidence and findings supporting issuance of the emergency temporary standard and its amendment and a discussion of its provisions are set forth in the aforementioned FEDERAL REGISTER publications. The emergency temporary standard was to have been effective on May 21, 1977. However, as a result of challenges to that standard, filed both in the Court of Appeals for the District of Columbia (*Industrial Union, AFL-CIO v. Bingham*, No. 77-1395) and in the Court of Appeals for the Fifth Circuit (*API v. OSHA*, No. 77-1516), a temporary restraining order was issued by the Fifth Circuit on May 20, 1977, and the standard never officially went into effect.

On May 27, 1977, OSHA published a proposed permanent standard to control occupational exposure to benzene (42 FR 27452). The emergency temporary standard and its preamble, which the new proposal supplemented, were incorporated in that proposal. The FEDERAL REGISTER document setting forth the proposal also contained a notice of hearing scheduling an informal public hearing to be held pursuant to section 6(b)(3) of the Act, and requesting the submission of written comments, data, views and arguments on all the issues raised by the proposed permanent standard and the emergency temporary standard. Subsequently, on June 24, 1977 (42 FR 32263), OSHA excluded from the scope of the benzene hearing and from the final permanent standard those activities related to the storage, transportation, distribution, dispensing and sale of gasoline as a fuel subsequent to its discharge from bulk terminals. OSHA explained in that notice its intention to assess the regulatory action to be taken to protect workers involved in these activities after conclusion of the deliberations of a joint EPA-NIOSH-OSHA Task Force.

The public hearings on the benzene proposal were held July 19 through August 10, 1977. A total of 95 individuals appeared at these hearings as witnesses. Among the witnesses were employers and employer associations from a variety of industries: petroleum refining, petrochemical, oil and gas production, aviation fueling; and coke ovens and coke by-products. In addition, representatives of the affected workforce, including a number of employees who have been exposed to benzene, unions, government agencies, public interest groups and other interested parties appeared. Furthermore, comments were received from representatives of other industries, such as analytical and research laboratories, paint manufacturing, construction, maritime, and rubber manufacturing and from users of pure benzene as well as users of benzene contaminated solvents. Public participation was representative of a large segment of the benzene users. The verbatim transcript of the hearings, as well as the numerous comments, exhibits and briefs submitted to OSHA before, during and after the hearings, are part of this rulemaking record, along with other relevant documents. The hearing record was originally scheduled to close on August 20, 1977 but, at the request of industry participants, the record was kept open until September 2, 1977 for the submission of additional evidence and until September 27, 1977 for the submission of briefs, summaries and arguments.

In conjunction with the development of the proposed standard, OSHA prepared a draft environmental impact statement. The draft environmental statement was published in the FEDERAL REGISTER (42 FR 27455). On June 17, 1977, the Council on Environmental Quality published a notice of availability of the benzene draft environmental impact statement (Ex 7). In addition to the 45 day comment period specified in 29 CFR 1999.4(g), the environmental impact of the proposed standard was also an issue for the benzene hearing as provided by 29 CFR 1999.4(h) and the notice of proposed rulemaking (42 FR 27452). A notice of availability of the final environmental impact statement for benzene was published on February 3, 1978 by EPA (43 FR 4674).

In addition to the draft environmental impact statement, OSHA prepared an economic and inflationary impact assessment of the proposed standard evaluating factors relevant under section 6(b) of the Act (29 U.S.C. 655 (b)(5), Secretary of Labor's Order 15-75 (40 FR 54484) and Executive Orders Nos. 11821 (39 FR 41501) and 11949 (42 FR 1017). The notice of the proposed standard indicated that the economic impact of this proposal was to be considered at the hearing (42 FR 27452) and certified that the economic and inflationary impact of the proposed standard has been carefully evaluated in accordance with Executive Orders 11821 and 11949.

This permanent benzene standard is based on a careful consideration of the entire record in this proceeding, including materials relied on in the emergency temporary standard, materials referenced in the proposal, and the record of the informal rulemaking hearing including the transcript, exhibits and pre-hearing and post-hearing written comments. Copies of the official list of hearing exhibits, comments, and notices of intent to appear at the hearing can be obtained from the Docket Office, Docket H-059, Room S6212, U.S. Department of Labor, 3rd Street and Constitution Avenue NW., Washington, D.C. 20210.

III. PERTINENT LEGAL AUTHORITY

The primary purpose of the Act is to assure, so far as possible, safe and healthful working conditions for every working man and woman. One means prescribed by Congress to achieve this goal is the authority vested in the Secretary of Labor to set mandatory safety and health standards.

Occupational safety and health standards provide notice of the requisite conduct or exposure level and provide a basis for assuring the existence of safe and healthful workplaces. The act provides that:

The Secretary, in promulgating standards dealing with toxic materials or harmful physical agents under this subsection, shall set the standard which most adequately assures, to the extent feasible, on the basis of the best available evidence, that no employee will suffer material impairment of health or functional capacity even if such employee has regular exposure to the hazard dealt with by such standard for the period of his working life. Development of standards under this subsection shall be based upon research, demonstrations, experiments, and such other information as may be appropriate. In addition to the attainment of the highest degree of health and safety protection for the employee, other considerations shall be the latest available scientific data in the field, the feasibility of the standards, and experience gained under this and other health and safety laws. (Section 6(b)(5).)

Sections 2(b) (5) and (6), (20), (21), (22), and (24) of the Act reflect Congress' recognition that conclusive medical or scientific evidence including causative factors, epidemiological studies or dose-response data may not exist for many toxic materials or harmful physical agents. Nevertheless, standards cannot be postponed because definitive medical or scientific evidence is not currently available. Indeed, standards need only be based on the best available evidence. The legislative history makes it clear that "it is not intended that the Secretary be paralyzed by debate surrounding diverse medical opinion." House Committee on Education and Labor, Report No. 91-1291, 91st Cong., 2d Session, p. 18 (1970). This Congressional judgment is supported by the courts which have reviewed standards promulgated under the Act. In sustaining the standard for occupational exposure to vinyl chloride (29 CFR 1910.1017), the U.S. Court of Appeals for the Second Circuit stated that "it remains the duty of the Secretary to act to protect the working man, and to act even in circumstances where existing methodology or research is deficient." "Society of the Plastics Industry Inc. v. Occupational Safety and Health Administration," 509 F. 2d 1301 (C.A. 2 1975). cert. den. 95 S. Ct. 1998, 4 L. Ed. 2d 482 (1975). A similar rationale was applied by the U.S. Court of Appeals for the District of Columbia Circuit in reviewing the standard for occupational exposure to asbestos (29 CFR 1910.1001). The Court stated that:

Some of the questions involved in the promulgation of these standards are on the frontiers of scientific knowledge, and consequently as to them insufficient data is presently available to make a fully informed factual determination. Decision-making must in that circumstance depend to a greater extent upon policy judgments and less upon purely factual judgments.

"Industrial Union Department, *AFL-CIO* v. *Hodgson*," 499 F. 2d 467, 474 (C.A.D.C. 1974).

In setting standards, the Secretary is expressly required to consider the feasibility of the proposed standards. Senate Committee on Labor and Public Welfare, S. Rep. No. 91-1282, 91st Cong., 2d Sess., p. 58 (1970). Nevertheless, considerations of technological feasibility are not limited to devices already developed and in use. Standards may require improvements in existing technologies or require the development of new technology. "*Society of the Plastic Industry, Inc.* v. *Occupational Safety and Health Administration,*" supra at 1309.

Where appropriate, the standards are required to include provisions for labels or other forms of warning to apprise employees of hazards, suitable protective equipment, control procedures, monitoring and measuring of employee exposure, employee access to the results of monitoring, and appropriate medical examination (section 6(b)(7)). Standards may also prescribe recordkeeping requirements where necessary or appropriate for enforcement of the Act or for developing information regarding occupational accidents and illnesses (section 8(c)). The permanent standard for benzene was developed on the basis of the above legal considerations.

IV. HEALTH EFFECTS

A. GENERAL

Inhalation is the primary route of entry of benzene in man. Benezene diffuses rapidly through the lungs and is quickly absorbed into the blood. The rate of absorption is greatest during the first five minutes and thereafter declines significantly. Benzene saturation of the circulating blood may reach as high as 70–80 percent saturation level within the first 30 minutes. However, relatively complete saturation of the blood may not be attained for two to three days.

The benzene absorbed by the circulating blood is distributed throughout the body where, because of its liposolubility, it tends to accumulate in various body organs in proportion to their fat content.

Upon removal from benzene exposure, the concentration of benzene in the expired breath follows an exponential decay curve, reflecting removal of benzene from various body compartments. Elimination via this route for relatively high concentrations has been estimated to range from 12 to 50 percent of the total amount of benzene absorbed in humans.

Most of the absorbed benzene remaining ultimately is metabolized by enzymes contained in the liver to derivatives which are more water soluble thereby facilitating their removal by the kidneys. A first intermediate in the biotransformation of benzene is believed to be benzene epoxide, a highly reactive chemical. This is one of several candidates—others: Hydroquinone and catechol, (Snyder Tr. 3229) suggested as the active agent responsible for benzene's hematotoxic effects. Phenol, and to a lesser extent, hydroquinone, pyrocatechol, and phenyl-mercapturic acid are the primary metabolites of benzene found in urine.

B. ACUTE EFFECTS

Exposures to high concentrations of benzene produce an almost immediate effect upon the central nervous system. Benzene concentrations near 20,000 ppm are fatal within minutes, with death occurring from acute circulatory failure or coma, with or without convulsions. Milder exposures produce a period of nervous excitation, euphoria, headache and nausea, followed by a period of depression which can result in cardiovascular collapse and/or unconsciousness. The occurrence of nonspecific nervous disturbances as an after-effect of acute exposures is dependent on duration of unconsciousness and/or severity of circulatory failure. Breathlessness, nervous irritability, and unsteadiness in walking have been observed to persist for a period of several weeks. Inhalation of still lower concentrations (250-500 ppm) yields signs and symptoms of mild poisoning, characterized by vertigo, drowsiness, headache, and nausea. Rapid recovery from these symptoms usually occurs following cessation of exposure.

These effects due to acute exposures to high concentrations of benzene have been recognized for many years and are well documented in classic toxicological textbooks and literature.

Direct contact with the liquid may cause erythema and blistering. Prolonged or repeated skin contact, even with small quantities of benzene, has been associated with the development of a dry, scaly dermatitis, or with secondary dermal infections.

C. CHRONIC EFFECTS

1. *Background.* The primary focus of this regulation is to minimize worker risk resulting from chronic exposure to low levels of benzene. These effects of benzene exposure in man have been recognized for approximately 80 years. As benzene attacks the hematopoietic (blood-forming) systems and especially the bone marrow, its toxicity is manifested primarily by alterations in the level of the formed elements in the circulating blood (red cells, white cells, and platelets). The degree of severity ranges from mild and transient episodes to severe and fatal disorders. The mechanism by which benzene produces its toxic effects, although under investigation, is still unknown. (Goldstein, Ex. 43.B, p. 132).

The adverse hematopoietic effects of benzene, including leukemia, have been documented in a variety of industries and occupations and include the rubber, shoe, rotogravure, painting, chemical processing, can manufacturing industries and more recently, the manufacture of natural rubber cast film. These studies range from single case reports, through cross-sectional studies to retrospective studies of morbidity-mortality among a defined cohort of workers industrially exposed to benzene. An important distinction among these investigations is that the cross-sectional method detects cases of mild benzene-induced hematotoxic effects in current employees who do not demonstrate signs of overt toxicity, whereas the retrospective method detects overt and fatal toxic effects subsequent to termination of employment.

OSHA is aware of the varying quality of the individually reported studies. Based on a review of the entire set of studies, taken as a whole, the accumulated evidence is conclusive that benzene exposure is causally related to the induction of leukemia (a cancer of the blood-forming system), various cytopenias (decreased levels of a formed element in the circulating blood), aplastic anemia (a non-functioning bone marrow) and to development of chromosomal aberrations.

The evidence supportive of this conclusion is derived from: (a) A high degree of association of blood dyscrasias with benzene exposure; (b) the apparent lack of a similar association with other known volatile chemicals in the same workplace; (c) outbreaks of hematotoxicity temporarily related to the introduction of benzene to an industry and conversely, a reduction in blood-related disease when other solvents are substituted for benzene; and (d) the experimental demonstration of marrow toxicity in animals solely exposed to benzene (Goldstein, Ex. 43B, p. 133).

The following studies are representative, although by no means all inclusive, of the published literature on the chronic effects of benzene exposure. These investigations do, however, illustrate the diversity and variability of the effects which dominate published reports. There are also several recent reviews and summaries concerning the hematological effects resulting from benzene exposure (See: Vigliani and Forni, Ex. 2-15; National Research Council, Ex. 2-4; NIOSH, Ex. 2-3, 2-5; NYU report; Ex. 43.B and ORC/Jandl, P.C. 34; Snyder and Kocsis, Ex. 2.B-288, and the International Workshop on the Toxicology of Benzene ("International Workshop") (Ex. 18).

2. *Non-Malignant Blood Disorders.*

a. *Human studies.* The most common effect resulting from chronic exposure to benzene is a decrease in the levels of erythrocytes (red blood cells), leukocytes (white blood cells) and thrombocytes (platelets) in the circulating blood. In simplified terms, a decline in red cells is termed anemia, a decrease in the level of white cells is leukopenia and a decline in the platelet count is called thrombocytopenia. Persons found to have depressed blood cell counts may or may not depending, in part on the severity of the decline, display overt physical symptoms. Anemia results in a decreased capacity of the blood to transport oxygen to various parts of the body, and persons so diagnosed may appear pale and weak and fatigue easily. However, the non-specific symptoms may develop gradually and not require medical attention until there are significant declines in red cell counts and blood hemoglobins. Chronic anemia may also result in physiological adjustments by the cardiovascular system and exacerbate difficulties in those with coexisting disease such as coronary insufficiency or chronic obstructive bronchopulmonary disease (Wintrobe, Ex 2A-107, p. 532). Since white cells provide a defense against many diseases, persons with leukopenia are prone to recurrent infections. Goldstein has written that "Infections are a dreaded complication of bone marrow toxicity and not uncommonly associated with a cause of death in benzene-induced pancytopenia"[2] (Ex 43B, p. 144). Thrombocytopenia results in an impaired clotting of the blood, and persons with this disorder may exhibit bleeding tendencies, such as easy bruising, nosebleeds, and hemorrhage.

[2] In the NYU review of "A Critical Evaluation of Benzene Toxicity", Goldstein uses the term pancytopenia in a general sense, defined as a decrease in the level of circulatory erythrocytes, granulocytes, and platelets. His rationale is that there is excellent evidence which suggests that all of these cell lines originate from a common precursor stem cell (Exhibit 43B, p. 135). While noting that aplastic anemia is, in a pure sense, an absence or a decrease in identifiable granulocyte, erythrocyte, and platelet precursors within the marrow itself, Goldstein feels that it is useful to include aplastic anemia or hypoplastic anemia under the category of pancytopenia. This is because in some human cases of pancytopenia induced by benzene and in some animal experiments, a hyperplasia of the bone marrow is observed; also there exists the possibility that sampling errors may affect attempts to quantitate bone marrow precursor cells, since only a small fraction of the marrow is observed by aspiration techniques.

However, Jandl feels that aplastic anemia is not a sufficiently explicit term to describe failure of "marrow to provide an adequate population of dividing blood cells for the 3 series of formed elements," and observes that the terminology "aplastic anemia" has been applied to states of chronic or non-acute marrow suppression, whether or not anemia was the most striking feature (ORC/Jandl PC 34, p. 83, Add. 3(i)). Other terms used synonymously have been "hypoplastic anemia, bone marrow failure, refractory anemia and aregenerative anemia." Based on the degree of severity exhibited, Jandl recognizes 2 phases of marrow failure:

Footnote continued on next page.

Pancytopenia and aplastic anemia are more serious conditions in which all 3 formed elements are depressed. These non-cancerous diseases may, in and of themselves, be fatal. An additional concern is that some or all of these disorders induced by benzene, may, if allowed to continue, either progress to or represent a preleukemia stage which may eventually evolve into a frank leukemia.

Among the early studies describing benzene toxicity was that of Selling (Ex 2-12). He observed a significant depression in the levels of circulating blood cells in workers employed where benzene was used as a solvent for rubber. Because of the depressed condition seen in the marrow of his patients and the results of extensive animal experiments (where he was able to produce both destructive and "regenerative" effects by subcutaneous injection of benzene), Selling suggested that the cause of the cytopenias observed in the workers was due to an aplasia of the marrow.

An important early milestone of the benzene literature was the 1922 report by Hamilton entitled "The Growing Menace of Benzene (Benzol) Poisoning in American Industry" (Ex 159.C).

The document attempted to alert the medical community to the dangers associated with chronic benzene poisoning which was less well known than acute toxicity. This was followed several years later by the reports of the National Safety Council (NSC) which reported the prevalence of known cases by chronic benzene poisoning. These results are summarized by Jandl as follows:

The magnitude of toxicity—primarily consisting of lowered blood cell counts—was shockingly high. Over half of the workers exposed for a year or more had abnormalities or cytopenias of the blood cells, a great proportion had pancytopenia, and a number of deaths were noted, characterized by preceding pallor, weakness, easy bruising, hemorrhage and often fulminant infections (ORC/Jandl, PC 34, p. 10-11).

More data relating occupational exposure to benzene to the occurrence of

Footnote continued from preceding page.

"[B]y convention, a diminution in the level of all blood cells produced in the marrow accompanied by evidence that bone marrow cellularity is deficient, but from which recovery occurs, usually, termed 'pancytopenia.' And by convention, a more sustained, more severe, more likely fatal suppression of the marrow is termed 'aplastic anemia.'" (P.C. 26B, p. 83, Add. 3(i)).

OSHA recognizes the usefulness and the technical reasons for Jandl's establishment of quantitative diagnostic criteria for various non-malignant blood disorders and his "reassignment" of diagnoses contained in the literature according to this classification scheme. However, for convenience sake, the terms "pancytopenia" and "aplastic anemia" are used interchangeably throughout the preamble except when discussing particular points contained in Jandl's review. In these instances, the terminology will reflect Jandl's more definitive nomenclature.

blood disorders was provided in the late 30's and mid-40's by the studies of Greenburg et al., Mallory et al., and those of Hardy and Elkins.

Greenburg and co-workers (Ex. 2-8) investigated the problem of benzene toxicity among 332 workers employed in three rotogravure processes in New York City. In addition to physical examinations, including medical and occupational histories and laboratory tests, workplace air samples, and chemical analysis of ink solvents was performed in an attempt to provide a correlation with medical findings. Exposure data in Plant A revealed that of 11 samples taken, 8 were in excess of 100 ppm, in Plant B, 14 of 24 samples showed levels above 100 ppm and in Plant C, 6 of 13 samples exceeded 100 ppm, but of the total 48 samples taken, 33 were below 200 ppm. The medical results revealed that clinically evident degrees of "poisoning" were seen in 130 workers of which 22 had severe poisoning and 6 required hospitalization. Five of the workers with the most severe hematologic pictures expressed no subjective complaints, and physical examinations were negative. It is noteworthy that the individual sensitivity to benzene varied greatly. Moreover, the authors noted that the benzene-related blood changes may be persistent and could continue to develop even after exposure had ceased.

In the same year, Mallory et al. (Ex. 2-9) reported autopsy findings in 19 workers exposed to benzene from 6 months to 12 years, but no data from the work environment was available. The author noted that significant changes were consistently found throughout the entire hematopoietic system including the marrow, liver, spleen, and lymph nodes. Of the 19 cases studied, 6 exhibited hypoplasia of the bone marrow, whereas 9 cases showed hyperplasia and 2 were diagnosed as leukemic (1 acute myeloid; 1 acute leukemic). The authors concluded that exposure to benzene under varying conditions provided diverse reactions and that individual variation was of great importance.

In 1948, Hardy and Elkins (Ex. 2-10) investigated an artificial leather plant in Massachusetts where in 1946, a man employed as a coater for 12 years became ill and died from benzene poisoning. Sixteen of the 52 workers at the plant exhibited a "remarkable deviation" in more than one blood component, but none were clinically ill. Benzene concentrations taken in 1938 ranged from 45 to 80 ppm, and thereafter repeated tests at the work station of the deceased subject revealed a concentration of about 60 ppm. Elkins stated "[D]uring World War II, however, the plant worked long hours, so that presumably his (the decedent's) exposure overall had been greater than would be indicated by the concentration of benzene found." (Elkins, Ex. 156.3, p. 7). However, on the basis of this case, Elkins and others revised the PEL in Massachusetts to 35 ppm. The authors also concluded that workers exposed to benzene should have routine complete blood examinations and stressed the importance of preemployment medical evaluations which should include a medical history and a hematologic baseline.

In 1956, Savilahti reported on 147 workers who used a rubber adhesive dissolved in benzene, and who were exposed to approximately 400 ppm benzene (Ex. 2-95). Blood tests demonstrated hematological deficiencies in 73% of those examined. Thrombocytopenia, the most common symptom, was observed in 62% of those tested, followed by anemia (35%) and leukopenia (32%). Deficiencies in all 3 cell types were observed in 21% of those examined and deficiencies in two cell-types in 14%. As a result of this investigation, the use of benzene as an adhesive solvent was discontinued.

In a nine-year follow-up to the Savilahti report, Hernberg et al. (Ex. 2A-252) reexamined 125 of these workers previously exposed to benzene who were free of such exposure since 1956. This is one of the few studies available which presents a longitudinal assessment of workers previously exposed to benzene for which follow-up exceeded 2 years. The results showed a greater tendency toward recovery of leukocytes than erythrocytes or platelets; a finding which is in agreement with other investigations. Among male workers, the mean erythrocyte levels remained depressed for 9 years and were significantly less than the values reported for the control group. Despite recovery in the mean level of the platelet counts during the nine-year interval, concentrations for both men and women remained significantly below that of a control group. Based on their findings, the authors stated that "The analysis also showed that the prognosis of the severe cases did not differ from that of the mild ones, provided the acute stage had been passed." (p. 204).

Worker exposure in the rubber coating industry to petroleum naphtha containing up to about 10% benzene was reported by Pagnotto et al. in 1961 (Ex. 159A). Several hemoglobin values less than the lower limit of normal were observed among 47 men given blood examinations. Benzene concentrations ranged from essentially zero to over 100 ppm, with a substantial number between 10 and 25 ppm. Their exposure to benzene ceased in 1965.

In 1973, blood tests were repeated on 18 of these workers who were still employed by the firm, the others having retired or left for other jobs (Pagnotto et al., Ex. 156.3.B). Selection bias may be inherent in this report as the retirees who were excluded may have been more likely to have had blood abnormalities. The blood picture of 14 out of 18 employees was characterized as "essentially normal" (Elkins, Ex. 156.3, p. 9); of the remaining 4, one was anemic, 2 were leukopenic and 1 was both anemic and exhibited leukocytosis (an elevated white blood cell count).

In 1969 Klinche et al. reported hematological damage at relatively low benzene concentrations (Ex. 2B-304). Eighteen roofers engaged in painting operations for an average of 17 years were investigated. Based upon a careful evaluation of the available results of hematological examinations, 6 were classified as suffering from early stages of benzene intoxication, 7 were suspected of chronic benzene intoxication and the remaining 5 were classified as having a normal blood picture. The paint solvent contained 0.12 percent benzene and benzene concentrations (as determined by Drager gas detector tubes) ranged from trace amounts to 15 ppm. Concentrations of toluene and xylene vapor up to 133 ppm were also found. It was not possible to quantitate past exposure to aromatic solvents. The authors concluded, however, that:

In the case of painted or coated areas of sufficient size, even small traces of these aromatics in the work material (benzene, toluene, and xylene) even in the open air, can cause respiratory air concentrations of the level of the MAC value (15 ppm for benzene). After the workers have been exposed for a sufficiently long time, one must assume that hematological damage will result.

In a recent survey, Aksoy et al. compared the hematological findings of 217 apparently healthy male workers who were exposed to as much as 210 ppm benzene when adhesives containing benzene were utilized (Ex. 2-11). The duration of exposure ranged from 3 months to 17 years. Hematological abnormalities were seen in 23.5% of the workers. Leukopenia, with or without thrombocytopenia, was the most common finding. In addition, relative to controls, benzene exposed workers were found to have a statistically significant reduction (p less than 0.001) in mean white cell and platelet counts. Aksoy and his colleagues reported a considerable variability in the 11 bone marrow specimens studied. The aspirates showed both hypocellularity and hypercellularity (a decrease and an increase in the number of cells present), a finding consistent with his previous reports. (See Aksoy et al. Ex. 2A-290). In 33 percent of the workers examined, the hemoglobin levels fell below the lower limit of normal. Aksoy was reluctant to ascribe the anemia to benzene exposure since it was correctable by iron therapy. However, Goldstein commenting on Aksoy's findings that "the mean corpuscular volume (MCV) was in the high normal to slightly elevated range (80–96 um^3) which is unusual for iron deficiency and more in keeping with a benzene effect." (Ex. 43B, p. 140).

In 1977, Dow reported that 2 of the 594 individuals occupationally exposed to benzene died of non-malignant blood disorders, compared to 0.2 deaths expected (DOW/OH/Ex. 154, Table 6). One worker died of pernicious anemia 4 years after retirement. His exposure was estimated to have averaged 30 ppm for approximately 184 months. The second employee, who died of aplastic anemia, had an estimated exposure of 5 ppm for approximately 98 months. Because of the small numbers involved, the significance of Dow's findings is subject to some question, but the trend of excess death is consistent with the evidence presented above.

An issue deserving discussion is the similarity and difference between benzene-induced non-malignant blood disorders and those resulting from other causes. In the NYU report, Goldstein noted that:

While the literature is inadequate to draw firm conclusions, the progression and outcome [fatality rate] of benzene-induced pancytopenias does not appear to differ substantially from that of published series of patients with idiopathic aplastic anemia (Ex. 43.B, p. 145).ª (Ex. 43.B, p. 148)

The NRC report comes to a similar conclusion:

[T]he available evidence does not indicate that the reported benzene-induced blood dyscrasias differ in any way from similar dyscrasias which are caused by other myelotoxic agents or for which the etiology is unknown (NRC, Ex. 2-4 p. 4).

On the other hand, Jandl contends that the high degree of reversibility of benzene-induced non-malignant blood disorders distinguishes these disorders from those caused by other agents which are idiopathic in nature. He reports that the chance for recovery in idiopathic aplastic anemia is 16%, that for drug-related aplastic anemia it is 33% and that based upon a review of 169 patients with benzene-induced aplasia reported between the years 1939–1975, the prognosis for recovery is 87%. (ORC/Jandl, Ex. 34, p. 37) Furthermore, he notes that milder forms of blood disorders resulting from benzene exposure appears to be 100% reversible (ORC/Jandl, PC 34, p. 39).

ª However, Goldstein does suggest three possible differences between benzene-induced pancytopenia and so-called idiopathic aplastic anemia: the first is the observance of marrow hyperplasia in benzene-induced disease; the second is the presence of lymphocytopenia; and the third is "relatively frequent but inconsistent" reports of red blood cell macrocytosis resulting from benzene exposure. (Ex. 43.B, p. 148)

OSHA is aware that literature indicates that many cases of nonmalignant blood disorders may be reversed. For example, Maugeri has stated:

the hemopathy due to benzol can, as in most cases studied by us, be cured if it is diagnosed immediately (Maugeri & Pollini Ex. 156. I p. 2).

Hernberg et al. in a follow-up study of 125 workers stated that: "the prognosis of these patients proved to be more favorable than might have been expected." (Ex. 2A-252, p. 209). Also, both Tabershaw and Aksoy have noted the reversibility of various cytopenias (Tr. 2547).

However, certain other investigators report that these disorders have persisted for as long as 12 years after cessation of benzene exposure. Goldwater and Tewksbury noted continued blood abnormalities in 10% of 108 men examined 24 months after exposure ended. (Ex. 141). Helmer observed abnormalities in 25% of 60 patients followed for 16 months, including 2 deaths. Rejcek and Rejskova found that 8 of 4500 workers manifested persistent leukopenia 12 years after exposure. (Cf. Hernberg, Ex. 2A-252, p. 204). Aksoy observed a further distinguishing feature of severe blood disorders resulting from benzene exposure. In over 100 cases of aplastic anemia classified as idiopathic or associated with etiological agents other than benzene, no cases of leukemia were observed (Aksoy, Ex. 60, p. 6A). This contrasts to the development of leukemia among some cases of aplastic anemia associated with benzene exposure.

As is apparent from the above discussion, there is no unanimity of opinion whether benzene-induced blood disorders persists, and are reversible after cessation of benzene exposure and whether such disorders differ from those resulting from other known or unknown causes.

OSHA recognizes that in many cases these disorders appear to be reversible, within the definitions of "normality" or "reversibility" used by the individual authors. However, it is not known whether for particular workers the blood indices actually return to pre-exposed or baseline valves. Because the original literature does not provide suitable long-term longitudinal analysis including adequate sample size, preexposure values nor baseline hematological values for comparative purposes on an individual basis, it is difficult to decide how much reliance can be placed on such studies. Moreover, the evidence is insufficient to determine whether or not individuals who have "recovered" are at greater risk to subsequent development of leukemia.

Finally, OSHA is unable to verify Jandl's quantitative estimates of recovery as the sources selected for his calculations are not adequately presented to permit the development of valid inferences. Consequently, the Agency does not believe it appropriate to conclude that these non-malignant blood disorders will in all cases, be reversed to the individual's normal values.

b. *Animal Studies.* One of the first studies describing experimentally induced benzene toxicity was published by Selling in 1916 (Ex. 2-12). He was able to produce both destructive and "regenerative" effects in rabbits by subcutaneous injections of commercially pure benzene in olive oil. Bone marrow specimens clearly demonstrated a constant, well defined aplasia associated with the disappearance of leukocytes in the peripheral blood. Later studies conducted by Gerarde demonstrated that a reversible hematopoietic injury in rats was produced at a concentration of 1 ml/kg administered by subcutaneous injection for 14 days (Ex. 2-64).

In an extensive review of experimental benzene intoxication (Ex. 43.B, Ch. V), Leong observed that the results from most of the inhalation studies reported up to Spring (1977) are difficult to interpret due to inadequacies in the experimental design. Many of the studies were conducted without proper control groups to allow comparison of hematologic values against experimental results. Factors such as age, sex, strain, hormone status and to a greater extent, emotional reactions of the animals and techniques used for blood sampling are known to affect hematological parameters. Because of these defects, many of the animal studies do not provide sufficiently definitive exposure-effect information.

In 1944, Svirbely and colleagues exposed rats to benzene vapors at a concentration of 1,000 ppm for 7 hours per day, 5 consecutive days a week for 28 weeks (Ex. 2A-201). Benzene-induced leukopenia was observed during the first half of the exposure period, followed by an increase in the white cell count during the latter part of the experiment. The significance of these results is unclear since similar shifts in the white cell count occurred in the non-exposed control group.

The hematological findings obtained in inhalation experiments at lower concentrations have demonstrated considerable variation. Guinea pigs and rats exposed to an average of 88 ppm benzene for 8 hours per day for 5 days a week demonstrated a mild leukopenia after 32 and 136 days of exposure, respectively (Cf Leong, Ex. 43.B, pp. 89-90). Deichman et al. reported in 1963, that benzene can effect hematological changes, at even lower concentrations (Ex. 2-16). Eight groups of rats were exposed to benzene vapors at average concentrations of 831, 65, 61, 47, 44, 31, 29, and 15 ppm, respectively. The duration of exposure was 5 hours per day, 4 days per week for periods ranging from 5 weeks to 7 months. Exposure to the 3 highest concentrations resulted in significant leukopenia after 2-4 weeks. A moderate, but definite leukopenia was reported in the groups exposed to 47 and 44 ppm after 5 to 8 weeks. No demonstrable alterations were observed at the 3 lowest concentrations during the 7 month experiment.

In a preliminary report submitted to OSHA prior to the rulemaking proceedings, Reinhart found no significant deviation of blood counts in test rats exposed to 388 ppm for 6 hours/day, 5 days a week for 12 consecutive weeks (Ex. 115. B.8). However, significantly depressed WBC counts were observed in mice, while in guinea pigs a progressive leukocytosis was observed which continued even into the post-exposure period. Reinhart used no control group, and the number of animals used per test group is considered to be inadequate.

Finally, Laskin and Goldstein prepared an abstract for the International Workshop on Toxicology of Benzene in Paris, November 9-11, 1976 describing their chronic benzene exposure studies (cf. Snyder, 156.2, p. 5; Ex. 178). For more than two years 150 Sprague-Dawley rats and 300 mice of 3 different strains were exposed to 100 or 300 ppm benzene 6 hours daily, 5 days per week. Anemia was observed in mice but not in rats. However, lymphocytopenia was observed in all animals without similar reduction in the granulocytes until the terminal stages of benzene poisoning. This study suggests that lymphocytopenia may be an early marker for benzene toxicity.

Based on the record, the extent of non-malignant blood disorders resulting from benzene exposure is uncertain. Based only on reports contained in the literature, Jandl has calculated a number of cases and deaths annually due to aplastic anemia in the United States. Based upon data derived from several consecutive case series over the past 15 years, Jandl estimates that a 2% "probable" and 2% "possible" benzene etiology for cases of aplastic anemia (presumably not including milder disorders such as pancytopenia). Using the annual figure of 420 cases of aplastic anemia in the United States, Jandl arrived at an approximation that "about 10 and possibly 20 patient die annually due to benzene exposure." (P.C. 26B, p. 40-1, 86-87).

Goldstein suggested several reasons why the true incidence of pancytopenia resulting from benzene exposure may be seriously underestimated (Ex. 43B, pp. 138-39):

(a) That mild cases of pancytopenia may not be discovered unless searched for since they often cause no overt symptoms.

(b) That since hematoxicity induced by benzene is a phenomenon well known to the medical community, such cases tend not to be recorded in the literature, and

(c) In cases of idiopathic aplastic anemia, possible relevant exposure to benzene may have been overlooked.

OSHA believes that the present data does not permit an accurate assessment of the morbidity or mortality for benzene-induced on nonmalignant blood disorders.

C. CONCLUSIONS

Industry participants have argued that the present permissible exposure limits (10 ppm TWA and 25 ppm ceiling) provide adequate protection against the non-malignant effects of chronic exposure to benzene. (AISI brief, PC 36, p. 51, 66; MCA brief, PC 35, p. 32).

Snyder has noted, "[i]n both animals and man the lowest level of exposure demonstrated to produced bone marrow toxicity was approximately 40 ppm." (Tr. 3224-25). Elkins believes that threshold-for injury to the blood-forming system is between 25 to 50 ppm (Tr. 3210); Tabershaw (Ex. 149A, p. 6) stated that cytopenia has not been demonstrated to occur below 25 ppm. ACGIH (Ex. 156.3D) and the International Workshop on Toxicology for Benzene, both recommended a 10 ppm TWA and 25 ppm ceiling. (Ex. 17).

Attention was drawn to the Dow morbidity study in which no abnormal blood findings were observed among 287 workers exposed to benzene for 1 to 20 years, at concentrations estimated to range from 2 to 30 ppm (Ex. 154).

The Pennzoil Company commented to OSHA that from the past nine years, 160 employees exposed to 1-5 ppm benzene have been monitored by blood test, with up to 25 years work experience for some there has been "no clinical evidence of medical problems associated with benzene". (Ex. 41.1, p. 2). However, no data were presented to allow an assessment of this statement. Others have reported similar views to OSHA again without formal presentation of data. (Joyner, Tr. 2286-87; Texaco, Ex. 41-4, p. 2, P.P.G. Industries, Ex. 6-41, p. 2 and Up John, Ex. 6-4, p. 1). It should be noted that these submissions were characterized by the American Iron and Steel Institute as largely "* * * anecdotal in nature" (Ex. 36, p. 64). In further support of its position, industry cites the 1974 NIOSH criteria document recommendation that a 10 ppm TWA as a "conservative limit" (Ex. 2-2, p. 73).

It is clear, there is no dispute that bone marrow toxicity can result from exposure to benzene above 25-40 ppm range. Furthermore, it is recognized that higher dosages produce increased response and that lower dosages are associated with reduced response. It is not clear, what adverse health effects occur below 20 ppm.

There are suggestions of hematological alterations (other than chromosomal aberrations) at concentrations below 25 ppm. Pagnotto et al. indicated that such effects occurred among workers employed in rubber coating industy exposed to concentrations for the most part between 6 and 25 ppm benzene (Ex 2A-116; Tr. pp. 360-361). Horiguchi has reported small changes in leukocyte function concentrations as low as 10 ppm, in the absence of any marrow change and with only slight variations in leukocyte count and leukocyt phagocytic activity among workers exposed to benzene in concentrations of 0.8-6 ppm (Ex 2A-161, p. 7). And there are also other reports of less apparent effects at airborne levels of 6-16 ppm in workers who do not exhibit characteristic hematological abnormalities. (Kahn and Muzyka, Ex 2A-260). It is not possible at this time to establish with any confidence a consistent dose-response relationship between benzene exposure and adverse health effects.

Where exposure information is available, it is generally area sampling and not personal exposure data that is reported. It is difficult to determine individual worker exposure based on area sampling. Compounding this difficulty are problems related to the fluctuating character of occupational exposure; the small amount of long-term data; and the fact that in small populations, workers with heightened sensitivity may not be encountered. At lower exposure levels, there is a paucity of data to provide a basis for definitive conclusions as to the extent of nonmalignant effects. In this regard Battle states that:

[E]ven a cursory study of the literature reveals a distressing lack of exposure-hematologic effect data particularly in the low benzene exposure area of less than 10 parts per million. (PC 26C, p. 2).

Moreover, the absence of pre-exposure hematological values for comparative purposes limits the usefulness of the few studies and reports which are available, because we cannot tell whether a particular worker studied with blood values at either extreme of the normal range is suffering adverse effects. Therefore, the available literature is insufficient to permit construction of dose-effect curve.

As indicated above, the quality of evidence available in the record is considered insufficient to permit meaningful conclusions concerning exposure to benzene at low levels and resulting health effects. Industry participants have cited the 10 ppm level established by the ACGIH as evidence that this level can be considered safe. However, in establishing TLV's, ACGIH recognizes that for some workers harmful health effects may result from exposure to the toxic substance at levels below the TLV. Therefore, the 10 ppm TLV for benzene is recognized by ACGIH as a level which does not protect all workers from material impairment of health.

Leukemia aside, with benzene we are dealing with an etiological agent well-documented to produce a variety of blood abnormalities, some of which are fatal. OSHA is aware of several studies reporting blood abnormalities at levels below 25 ppm, at levels perhaps as low as 10 ppm. Because of these considerations, and the wide range of human sensitivity to benzene, OSHA cannot determine a minimum effect level for benzene and, furthermore, cannot conclude that 10 ppm provides sufficient protection against non-neoplastic effects to all workers. OSHA recognizes that prudent public health policy, and established toxicological principles necessitates setting the permissible exposure limit sufficiently below the levels at which adverse effects have been observed to assure adequate protection for all exposed employees. It is customary to use a safety factor of 10-100 or greater depending on the seriousness of the toxic effects and the nature of the data being relied upon. That is, the lowest levels at which effects had been observed would be reduced by the safety factor chosen in establishing the exposure limit. Taking this approach in the case of benzene would lead to a permissible exposure limit substantially less than 10 ppm without regard to the issue of leukemia and the view that no safe level can be established for the carcinogenic risk.

IV. Leukemia

A. Definition and Description of the Disease

Robbins defines leukemia as:

Leukemia may best be considered as a neoplasm (cancer) of the white blood cells and is so classified in the International Lists of Causes of Death. It is characterized chiefly by: The appearance of abnormal, immature white cells in the circulating blood; diffuse and almost total replacement of the bone marrow with the leukemic cells; and widespread infiltrates of the liver, spleen, and other tissues, analogous to metastatic dissemination of solid tissue cancer. (Ex. 2-24, p. 728).

Once leukemia is diagnosed there is virtually no chance of recovery. There are different categories of leukemia, depending on the duration of the disease, I.E., acute or chronic; an increase or non-increase in the number of abnormal cells, i.e., leukemia or aleukemia; and the cell type involved, i.e., myelord, monocytic or lymphoid. Goldstein also emphasizes that, "there are other differences between various subtypes of leukemia in terms of incidence, clinical course, prognosis and presumably etiological mechanism" (Ex 43B, p. 156).

The most prevalent subtype of leukemia in adults, and the type most commonly associated with benzene is acute myelogenous leukemia (AML). This disease is variously known as acute myeloid leukemia, acute granulocytic leukemia, and acute myeloblastic leukemia (ORC/Jandl, PC 34). There are several variants of AML, probably related to the pluripotential of the precursor cell and include erythroleukemia, (DiGuglielmo's Syndrome) and acute monocytic leukemia (myelomonocytic, or monomyelocytic). (ORC/Jandl, PC 34, p. 71).

Despite advances in leukemia therapy and the fact that half the adults diagnosed with AML enter into a remission (generally averaging 6 to 8 months) the prognosis of this disease remains poor (ORC/Jandl, P.C. 34, p. 73). Those who enter into remission have life expectancies averaging from 12 to 18 months, and those who fail to respond to therapy have a 50 percent survival rate of only 3 to 6 months. Aksoy noted that in his experience, for cases of benzene-induced leukemias, the period of survival after discovery was short, usually less than 6 months (Tr-275).

Myeloproliferative disorders, including chronic myelogenous leukemia (CML) have only occasionally been attributed to benzene. Lymphocytic leukemia is also subdivided into acute and chronic subtypes. The acute form (ALL), the type commonly seen in children, has only rarely been associated with benzene exposure. Chronic lymphocytic leukemia (CLL), a slowly progressive disease, and often not a cause of death, generally occurs late in life. There have been several reports associating CLL with benzene, most particularly in several reports of French origin. The evidence as to chronic lymphocytic leukemia, which will be discussed in detail later, does not appear to be conclusive.

B. Studies—Human

Evidence in the record clearly demonstrates that benzene is a human leukemogen. Aksoy testified that the first case of leukemia due to benzene exposure was that published by Noir and Claude in France in 1897. (Aksoy, Tr 154). In 1939, Mallory, et al. described two cases of leukemia resulting from chronic exposure to benzene which supplemented ten previously reported in 1935 by Penati and Vigliani. (Mallory, Ex. 2-9), Mallory et al. stated:

Certainly no more favorable conditions for the development of neoplasm can be imagined than prolonged and intense stimulation of reproductive activity, and simultaneous arrest of maturation.

The evidence that chronic exposure to benzene produces leukemia in human beings is still incomplete but it is accumulating at a rate and to a volume which command serious consideration. (Ex. 2-9, p. 365).

In 1965, in a partial review of the literature, Browning tabulated 61 cases of leukemia among individuals having reported prior exposure to benzene (Ex 2-31). The majority (40) were of the myeloid series—including 12 cases of erythroleukemia, 7 were classified as lymphatic and the remaining 14 grouped under the heading of "aleukemic leukemia" (a type characterized by a decline in white blood cell number).

Vigliani and Saita, in their review of 47 individuals suffering from benzene hematoxicity between the period of 1942 and 1963, presented clinical and laboratory accounts of six additional cases, all of whom were diagnosed as having haemocytoblastic leukemia (Ex 2-27). The duration of exposure of these individuals to resins, inks, varnishes, or glues containing varying concentrations of benzene ranged up to 19 years. Data on the concentrations of benzene in the workplace environment were extremely limited or non-existent. Occupational histories and medical status prior to the final diagnoses were not available, but the authors stated that "attribution of the cases (of leukemia cited) to the exposure cannot be doubted." During the years 1962-63, Vigliani and Saita noted a sharp rise in the diagnosis of leukemia among individuals having reported prior exposure to benzene which coincided with an increase in the number of newly diagnosed benzene poisoning cases. The risk of acute leukemia for the workers exposed to benzene in Milan and Pavia was estimated to be about 20 times greater than the risk for the general adult population. (Ex. 2-27). In 1975, Vigliani and Forni observed that in the rotogravure industry no new cases of aplastic anemia nor of leukemia were found among workers exposed solely to toluene after this solvent was substituted for benzene in 1964. Although there have been some reports of hematological disorders associated with toluene exposure (See Girard et al., Ex. 2B-283), the finding of Vigliani is in concert with the conclusion of Brair that, toluene and benzene are "* * * quite different chemical entities from toxicological and pharmacological points of view" and that, "toluene free from benzene can be safely used in industry as a suitable solvent in replacement of benzene." (Ex. 39-5, p. 7). Vigliani and coworkers also observed that workers exposed to toluene did not exhibit chromosomal aberrations, a "finding" observed by these authors in employees who worked with benzene.

In 1972, Aksoy et al. reported the deaths of 4 Turkish shoemakers resulting from their exposure to benzene for periods ranging from 6 to 14 years (Ex 2-29). At the time of the study, air concentrations were found to be between 150-210 ppm of benzene. Previous occupational exposure data were not provided. Two of the four patients developed acute leukemia approximately two and three years after the occurrence of aplastic anemia, although the other two did not. These case reports are supplemented by surveys of newly diagnosed cases in a medical referral area. Aksoy summarized and updated his findings at the benzene rulemaking (Ex 60). He pointed out that from 1967 to 1973, he had observed 26 patients with acute leukemia or preleukemia* associated with benzene exposure. Based on a population of 28,500 shoeworkers in the area, Aksoy estimated the incidence of leukemia to be twice the prevailing value of 6/100,000 observed in the general population of Western nations. From September 1973 to 1975, 14 additional cases were admitted to the hematology departments in Istanbul and Cerrphasa Medical Schools making a total of 40 in all. Thirty-four of these patients were among the groups of shoeworkers described previously. Aksoy either examined the cases personally or reviewed the charts.

API has criticized Aksoy's use of the baseline figure of 6/100,000, citing a statement from Clemmenson, that the incidence rate for a standard European population is 8 to 14/100,000 (API Ex. pp. 50-51). However, Aksoy has testified that the two-fold excess estimate among benzene/exposed shoeworkers may be underestimated since the incidence of leukemia among the Turkish population is only 3.0 per 100,000. Moreover, Aksoy stated that "undoubtedly there were other additional patients among shoeworkers who were not included in our study" (Ex. 61, p. 2). Aksoy's findings of a decline in leukemia cases during the past several years after other solvents were substituted for benzene (Aksoy, Ex. 61, 99. pp. 2-3) further provides, in an inverse manner, additional supportive evidence for the causal relationship between benzene and leukemia. This finding reinforces the results of vigliani discussed above.

In his testimony, Marvin Sakol, an Akron hematologist, described what he considered an epidemic of a rare form of leukemia among workers employed in a small department at a local industry (Ex. 61). Over a nine year period, beginning in 1954, he observed nine cases of acute myeloblastic leukemia, including "at least four and probably all nine" with DiGuglielmo leukemia (erythroleukemia). Due to the rarity of erythroleukemia Vigliani noted that the 20 or more cases attributable to benzene, reported in the literature seem to be significant (Ex. 2-49). Diagnoses from slides in some of Sakol's cases were confirmed personally by DiGuglielmo. He learned with much difficulty that all of these cases were exposed to benzene while employed in a Pliofilm operation at a plant subsequently studied by Infante et al. Sakol testified that he had knowledge of additional cases of benzene-related leukemia which have not been included in the NIOSH study. These cases may never be recorded officially because the families refused permission to release the names of the decedents for fear of loss of compensation and job. This testimony indicates an underestimation of the leukemic risk among Pliofilm workers.

Based on the hypothesis that the risk of leukemia was higher among workers who were exposed to benzene and medical X-rays, Ishimaru et al. conducted a retrospective epidemiological investigation examining the relationship between occupation and environmental factors, other than A-bomb exposure, and the incidence of leukemia in Nagasaki and Hiroshima between 1945 and 1967 (Ex. 2-33). Fifteen occupations were selected in which there had been exposure to either medical X-rays or solvents, especially benzene and its derivatives. This case-control study compared all cases diagnosed as definite or probable leukemias between 1945 and 1967 and residing at the time of the onset of the disease, in Hiroshima or Nagasaki. Controls were matched for city, sex, date of birth (±30 months), distance from the atomic bomb explosion, and residence in either city at the onset of disease. Four hundred ninety-two leukemia cases were identified, and matched controls were obtained for 413. Three hundred and three adult cases with the onset of leukemia at age 15 years or over and their controls were compared. The risk of leukemia was found to be significantly higher (about 2.5 times greater) among those with a history of employment in occupations in which various volatile solvents were used as compared to those without. The relative risk was 1.8 times higher for chronic leukemia and 2.9 times higher for acute. Eighteen of the leukemia cases associated with solvents were located in distant and non-exposed radiation areas and were considered too far from the A-bomb explosion for radiation to have enhanced the increased risk. Accepting the source of error inherent in the method which was used to collect the data, the results of this study nonetheless reinforce the observation that an increase in leukemia existed in that portion of the population exposed to radiation and employed in an occupation where solvents especially benzene, were used.

---

*The criteria used for the diagnosis of preleukemia were similar to those of Wintrobe: "[P]releukemia seems a reasonable term * * * when this diagnosis of AML may be suspected but cannot be made with any confidence." (Ex. 2-107, p. 1477).

In 1970, Girard et al. published an epidemiological study (Ex. 2B-283) undertaken to determine previous exposure to benzene or toluene in 401 patients suffering from malignant hemopathies. Extensive interviews about their work environment and chemical analyses of solvents were utilized to estimate exposure. Their findings demonstrated that compared to a control group admitted for non-hematological diseases there was a statistically significant greater frequency of past exposures to benzene or toluene among subjects with aplastic anemia, chronic lymphocytic leukemia and acute leukemia.

In April 1977, NIOSH submitted to OSHA a preliminary report by Infante et al. of a study of leukemia among benzene-exposed workers employed in the production of natural rubber cast film (Pliofilm) (Ex. 2-57, Ex. 2A-271). The study included all white males assigned to the Pliofilm production area who were hourly employees and who at any time between January 1, 1940 and December 31, 1949 had direct exposure to benzene at the Goodyear Akron and St. Mary's plants. Only the men employed in a section having known potential exposure to benzene ("wet side" in industry terminology (Rinsky, Tr. 818-820)) were included in the cohort. Men employed in the Pliofilm operations, but not in production jobs (so called "dry side" workers) were not included in the study. Follow-up of vital status was attempted from termination of employment to June 30, 1975 and was achieved for 75% of the total of 748 benzene-exposed workers. So as not to overestimate the true risk of lymphatic and hematopoietic malignancies associated with benzene exposure, the 25% of the cohort for whom vital status was not determined were assumed to be alive. Little or no quantitative exposure data existed for these plants.

Causes of death were determined from death certificates, and person-years at risk were determined by a modified life-table method. Person-years of observation and causes of death were determined from January 1, 1950 to December 31, 1975. Two control populations were used to generate the numbers of expected deaths in the study. The first comprised the U.S. white male population, while the second was white males employed in an Ohio fibrous glass production facility during the 1940's and who had achieved five years of employment by June 1, 1972.

The most striking finding was the observation of a statistically significant (p less than 0.002) five-fold increased risk of dying of leukemia compared to the U.S. male rates (7 deaths observed vs. 1.38 deaths expected, p less than 0.002), as well as to the fibrous glass workers (7 deaths observed, 1.48 deaths expected, p less than 0.002).

Criticisms of various aspects of this study were raised by participants to the rulemaking. The issue of what level of benzene the workers in the Infante et al. cohort were exposed to was the most discussed area. Participants criticized the information available in the Infante study. The study referred to a November 1946 report of the Industrial Commission of Ohio which indicated that: "Tests were made with benzol detectors and the results indicate that concentrations have been reduced to a safe level, and in most instances range from 0 to 10 or 15 parts per million." Comments at the hearing demonstrated that there were area exposures during this study period exceeding these levels, at times reaching values of hundreds of parts per million. Since no personal monitoring data are available, any conclusion regarding the actual individual time-weighted average exposure is speculative. Because of the lack of definitive exposure data, OSHA cannot derive any conclusions linking the excess leukemia risk observed with any specific exposure level.

The study was also criticized for combining the two separate plants in the same analysis. It was suggested that an independent analysis of the two plants would produce different results, thereby suggesting that factors other than similar exposure to a single agent (benzene) contribute to the excess leukemia. (Lamm, Tr. 2538). OSHA believes that the use of non-age-adjusted leukemia death rates, as utilized in Lamm's analysis, is inappropriate for the assessment of an event associated with age. Therefore, OSHA believes that this criticism is unsupportable on this basis. Moreover, the cohorts were combined based on the similarity of the processes in the two plants (Tr. 842). Also, because of the small number of subjects in the study cohort, the authors stated that it would be inappropriate to analyze the data separately for the two plants. (Infante, Tr. 935, 796-97).

A question has been raised by industry participants as to the validity of excluding the "dry side" (Pliofilm finishing) workers from the study cohort (Lamm, Tr. 2638-29; API, PC 33, p. 42-43). As explained by one of the authors, these workers were never intended for inclusion in the cohort following discussion with company personnel indicating there was no benzene exposure on the dry side (Infante, Tr. 929-30). Testimony was also offered that if "dry side" employees had job operations which kept them predominantly on the wet side of the plant, then those employes were included in the cohort (Tr. 931). Limited data subsequently released by the University of North Carolina at the OSHA hearing indicated that there may have been some benzene exposure (0, 11.8 20 ppm) on the "dry side". However, there were only 3 sample points and no details were given as to sampling locations, duration of sampling, or definition of departments, to permit interpretation of this information (Ex. 187B.6, Table 3, p. 10). Infante testified that a cursory examination of "dry side" workers revealed no leukemias. However, Infante stated, "* * * there really are not enough people who are employed on the dry side who never went into the wet side to make a meaningful conclusion about that (leukemia deaths from exposure among workers on the dry side)" (Tr. 930). Based upon the above considerations, it appears that the "dry side" workers should not have been included as was the case.

A question was raised as to whether the study demonstrated the true risk of leukemia among workers exposed to benzene (Tr. 311). Sakol referenced two cases of leukemia at the same plant which he personally diagnosed, which were not so certified on death certificates. He cited these cases in support of the substantial underascertainment of leukemia mortality in the study of Infante et. al., due to their sole reliance on death certificates.

The use of the Connecticut Tumor Registry data to determine the expected number of type-specific leukemia deaths for the Infante cohort was criticized. Jandl calculates that the tenfold excess risk of myelogenous and monocytic leukemia found by Infante et. al. may have been off by at most a factor of two. (ORC/Jandl, PC 34, p. 77). (Also, there is a counter-balancing underestimate of the risk because at the time of the report, only 75 percent followup had been completed and the remaining 20 percent of the study population were all regarded in the statistics as being alive at the end of the study period. Both the stengths and weaknesses of the Infante et. al. study have been carefully evaluated by this Agency. OSHA believes this study is one of the more definitive investigations in the benzene literature and provides additional evidence supportive of earlier studies which implicate benzene exposure and the subsequent development of leukemia.

An opposite result was obtained in a Tabershaw-Cooper Associates Inc. mortality study of petroleum refinery workers (Ex. 2-5a). The study population was a sample of workers from the 251 U.S. refineries operating in 1971. The authors attempted to select a sample of 17 U.S. refineries to give a representative distribution with respect to geographic area, ownership, and size. Hourly workers with more than one year of work experience between January 1, 1962, and December

31, 1971, were studied. All workers were classified as to their potential hydrocarbon exposure. The authors reported that better than 94% follow-up was achieved. An excess mortality from lymphoma was observed although it did not achieve statistical significance.

Subsequently, the study was expanded to include 5,145 workers who had been hired on or before 1943 and who were working sometime during January 1, 1962, through December 31, 1971. No excess mortality from leukemia or lymphomas was observed in this subpopulation having 20 or more years since first hydrocarbon exposure. Gaffey, one of the authors of the study, stated that the study had been commissioned not specifically to study the effects of benzene, and in fact he could not delineate what proportion of the study population had ever been exposed to benzene (Tr. 1343-1344). Moreover, as adduced during the hearing, Gaffey acknowledged that there was no industrial hygiene assessment of the work environment (Tr. 1251) to provide the basis for identifying which workers may have been exposed to benzene.

Because this study was designed to assess potential effects of "hydrocarbon" and not benzene exposure and since no group of workers exposed to benzene was clearly identified, this study provides no relevant information as to the leukemogenic risk of exposure to benzene. Therefore, OSHA does not view this study as providing negative proof that benzene does not cause leukemia.

Dow submitted a study which addressed the incidence and mortality of leukemia among benzene-exposed workers (Ex. 154). The mortality experience for all employees occupationally exposed to benzene from 1940 to 1973 was studied. Expected deaths were calculated using age-cause-specific mortality rates for the corresponding U.S. white male population from 1942 to 1972. Each of the 594 workers were classified in approximate benzene exposure categories. A rough cumulative dosage was calculated. When consideration was given to the incidence of myelocytic leukemia, the authors concluded that a statistically significant excess was demonstrated (3 deaths observed vs. 0.8 expected, p less than 0.05) (Ex. 154, p. 12). Mortality data were analyzed by production areas and according to cumulative dosage and length of occupational exposure. Those people in the cohort who had exposures to known carcinogens were excluded from analyses dealing with dose-response in relation to benzene exposure.

An increase in leukemia incidence was observed in the population where 2 deaths were observed vs. 1.0 expected. An additional death resulting from myelogenous leukemia was identified among the study cohort subsequent to the earlier published report. Based upon their findings the authors stated that:

> In these cases, varied work histories, low levels of potential benzene exposure relative to other employees in the cohort, and the lack of information in regard to total medical history made a retrospective assessment of the possible relationship to benzene exposure very judgmental.
> (Ott, Ex. 154, p. 12.)

However, based upon the recent finding of the third death due to leukemia, Dow lowered its internal ceiling limit from 25 ppm to 10 ppm (Ex. 82).

In addition to the 3 deaths reported due to leukemia, 2 deaths attributable to non-malignant blood disorders were also reported: aplastic anemia and pernicious anemia. (OH, Ex. 156, p. 9.) Deaths attributed to aplastic anemia without examination of bone marrow aspirates, may in actuality, be due to leukemia (Vigliani Ex. 2-15, p. 123), thereby potentially resulting in an underestimation of leukemia risk among the benzene-exposed cohort.

OSHA recognizes that the decedents were for the most part exposed to relatively low concentrations of benzene, and no leukemia deaths occurred in the higher exposure groups. However, because of the small population size as well as the possibility of sensitivity of those individuals developing leukemia, it cannot be concluded that these deaths are not caused by benzene exposure.

OSHA does conclude that the findings of this study are consistent with the findings of many studies that there is an excess leukemia risk among benzene exposed employees.

Allied Chemical Corp. submitted a post-hearing comment containing mortality data for employees with potential benzene exposure (P.C. 22). Death certificates were obtained for all workers dying between 1961 and mid-1977. Rough exposure information was derived by back extrapolation, limited exposure data, and subjective reports of employees regarding signs and symptoms. Based on these estimates of benzene exposure, there was no observed cluster of leukemia in operations having benzene exposure. Because this study was presented to OSHA after the hearing closed, more specific details concerning the estimate of exposure, the method utilized for obtaining estimates of person-years at risk, or how the death certificates were coded were not obtained.

In 1974, Thorpe reported that the incidence of leukemia, among a population of 38,000 workers exposed to low levels of benzene over a ten year period (1962-1972), was not statistically different from that expected on the basis of the general population (Ex. 2-34). However, the study has been criticized for the relaxed casefinding techniques and analytical methods (Brown, Ex. 2A-35). Thorpe himself acknowledged the following serious methodologic deficiencies:

1. Validity of leukemic diagnoses.
2. Quantitative determination of exposure levels.
3. Adequacy of retiree follow-up.
4. Complete occupational histories.

As in the case of the Tabershaw/Cooper study, OSHA cannot separate the benzene-exposed workers from those with little or no exposure. This difficulty, compounded by the 4 deficiencies listed above, also precludes reliance upon the finding of the Thorpe study.

University of Pittsburgh investigators (Lloyd, Redmond, Ex. 113) have published a notable series of epidemiological studies of steelworkers mortality. AISI (P.C. 36, pp. 59-60) cited the above studies as evidence of no leukemia risk among benzene-exposed coke oven workers. Most notably, the industry cited the fact that byproduct plant coke oven workers were not found to have been at excess risk (P.C. 36, p. 60) and recommended that coke ovens be excluded from the benzene standard (P.C. 36, p. 100).

However, as Bickmore stated:

> * * * In all 35, the expected, the incident (SIC) of mortality and expectation is so low as to preclude any meaningful statistical calculations by the professionals that made the study. (Tr. 3314)

These coke oven worker studies were designed to assess worker risk of exposure to coke oven emissions (Ex. 113 p. 1) (P.C. 27C p. 1, P.C. 27D p. 1). Given the small numbers of observed and expected rates of leukemia, and in the absence of a study designed to assess coke oven worker exposure to benzene, OSHA believes it inappropriate to rely on these studies as evidence that there is not a leukemia risk of exposure to benzene.

During the hearing, Stallones discussed the unpublished study of 3,600 Shell Oil Company workers with "potential" benzene exposure (Ex. 115C.2). Among this population he testified that he observed no excessive leukemia mortality in the three year period ending in 1976. Stallones could not determine how many workers were exposed to benzene. In fact, clerks and officer workers were included in the study group. Again, failure to correctly identify or define a benzene-exposed cohort limits the usefulness of this study.

A series of published and unpublished epidemiologic studies investigating the mortality experience of rubber workers have been entered into the hearing record. (McMichaels et al. Ex. 2-36, -37, 2B-295); Monsan and Nakano (Ex. 2-83); Fox et al. (Ex. 2-61); Andjelkovic (Ex. 2-54); Occupational Health Studies Group, UNC (Ex. 187.B). These studies have, in general, shown 'excess of lymphoma and leukemia among rubber workers. The leukemia was predominantly lymphatic, a cell type not commonly associated with benzene. The proportion of rubber workers actually exposed to benzene in these studies is undetermined. The presence of other chemical exposures raises the question of an alternative causative agent for the excess of lymphatic leukemia and lymphoma observed among rubber workers. In view of these considerations, these studies provide no additional evidence for the leukemogenic potential of benzene.

There has been discussion whether benzene is a *primary carcinogen* (can directly effect a neoplastic alteration without host-mediated activation); a *secondary carcinogen* (chemicals which may increase susceptibility of cells, to a primary carcinogen); a *pro-carcinogen* (requires biochemical alteration prior to effect); or a *co-carcinogen* (which requires the interaction of another chemical to elicit a neoplastic response). (Olson, Tr. 2892; NRC, Ex. 2-4, p. ii; API/NPRA Brief, P.C. 33, p. 86-87; Aksoy, Tr. 177; Furst, Tr. 1744; Stockinger; P.C. 32-H, 53); Weisburger, P.C. 32-I; Tabershaw Tr. 2549, 2545). The evidence is, at present, insufficient to choose among these alternatives. Namely, what is apparent, however, is the effect of benzene exposure—a significantly increased risk of death from leukemia.

Types of leukemia clearly associated with benzene exposure include acute myelogenous leukemia (AML) and its variants. For example, Aksoy testified that among 40 cases of leukemia among benzene-exposed workers in Turkey, AML and its variants were the most frequently observed form (Ex. 60, p. 7). In the experience of Vigliani and his coworkers, only acute or subacute myelogenous leukemias were observed (Ex. 2-49). The predominant cell type found by Infante (Ex. 2A-271) and others were also myelogenous in character. This well-established association between benzene and AML is in concert with the well-known toxic effects of benzene on bone marrow stem cells (Goldstein, Ex. 43B, p. 162).

In contrast to the definitive relationship between benzene and the induction of AML and its variants, there is considerable scientific debate concerning the association between benzene exposure and the development of chronic forms of leukemia (CLL and ML). Jandl has stated that "* * * lymphoid leukemia is not a feature of benzene toxicity." (ORC/Jandl, P.C. 34, p. 59), and Lamm ruled out chronic leukemia, not as a "possibility", but as a "probability" (Tr. 2558). Aksoy found no cases of CML among the workers he studied. (Ex. 60, p. 6-F),

and no cases of CLL were seen by the Italian investigators. A different result has been reported by several French studies which have reported a relatively high incidence of CLL. In some instances, there were more cases of CLL than AML. They also report a higher incidence of CML than would be expected based on Aksoy's and Vigliani's findings. Supporting the high incidence of CLL in the French studies is the report by Tarreef which describes 16 cases of leukemia which resulted from long-term exposure to benzene, 3 diagnosed as CLL (Ex. 2-28).

The evidence is inconclusive as to the relationship between benzene exposure and the development of forms of leukemia other than AML and its variants. Vigliani stated (1976) that, although he did not observe chronic leukemias in his series, "there is no *a priori* reason for not accepting them as benzene-induced leukemias * * *" (Ex. 2-49, p. 148). OSHA believes at this time that it is prudent to recognize the possibility that benzene exposure may be related to chronic forms of leukemia.

One of the major issues of this rulemaking is whether a blood disorder is a necessary precursor for the subsequent development of leukemia. For example, in 1974 NIOSH speculated as follows:

It can be postulated that bone marrow changes and blood dyscrasias would precede leukemia if induced by benzene, so that if these changes were prevented leukemia should not result (Ex. 32B).

Aksoy has testified that benzene leukemias usually develop after a previsous pancytopenic period. (Tr. 198) Maugeri, reports that, "the leukemias in our cases were always of the acute, aleukemic slow developing type, and they were always preceded by [an] aplastic condition" (Ex. 39-10, p. 2). Snyder remarked that he was unaware of benzene leukemias without previous demonstration of some kind of bone marrow damage (Tr. 3226) and Goldstein has noted that, "there does not appear to be any proven cases of leukemia in the absence of previous pancytopenia" (Ex. 43B, p. 165). But Goldstein cautions that this interpretation is, "open to speculation, especially in view of the paucity of routine laboratory data preceding the onset of leukemia" (Ex. 43B, p. 165).

As observed by Goldstein, since the mechanism by which benzene induces leukemia has not been elucidated, it is possible that leukemia develops, not in response to the pancytopenic effects of benzene, but rather to the direct carcinogenic effect on the marrow hematopoietic stem cells not necessarily accompanied by any other evidence of marrow effect (Ex. 75, p. 3). In such events, protection against non-neoplastic blood disorders would not rule out subsequent development of leukemia.

Vigliani, in discussing the evidence causally relating leukemia to benzene exposure observed that such leukemia may occur, "* * * directly or following aplastic changes in the bone marrow * * *" (Ex. 2-49, p. 148). In response to a question concerning whether aplastic anemia is a necessary precursor for leukemia, Lamm responded in the negative (Tr. 2562). Also, Browning in 1965 observed that, "benzene leukemia is frequently superimposed upon a condition of aplastic anemia, *but it can develop without a preceding peripheral blood picture characteristic of bone marrow aplastia* (emphasis added) (Ex. 2-31).

Finally another industry participant was reluctant to conclude that cytopenias "always" precede leukemia (Tabershaw, Tr. 2546). The evidence is not conclusive. There has been no scientific study of leukemia and preceding pancytopenia resulting from low levels of benzene exposure. The current lack of a reproducible benzene leukemia animal model forecloses the use of experimental evidence. There is no agreement on prospective criteria for benzene-induced blood abnormalities. Therefore, the statement that non-malignant pathological changes always precede leukemia is essentially based on retrospective analysis. The fact that non-malignant blood changes may occur shortly (within a few months) after first exposure to benzene, while leukemia may appear as long as 10 years or more after initial exposure suggests that separate etiologic pathways may be involved.

A corollary issue related to preceding blood disorders is whether recovery from such abnormalities eliminates the risk of leukemia deaths attributable to benzene exposure. Rosen has stated that: "[w]hether reversing the early toxicity can prevent the development of more serious blood dyscrasias in some cases is open to question, but it is a definite possibility" (Ex. P.C. 26A, p. 4).

Jandl has written: "In carefully studying other reports of long latent periods, or of any period between an initial aplasia and a terminal leukemia in which the patient was hematologically normal, I failed to find any well-documented example of complete recovery followed years later (despite abstinence from benzene exposure) by acute myelogenous leukemia" (ORC/Jandl, P.C. 34, p. 24).

Because of the lack of information with a suitably identified cohort followed for a sufficient number of years, it is impossible to make a determination whether or not these "recovered" workers are actually at increased risk of dying of leukemia. This uncertainty has been expressed in the conclusions in the Dow Chemical Company morbidity studies:

The question that remains unanswered, because of the relatively small sample size of this study, is what is the probability that serious effects of benzene (such as leukemia) will appear years after first exposure and after a period of health surveillance during which no hematological abnormalities are noted (Dow/Fishbeck, Ex. 154, p. 19).

OSHA agrees with the caveat expressed by Dow, and therefore the benzene regulation is not predicated on reversal of blood abnormalities protecting against the development of leukemia.

Benzene-associated leukemia is frequently characterized by a long delay between initial exposure and the onset of disease. Sellyei and Kelemen report a case of subacute granulocytic leukemia in a patient 8½ years after starting to work with benzene (Ex. 2B-258). Hernberg reported a case of a 10-year period between initial exposure and the onset of leukemia in a woman with normal interval blood counts who was exposed to benzene for a total of 6 months (Ex. 217-152). In 1974, a case of acute leukemia was observed in a woman who, following a benzene-induced anemia, was reported to have an exposure-free period of 14 years with almost normal blood counts (Vigliani and Saita, Ex. 2-50, p. 214). Infante et al. report a range of 2-21 years between initial exposure and death; all but one exceeded 10 years (Ex. 2A-272). Goldstein hypothesizes that if benzene-induced hematologic damage is expressed by an * * *

aberration in the bone marrow producing a stem cell that is liable to a further mutagenic event, such an aberration could conceivably occur many years after the original insult. (Ex. 43B, p. 166).

The delay in the expression of overt disease is a characteristic of many other carcinogenic compounds. With regard to benzene, it has been documented that there is a relatively rapid decline in the incidence of leukemias following cessation of worker exposure to benzene. Aksoy has testified that after the substitution of other solvents for benzene, there has been a decline in the annual number of workers diagnosed with benzene-leukemia (Ex. 60, p. 3). Vigliani and Forni reported that there was a sharp decrease in cases of chronic benzene poisoning after the use of benzene as a solvent was banned (Ex. 2-50, p. 214). These observations appear to be paradoxical in light of the latency periods also reported. However, Forni and Vigliani have noted that, occasional cases of acute leukemia with a long latency period may still be observed (Ex.2-50, p. 215) and Aksoy, while reporting that no additional cases of leukemia among shoeworkers were diagnosed in 1976, noted that this did not preclude the development of additional cases in the future (Tr. 169).* The question of latency periods for benzene-induced leukemias has not been resolved and requires further investigation.

Another area of discussion is the question of identifying individuals who may be particularly sensitive to the hematotoxic effects (including the development of leukemia) of benzene. Several witnesses including industry participants generally expressed the view that there probably were differences in sensitivity (Sakol Tr. 228; Shaw Tr. 412; Snyder Ex. 54.2, p. 25, Thorpe, Eckard Tr. 2052; Furst Tr. 1791-42). It appears from the evidence that there are individuals in given populations who may, because of genetic factors and/or concurrent or prior exposure to other environmental agents, be especially sensitive to benzene-induced blood disorders.

Aksoy and his colleagues reported a possible familial link in benzene-induced leukemias in 5 of 40 workers indicating a possible genetic predisposition to benzene's harmful effects (Ex. 60, pp. 8-10). Also, Jandl on the basis of tentative data has attempted to identify a subpopulation with risk factors predisposing to increased sensitivity to benzene (ORC/Jandl, PC 34, pp. 43-44). These include:

a. Those between the ages of 16 and 30;

b. Those with prior exposure to benzene or exposed to radiation in excess of that required for diagnostic purposes, and;

c. Those with evidence of a history of other less severe myeloproliferative disorders such as remitted aplastic anemia, certain abnormal circulating blood cells, or those with persistent abnormalities in one or more blood cell types.

Other predisposing factors may include exposure to drugs or alcohol which may act to modify the metabolism of benzene. These postulated effects are not consistently observed or universally accepted by investigators. In his testimony, Shaw stated that there was no way to determine who is susceptible to benzene and who is not (Tr. 420). And despite "well substantiated information" concerning the increased sensitivity of certain groups to benzene, the International Workshop was unable to develop and propose specific guidelines on this matter (Ex. 17, p. 7). These sensitive individuals, interspersed among the working population, are certainly among those who are at highest risk of material impairment from benzene exposure. The size of this group is unknown. It is the purpose of this regulation (especially the permissible exposure level and detailed medical surveillance protocol) to not only minimize harmful effects of benzene exposure, but also provide early diagnosis and treatment should such effects occur.

c. *Animal studies.* The wide variety of non-malignant hematological disorders observed in humans exposed to benzene which range from simple anemia, and leukopenia to aplastic anemia have been experimentally induced in animals. However, attempts to demonstrate the development of leukemia in animals exposed to benzene has met with less success. Until recently, a study by Lignac in 1932 was the only animal study known to OSHA in which leukemia has been observed in animals exposed to benzene. (Ex. 2-38). Fifty-four mice (28 females; 26 males) were given subcutaneous injections of benzene (0.001 ml in 0.1 ml of olive oil) for 17 to 21 weeks. Nine mice were initially excluded following intercurrent infection and an additional 12 were lost through atrophy of various organs, especially the spleen. Lignac attributed these 12 deaths to the size to the dose of benzene. Eight of the remaining 44 mice developed leukemia or Kundrat's lymphosarcoma and died 4 to 11 months after receiving the first injection. The absence of concurrent controls makes interpretation of the results problematic and uncertain. Failure to identify the mouse strain studied has frustrated efforts to independently confirm the findings.

Other studies have failed to reproduce Lignac's results. Amiel in 1960 utilized four inbred strains of mice and subjected them to the same experimental program outlined in Lignac's study. (Ex. 2-39). No leukemic or aplastic hemopathies were observed. Ward et al. administered benzene subcutaneously to a strain of mice which is known to be responsive to leukemogenic agents. (Ex. 2-40). A slight increase in the percentage of granulocytic leukemias was observed in the benzene-treated mice as compared with the controls; however, the authors viewed the increase as not statistically significant.

In a letter transmitted Aug. 9, 1977 to Eula Bingham, Asst. Secretary of Labor, Nelson described preliminary results of an inhalation experiment using rats and mice (Ex. 178). Animals were exposed to benzene vapors 6 hours per day, 5 days per week for up to 2 years. Two possible leukemias were observed in a group of 40 strain CD mice exposed to 300 ppm. In one animal, elevated WBC counts were observed 193 days after exposure, with death following 3 days later. The autopsy findings were consistent with CML, a disease not known to arise spontaneously in this strain. The second mouse exhibited the presence of abnormal blast forms in the peripheral blood after 211 days of exposure and died 4 days later. Whether this finding indicates AML, a disease previously observed in this strain, or acute lymphoblastic or stem cell leukemia (spontaneous incidence about 4 per-

---

*A short latency resulting from high exposures is another possible explanation.

cent) had not been resolved at the time the report was submitted.

One of the 40 rats exposed to 100 ppm benzene developed an elevated WBC count after 240 days of exposure with the cell counts continuing to increase slowly. Peripheral blood smears showed some immature neutrophils. This finding is compatible with a diagnosis of CML, a disorder which is relatively rare for the rat. However, a nonleukemic leukemoid process could not be excluded.

Nelson characterized the results of the study as:

* * * slender evidence of the production leukemia. Nevertheless, despite the shakiness, we believe these results to be extremely suggestive and provide an urgent basis for intensive further study (Ex. 178).

D. *Conclusions.* Leukemia is a serious world-wide problem. In this country alone, approximately 20,000 adults die annually from this disease, and about 12,000 individuals develop and die from AML each year (ORC/Jandl, P.C. 34, p. 72). However, based on the published literature the number of AML deaths attributable to benzene exposure is unknown. In this regard, Goldstein states: "The incidence of leukemia appears to be a small fraction of all benzene-induced hematotoxicity." (Ex. 54.B, p. 160). One reasonably complete 9 year follow-up study of 147 workers exposed to benzene, of which 109 displayed some degree of cytopenia, 1 case of leukemia was observed (Hernberg et al., Ex. 2-252). Dr. Jandl observes that, "* * * few patients with proven cases of acute myelogenous leukemia can be discovered in the American literature in the past 10 years * * *." However, the unusual number of erythroleukemias observed by Sakol at the Akron Pliofilm operation (Sakol, Ex. 61) indicates that there may be a severe underreporting of benzene-induced leukemias. Also, the relationship between initial leukemia and exposure to benzene may be less apparent than the benzene-induced cytopenias due to the substantial delay between exposure and manifestation of symptoms (Goldstein, Ex. 43.B, p. 160). Moreover, in many such instances the victim may be unaware of earlier exposure.

When case reports are derived from patients who are sufficiently sick to seek medical attention, the relative prevalence of leukemia is reported to be much higher. For example, in the 1976 report by Vigliani and Forni in which a total of 83 compensation cases of benzene toxicity were observed, leukemia accounted for 19 (Ex. 2-15).

Vigliani has estimated based upon available literature, that the number of cases of leukemia attributed to benzene exposure is "at least 150" (Ex. 2-49, p. 144). The document, "Benzene in the Work Environment", used to establish the German Occupational Standard, summarized over 250 reported cases of leukemia in individuals who had been chronically exposed to benzene. (Ex. 2-58). Jandl states there have been at least 250 cases of AML in the United States which are believed to be the result of exposure to benzene. (ORC/Jandl, PC 34, p. 103). It is impossible to extrapolate these published findings to either an absolute total or a relative incidence. Beyond the likely underreporting of cases, the incidence of benzene leukemia may be greater than previously recognized since many diagnoses were made without results of marrow aspirates. Thus, peripheral pancytopenia may have been inaccurately labelled aplastic anemia when marrow examination might well have revealed acute myeloblastic leukemia, leukopenic or aleukemic leukemia, a type characterized by a decline in circulating white cells. (Vigliani, Ex. 2-49, pp. 143-4).

The evidence in the record conclusively establishes that benzene is a human carcinogen. The determination of benzene's leukemogenicity is derived from the evaluation of all the evidence in totality and is not based on any one particular study. OSHA recognizes, as indicated above that individual reports vary considerably in quality, and that some investigations have significant methodological deficiencies. While recognizing the strengths and weaknesses in individual studies, OSHA nevertheless concludes that the benzene record as a whole clearly establishes a causal relationship between benzene and leukemia.

The evidence indicating that relationship has been expanding and the international scientific community has increasingly acknowledged that worker exposure to benzene is associated with an increased risk for the induction of leukemia, in fact, today there is little dispute that benzene is a human leukemogen.

For example, in 1976 NIOSH stated:

NIOSH considers the accumulated evidence from clinical as well as epidemiologic data to be conclusive at this time that benzene is leukemogenic. Becauses it causes progressive, malignant disease of the blood-forming organs, NIOSH recommends, that for regulatory purposes, benzene be considered carcinogenic in man (Ex. 2-5).

In a paper presented at the International Workshop, Brair wrote that:

It is now generally recognized that benzene leukemia is a nosological entity. (Ex. 34-5, p. 4).

Vigliani has observed:

[F]rom the vast literature accumulating over the last forty years, the conclusion can be drawn that the clinical as well as epidemiological data are indicative of a strong leukemogenic action in man * * * (Ex. 2-49, p. 148).

Aksoy testified that:

[T]here is overwhelming scientific evidence that benzene can cause leukemia in man (Tr-144).

Other witnesses and commentators were in basic agreement with this conclusion: (Goldstein, TR-351, 389; Sakol, TR-316; Jandl, P. C. 26B. Addendum V, P. V, Tabershaw Tr 2545).

In 1971, the Senate of the Deutsche Forschungsgemeinschaft for the Examination of Hazardous Industrial Materials indicted benzene as a human carcinogen (See NIOSH Ex. 84A, p. 1). Two years later, Eckardt, then Medical Director of Exxon Corp., stated,

[t]he accumulated evidence in the literature leads to the inevitable [sic] conclusion that benzene is a leukemogenic agent, particularly in cases that have previously displayed a panmyelopathy. Although this had long been suspected, the data reported in the literature was not sufficiently convincing to establish the leukemogenic nature of benzene. However, the more recent observations seem to establish the association beyond a doubt (Ex. 84, B.2., p. 7).

Industry participants in the benzene rule-making did not, for the most part, challenge benzene's leukemogenicity. Rather, they argued that there is a threshold level for benzene and that, consequently, benzene is not a leukemogen at low levels. In support of this argument, industry cited several epidemiological studies purporting to demonstrate that an excess risk of leukemia is not seen in workers exposed to benzene at levels lower than 10 ppm. Although the epidemiological method can provide strong evidence of a causal relationship between exposure and disease in the case of positive findings, it is by its very nature relatively crude and an insensitive measure. In the case of negative findings, the results are especially difficult to interpret for several reasons. One methodological artifact commonly observed is the failure to adequately define the composition of the study cohort. The fundamental premise of any epidemiological investigation of the potential relationship of occupational exposure to an agent and observed health effects is, in OSHA's view, a careful definition of an exposed cohort. As was pointed out above, many of the benzene studies cited by industry suffer from this design defect. Other factors, such as incomplete followup of workers who may exit the cohort, and uncertainly as to the actual cause of death lead to further dilution of the reported results. The presence of a latency period exacerbates this problem: The long delay between exposure and manifestation of symptoms, compounded by the high mobility of our society make it medically and scientifically extremely difficult to identify workplace carcinogens. An additional difficulty arises from the apparent variability in individual sensitivity. Epidemiological study populations tend to be small and reaction to an etiological agent may not appear in the cohort, due to varying degrees of sensitivity. On the other hand, positive retrospective investigations suffer from these artifacts to a much lesser degree. For these reasons, it is OSHA's policy when evaluating negative studies, to hold them to higher standards of methodological accuracy.

As stated above, the positive studies on benzene demonstrate the causal relationship of benzene to the induction of leukemia. Although these studies, for the most part involve high exposure levels, it is OSHA's view that once the carcinogenicity of a substance has been established qualitatively, any exposure must be considered to be attended by risk when considering any given population. OSHA therefore believes that occupational exposure to benzene at low levels poses a carcinogenic risk to workers. The Agency, however, recognizes that not all individuals exposed to benzene will suffer harmful effects. The benzene record establishes that there may be individual effect levels. However, there is no way of ascertaining which workers in any exposed population are susceptible and to what levels.

OSHA further recognizes that determinations of carcinogenicity are normally based on animal studies, and when available, human evidence. OSHA acknowledges that, at the present time, there is no unequivocal animal model demonstrating the induction of leukemia. However, the lack of an animal model does not diminish the conclusive nature of human evidence demonstrating benzene's leukemogenicity.

OSHA acknowledges that there are many unresolved issues concerning the relationship between benzene exposure and leukemia:

Whether benzene is a co-carcinogen, a procarcinogen, etc.

Whether benzene exposure is causally related to induction of forms of leukemia other than AML and its variants.

Whether blood dyscrasias, such as aplastic anemia always precede benzene leukemia.

To what extent are leukemia deaths attributable to occupational exposure to benzene.

What factors identify a sensitive population.

These questions are on the frontier of scientific and medical knowledge and, given the mandate of the Act, OSHA cannot wait for answers while workers are exposed to this life-threatening substance. There is no doubt that benzene is a carcinogen and must, for the protection and safety of workers, be regulated as such. Given the inability to demonstrate a threshold or establish a safe level, it is appropriate that OSHA prescribe that the permissible exposure to benzene be reduced to the lowest level feasible.

4. *Chromosome studies.* OSHA has examined both the original literature of benzene-induced chromosomal changes and the material contained in recent reviews of the subject (NRC Report, Ex. 2-4, section entitled: Chromosome Effects; Forni, Ex. 156.H and; Wolman, NYU Report Ex. 43.B, Chapter VI). Evidence derived from human studies together with similar results obtained from experimental investigation clearly demonstrates that benzene can induce visible damage to chromosomes in lymphocytes and blood-forming cells. For example, Tyroler noted that: "I think there is rather convincing evidence that the chromosomal aberrations are associated with exposure to benzene." (Tyroler, Tr. 3100; also see International Workshop, Ex. 17, p. 6). The effects may be manifested as numerical alterations and/or structural rearrangements of the chromosomal material and include additions or deletions of chromosomes, segments of whole chromosomes or chromosome sets, in addition to exchanges which result in morphologically aberrant chromosomes (Wolman, NYU Report, Ex. 43.B, p.—). Forni has observed that the aberrations produced by benzene exposure are non-specific and are similar to those produced by ionizing radiation. (Ex. 156.H). To what extent these events are detrimental to human health is not known.

a. HUMAN STUDIES

There are numerous reports on chromosomal evaluations in worker populations both with and without clinical symptomatology resulting from exposure. In general, these studies reveal that there are statistically significant increases in chromosomal damage in those occupationally exposed to benzene. The chromosome alterations have been classified as either unstable changes (i.e., fragments, dicentric, tricentric and ring chromosomes) or stable changes (i.e. abnormal monocentric chromosomes due to deletions, translocations, inversions and trisomies). Aneuploidy (abnormal chromosome number) and/or polyploidy (a multiple chromosomal set) have also been observed. The early reports focused on the examination of workers exhibiting benzene hemopathy (a disease of the blood). As early as 1964 Pollini and Columbi published such a study. (Ex. 2-17). Examination of cultured bone marrow cells and peripheral lymphocytes showed increased frequencies of aneuploid cells and structural aberrations. A second study by the same group (Ex. 2-4, reference 50) of 4 patients with temporary or progressive blood dyscrasias revealed that the incidence of heteroploid (abnormal number) chromosomal patterns ranged around 70% both in the blood and in the marrow of each subject. The authors were, however, unable to establish a correlation between the duration and degree of exposure and either the frequency of chromosome aberration or the degree of toxicity. Other cytogenetic studies of subjects with benzene-related hemopathies have yielded similar findings (e.g. Forni and Moreo, Ex. 2-18 and 2-19, Sellyei and Keleman, Ex. 2A-258; Aksoy et al., Ex. 2-57; and Erdogan and Aksoy, Ex. 2A-192). Variables which make these case reports difficult to interpret and compare include: differential diagnoses, the use of lymphocytes artificially stimulated to grow in some cases and bone marrow cells in others; the occasional lack of "normal" baseline chromosomal breakage frequency; and the occurrence in some cases of pre-existing familial chromosome aberrations (Wolman, Ex. 43-B, p. 129). In spite of the limitations of these reports, Wolman has stated that some trends have been observed—additional chromosomes have been identified in several reports, tetraploidy (4 sets of chromosomes) or polyploidy were seen in some cases and in many instances an increased frequency of chromosome breakage was reported but not well documented. Wolman states that:

the clearest picture of the relationship between benzene exposure and chromosomal changes emerges, not from experimental studies or reports of human disease, but from studies of occupationally exposed workers (Ex. 43.B, p. 130).

The National Research Council (Ex. 2-4) considers the 1969 study by Vigliani and Forni (Ex. 2-20) to be one of the most systematic of this type reported. The results of cytogenetic analysis of 25 subjects who had recovered from benzene hemopathy were compared to the findings of controls matched for sex and age. In most cases, increased ratios of both stable and unstable chromosome aberrations were still present several years after cessation of exposure to benzene and/or recovery from poisoning, whereas the hematological analysis revealed normal blood counts in most cases. Follow-up cytological analyses of these subjects showed an overall decrease in unstable changes, and generally, a persistence or an increase in stable alterations. Other cytogenetic surveys of workers industrially exposed to benzene are: (Forni, Ex. 2-20; Girard et al., Ex. 2-40; Hartwich and Schwanitz, Ex. 217-129; and Vigliani and Forni, Ex. 2-14).

OSHA is also aware of several cytogenetic studies of workers chronically exposed to airborne concentrations of benzene probably less than 25 ppm. Several indicate an increase in chromosomal damage (Hartwich and Schwanitz, Ex. 129; Girard et al., Ex. 2-65 or 2-283, and Berlin et al. Ex. 2A-218) while others present negative findings (Forni, Ex. 156.H; Tough et al., Ex. 2-21B, and Burgett Ex. 2A-226).

b. EXPERIMENTAL STUDIES

The quantity of data available from experimental studies is less than that reported for exposed workers, but it provides additional corroborative evidence that chromosomal aberrations can be induced by benzene. An in vitro study utilizing cultured human lymphocytes, incubated with benzene, revealed that the frequency of chromatid-type aberrations increased with increasing dosages of benzene. It was further observed that the yield of dicentric and ring chromosomes induced by benzene damage was synergistically related to treatment with radiation. A high incidence of chromosomal damage has also been observed in cultured human leukocytes and Hela cells incubated with benzene (Koizumi et al., Ex. 2B-298). In 1970, Philip and Jensen reported that rats acutely intoxicated with subcutaneous injections of benzene displayed increased rates of chromatid breakage in direct bone marrow preparations 12 and 24 hours after injection, followed by a return to normal cytogenetic findings at 36 hours (Ex. 2-23). It has also been reported that rabbits made pancytopenic by chronic exposure to benzene via injection exhibited a high frequency of both chromatid and chromosome aberrations, in uncultured bone marrow cells (Ex. 2-22). After reviewing the animal evidence, OSHA concurs with Wolman's following conclusion:

[T]hat although it appears that exposure times in different species can induce increases in chromosome aberrations, there is no clear evidence for a dose-dependent response to benzene exposure. (Ex. 43B).

C. DISCUSSION

While the record clearly demonstrated that benzene causes chromosomal aberrations, it is equally clear that there is no unanimity of opinion as to what the benzene-induced chromosome aberrations mean in terms of demonstrable health effects, especially the relationship of chromosome damage to the induction of leukemia. For example, the International Workshop concluded: "The implication of the finding of increased chromosome aberrations for the occurrence of benzene leukemia is still not clear." (Ex. 17, p. 6).

Whether these aberration are, in some instances, (1) causally-related to the development of leukemia, or (2) are secondary to the neoplastic state and therefore are epiphenomena of metabolic alterations resulting from abnormal growth is not known (Sandberg, Ex. 2-102).

There are several theoretical considerations and some evidence supportive of the first alternative: Some investigators view chromosomal damage as a mutational event or an event which increases the probability of a mutation (e.g. Wolman, Ex. 43B, p. 126). As the induction or the maintenance of a neoplastic (cancerous) transformation results most probably from a somatic mutation, damage to chromosomal material may enhance the opportunity for such a cellular change to occur. Other interrelated hypotheses in support of a causal relationship between chromosomal damage and leukemia are:

(1) That benzene might induce various types of chromosomal aberrations and leukemia might evolve in those cases where a potential leukemia clone with a selective advantage develops in response to benzene exposure (Vigliani and Forni, Ex. 2-14).

(2) That cells exhibiting chromosomal instability or imbalance may be more susceptible to transformation by a leukemogenic virus (International Workshop, Ex. 17, p. 6) or;

(3) That chromosomal aberrations may lead to decreased immunological surveillance, and that in such instances, leukemia can develop if abnormal cell clones (some of which may be neoplastic) are not eliminated. In this context, Forni and Moreau reported a case of a benzene-induced leukemia in which both bone marrow and peripheral lymphocytes exed 47 chromosomes (1 more than normal) with the same karyotype, suggestive of clone formation. (Ex. 2-18).

There are also several investigators who believe that at this time, there is insufficient evidence to establish a causal link between visible chromosomal damage and leukemia (Ex. 156.2). For example, Snyder has stated that:

while exposure to benzene in sufficient concentration and for extended periods of time has been linked with both chromosomal aberrations and leukemia in humans, benzene-induced chromosome damage has been found in individuals who have not exhibited leukemia * * * (p. 22).

Additional evidence in favor of the non-causal relationship nature of the relationship between chromosomal damage and leukemia was presented by Jandl. He observes that exposure to several types of agents, not known to possess leukemogenic action, induce chromosomal abnormalities which may persist for months.

Sandberg cautions that no consistent karyotypic pattern has emerged for any type of cancer or leukemia and in about half of all cases of acute leukemia, some cases of chronic myelogenous leukemia and in almost all cases of chronic lymphocytic leukemia, no visible chromosomal alterations occur (Ex. 2-102). And also, Elkins feels that the evidence fails to establish a causal link between chromosome alterations and leukemia (Ex. 156C, p. 22-23).

Whether these gross alterations of chromosomes can be viewed as toxic or mutational events depends on the fate of the affected cell. If the alteration in the chromosomal material results in an inhibition of further cellular division, then in terms of its reproductive potential, the cell is dead and the damage inflicted may be classified as a toxic event. However, if the damage does not interfere with the reproductive ability of the cell, and the alteration is replicated, this may constitute a persistent gross mutation, "* * * a structural change in the genome which presumably alters cell function". (Wolman, Ex. 43.B, p. 126.) In addition to structural and numerical alterations, benzene has been observed to induce chromosomal breaks:

While breaks may be repaired and are not necessarily mutational events (in the sense of being inherited), each occurrence increases the probability of a structural aberration and, therefore of a mutation (Wolman, Ex. 43.B, chapter VI).

In viewing benzene as an agent potentially capable of interacting with the genetic material, several points should be made:

(1) All of the studies demonstrating benzene's "clastogenic" or chromosomal-damaging effect (Shaw, Ex. 2-94) have apparently reported the effects in somatic cells only. In view of the wide distribution of benzene in the body, it is not unreasonable, however, to assume that given somatic chromosomal damage resulting from benzene exposure, that germinal tissue may also be affected. In her review of clastogenic substances, Shaw has stated:

Because changes in the genes and chromosomes do not usually produce an immediate health hazard, they may go undetected for a lifetime or even for several generations. Yet, the human gene pool can become insiduously polluted. (Ex. 2-94, p. 409).

(2) Despite the considerable evidence demonstrating benzene's ability to inflict visible damage to chromosomes, benzene has not manifested mutagenic activity when tested in various microbial assay systems (Kraybill, Ex. 84. B 19, p. 314).

(3) The finding of gross chromosomal damage in bone marrow cells clearly demonstrates that despite competing detoxification reactions (see: Olson, Ex. 149, p. 12) benzene, or a reactive metabolite, is able to overwhelm protective defense mechanisms and enter the nucleus of hematopoietic cells.

In summary, the evidence clearly demonstrates that benzene is capable of causing significant increases in chromosomal aberrations in somatic blood cells in the absence of clinical and/or hematological symptomatology. It is also clear that for man, no quantitative dose-response relationship has been established for these effects. It is OSHA's interpretation of these findings that chromosomal damage represents an adverse biological event of serious concern which may pose or reflect a potential health risk and as such, must be considered in the larger purview of adverse health effects associated with benzene.

V. ECONOMIC CONSIDERATIONS

In setting standards for toxic substances, the Secretary is required by section 6(b)(5) of the Act to give due regard to the question of feasibility. Section 6(b)(5) mandates that final standards be set which most adequately assure employee safety and health "to the extent feasible, on the basis of the best available evidence" and further requires that, in the development of occupational safety and health standards, "considerations shall be the latest available scientific data in the field, the feasibility of the standards, and experience gained under this and other health and safety laws." While the precise meaning of feasibility is not clear from the Act, it is OSHA's view that the term may include the economic ramifications of requirements imposed by standards. The determination that OSHA has the authority to consider economic feasibility factors in developing standards has been endorsed by the courts. *Industrial Union Dept., AFL-CIO v. Hodgson*, 499 F. 2d 467 (C.A.D.C., 1974); *AFL-CIO v. Brennan*, 530 F. 2d 109 (C.A. 3, 1975). As pointed out by the D.C. Circuit Court of Appeals, Congress did not intend the Secretary to promulgate standards which drive entire industries or large numbers of employers out of business. On the other hand, "standards may be economically feasible even though, from the standpoint of employers, they are financially burdensome and affect profit margins adversely; further, the Court said, the concept of economic feasibility does not "necessarily guarantee the continued existence of individual employers." *Industrial Union Dept., AFL-CIO v. Hodgson*, supra, at page 478.

In accordance with the Secretary's position, it has long been OSHA's practice to analyze the economic feasibility of proposed standards where significant economic impact on employers covered by the proposals seem likely, to make such analysis available to affected parties for comment and subsequent hearing prior to issuance of final rules, and to invite the submission of other information on the economic impact and feasibility of proposed standards. In developing a final standard, therefore, OSHA evaluates the economic feasibility of the final standard on the basis of the information developed by its own studies of the proposal and submissions by the public during rulemaking.

To assess the economic feasibility of the proposed standard for benzene, OSHA undertook an extensive study of the proposal's economic impact on various affected industries. This study was conducted for OSHA by Arthur D. Little, Inc. (ADL). (Ex. 5A, 5B). The ADL study has provided basic economic data for the evaluation of the economic impact of the permanent benzene standard on the major affected industries. Additional information for this purpose was obtained through OSHA's analysis and consideration of all other economic data, comments, arguments and testimony submitted at the benzene hearing, in pre-hearing comments and in post-hearing comments and briefs. On the basis of the best available evidence, therefore, OSHA has determined, as explained in detail below, that the permanent benzene standard is economically feasible.

COMPLIANCE COSTS

Estimates of total costs of compliance with the permanent benzene standard for the major affected industries studies are as follows: First year operating costs for all industries combined are estimated by OSHA to be approximately in the range of $187 million to $205 million recurring annual costs are estimated at approximately $34 million and investment in engineering controls is expected to be approximately $266 million. Estimates of these costs for various individual industry sectors are analyzed below. As that analysis reveals, the greatest economic impact of the standard falls on the larger and more stable industries, such as petroleum refining and petrochemical production, which can readily absorb the costs or shift them forward to consumers. No testimony was offered by these industries that the proposed standard for benzene would imperil their existence. Even the higher projection of some costs by the American Petroleum Institute and other participants would not raise any serious question concerning the economic feasibility of the standard or the ability of the regulated industries to bear the additional costs.

The benzene standard, in its final form will require all industries that produce or use benzene, petrochemicals, products and services involving the use of solvents derived from petroleum, as well as production of crude and refined petroleum products, and primary distribution of gasoline, to undertake an initial determination of the extent to which their employees are exposed to benzene. The results of the initial exposure measurements will determine the types of activities each firm will be required to take to comply with the provisions of this standard. Where exposure measurements are below the action level (0.5 ppm), firms will have to provide information and training in the hazards related to benzene to their employees, comply with the labelling requirements, and retain the records of initial measurements. These compliance activities account for the entire first year operating costs for many of these industries. Firms with exposure levels above the action level will incur additional first year costs for monitoring and medical examination programs. Finally, those firms with exposure levels above the permissible exposure level will have additional first year costs for installation of appropriate engineering and work practice controls, and for respirators. It is assumed that engineering controls are all installed in the first year and reduce exposure levels to below the permissible exposure limit. However, those firms with continuing exposure measurements above the action level will have recurring annual costs for monitoring and medical surveillance. Regardless of the airborne exposure measurements, employees in operations with potential exposure to eye or repeated skin contact will have to be provided with appropriate personal protective equipment. Thus, some firms and industries will have higher and more sustained cost burdens than others.

OSHA's estimates of costs for compliance with the final standard differ from estimates of the proposal's compliance costs in part because the standard in its final form differs from the proposal in areas which significantly impact on the cost of compliance. ADL estimated the proposal's compliance costs for 20 industry sectors engaged in benzene production (petroleum and coke), petroleum refining, chemical processing, benzene transportation, and other industries, such as rubber manufacturing and laboratories (ADL, Vol. 1, Ch 5). ADL provided detailed cost estimates for each industry and for each compliance activity required by the proposal. Total costs for all surveyed industries were estimated by ADL as follows: First year operating costs were approximately $124 million, recurring annual costs were approximately $74 million, and costs for implementation of engineering controls were approximately $267 million. In developing each of these estimates, ADL used the control with the least cost as the basis of their calculations. In other words, where alternative methods of compliance are available to employers in an industry, and indeed may be more attractive to employers, the higher cost method was considered an optional process improvement, and only that portion of its total cost required to produce compliance was allocated to compliance with the proposed regulation (ADL, Vol 1, pp. 5-3). Additionally, ADL assumed compliance with the previous OSHA regulation (29 CFR 1910.1000, Table Z-2) requiring an exposure limit of 10 ppm TWA and a ceiling limit of 25 ppm, thus assessing to the benzene proposal only the incremental cost associated with moving into compliance with the proposed permissible exposure level from the previous exposure limit.

The ADL costs covered 24,312 facilities and 196,875 estimated exposed employees (ADL, Ex. 5A, Table 5-1, p. 5-2). Costs were estimated for the required compliance activities of monitoring, medical surveillance, respirators, personal protective equipment, training, signs and labels, and recordkeeping. Distribution of costs varied from sector to sector depending on such variables as the number of exposed workers who would require monitoring and medical surveillance and the estimated effectiveness of engineering and work practice controls. OSHA has examined the methods used by ADL to determine the costs for each provision of the standard and, in the absence of substantially contradictory testimony and alternative methods from affected industries, the approach of ADL has been largely adopted by OSHA in estimating the costs of compliance.

Based on all the evidence on the record, OSHA has estimated the cost of compliance with the benzene permanent standard for approximately 157,000 facilities and approximately 629,000 exposed employees.

Monitoring

As a basis for OSHA's estimates of compliance costs, OSHA has utilized for many industries ADL's sampling protocol for determining monitoring costs. ADL's protocol was designed to provide an initial exposure profile and characteristic sampling program.

ADL's representative explained that the protocol was intended to be representative, but that it might not fit each plant exactly and costs might be higher if the number of exposed workers in an industry, or those exposed above the permissible limit, were higher. (See Tr. 477-484, 522-526, and 592-595.) The derivation of the sampling base further explained in ADL's post-hearing submission as follows:

> The vast majority of facilities which must comply with this regulation are typified by twenty-four hour per day continuous processes representative of petrochemical operations. It was proposed to initially sample 50 percent of all exposed employees. However, the three shift nature of these operations and the commonality between worker exposures by job category between shifts indicated that sampling results from one shift could be properly interpreted as being representative for all shifts. Therefore only ⅓ of the exposed workers should be sampled. Multiplying ⅓ × ½ equals ⅙ of the exposed employees who must be monitored. (Post-Hearing 15E)

OSHA recognizes that these assumptions may not portray exactly the monitoring pattern of employers within other industries. It is OSHA's view, however, that the monitoring scheme as described is a reasonable basis for estimating the cost of compliance with the monitoring requirements of the final standard.

In estimating the percentage of employees within each exposure category—i.e., above the permissible exposure level, between the action level and the permissible exposure limit, and below the action level, OSHA has used varying assumptions based on exposure data, where available. For the major portion of the covered industries, OSHA has accepted ADL's estimate that 20% of the exposed employees are above the permissible exposure limit. While there may be some variations in actual percentages at individual plants, industry witnesses offered no alternative approaches to estimating the number of exposed employees above the permissible exposure limit. Determination of the percentage of employees in each of the two exposure categories below the permissible exposure limit was based to the extent possible on record evidence of exposure. Where such evidence was not available, it was assumed for cost calculation purposes that one-half of the employees were exposed below the action level and one-half were exposed between the action level and the permissible exposure limit. Use of this distribution of exposed workers, however, does not imply that OSHA believes it necessarily represents actual exposure profiles for all industries; it has been used only to provide some basis for estimating approximate compliance costs and assessing the economic impact of the standard on these industries. OSHA has therefore, relied on the best available evidence as to likely exposure patterns in determining that its sampling protocol reflects typical industry conditions, provides a uniform estimation method enabling inter-industry comparisons to be made, and yields reasonable approximations of the cost impact of the monitoring program.

OSHA has utilized the ADL methodology for calculating monitoring costs. There are, however, significant differences in total monitoring costs as a result of changes in the monitoring requirements of the final standard. The elimination of the proposed percentage exemption expands the number of employers covered by the standard and the number of employees affected. Cost estimates for compliance by employers using liquid mixtures containing less than 1% benzene were not given in the ADL study. However, inclusion of an action level in the standard, below which no regular monitoring or medical surveillance is required, should minimize the economic impact of the permanent standard. This is especially true since the proposal would have lowered the percentage exclusion to 0.1% after one year, and would have imposed periodic monitoring and medical surveillance of employees in these operations with any amount of exposure to benzene. Furthermore, it is anticipated that a significant number of operations utilizing mixtures with less than 1% benzene will be below the action level, in which case only initial monitoring must be conducted.

The introduction of an action level decreases for individual employers the cost of compliance for the monitoring activities. The proposal required quarterly monitoring where exposures were at or below 1 ppm. The final standard requires initial measurement to determine exposures but requires no further monitoring where exposures are below 0.5 ppm (unless a redetermination of exposures is necessitated by specific changes). As a result of this change in the monitoring requirements, some employers, who would have been required to monitor employees quarterly, will now only need to conduct initial measurements. Initial measurements it is estimated will cost approximately $109 million for all industries combined. However, the actual cost will be substantially below this, inasmuch as the standard allows employers who have conducted initial monitoring, pursuant to the benzene guidelines or the emergency temporary standard, to utilize those measurements in order to initially determine employee exposure; and the benzene record indicates that many employers have conducted these measurements.

Medical Surveillance

The final standard, unlike the proposal, requires medical surveillance only of those employees whose exposure is at or above the action level. Moreover, the standard provides for twice yearly medical exams rather than quarterly exams. ADL has estimated the annual cost for routine medical surveillance as approximately $207 per employee annually, (Vol. II, App. C). This figure included employer and employee time, physician charges and laboratory fees. (Vol. II, C-7).

Since routine medical surveillance is required by the permanent standard semi-annually rather than quarterly as contemplated by the proposal, OSHA has adjusted the ADL medical cost per employee to reflect the reduction in the number of periodic exams and to remove from the periodic exams the cost of the reticulocyte count and serum bilirubin, which the permanent standard requires only as part of the initial baseline exam. OSHA has, therefore, estimated the cost of the first year medical surveillance, including the work history, as approximately $105.00 per employee and the recurring annual cost of medical surveillance is estimated as approximately $78 per employee.

The final standard provides for medical surveillance in addition to semi-annual routine blood testing. Where routine medical surveillance reveals an abnormal blood picture, the employer is required to refer the employee to a hematologist. Where emergency situations occur, the employer is required to provide a urinary phenol test and, if urinary phenols are elevated, a repeat complete blood count (CBC); if the CBC reveals abnormalities then the employee must be examined by a hematologist. The cost of these non-routine examinations was not estimated by ADL since they would not have been required by the proposal. There is, of course, no certain way of determining the number of employees who will need examination by a hematologist. To calculate these additional costs, therefore, OSHA has assumed that 5% of all employees will have abnormal blood pictures and will be re-referred to a hematologist, and that the added cost will be $100 per referral.

OTHER COSTS

Total costs for other compliance activities, such as engineering controls, respirators, training, and signs and labels have also been assessed. These costs are based essentially on the cost factors furnished by ADL since the ADL costs were generally accepted by industry participants.

Capital costs for engineering controls have been determined for industries with operations above the permissible exposure limit on the basis of the type of engineering controls available for the individual industry. Many of the industries covered by the benzene standard involve the storage and movement within a closed system of liquids containing benzene in various concentrations. Emissions occur at such points as pumps, pipe line joints, compressors, sampling points, and gauging stations. (Vol. 1, 4-35). Engineering controls for these systems consists of replacing worn pumps and compressors with equipment specifically designed to minimize emissions, replacing gaskets to ensure tight fit of joints, welding joints, and the installation of automative gauging and closed loop sampling devices. Specific units of these systems have been costed by ADL and aggregate costs shown below are based on an estimated number of these units which are required for a typical plant. To determine respirator costs, OSHA has used ADL's cost per employee for respirators. (Vol. II C-8, 9). For personal protective equipment, which is required by the standard for employees who may be exposed to eye or repeated dermal contact with benzene, the cost of $13.16 per employee per year has been used. This cost includes the cost of gloves, apron, faceshield, and goggles (Vol. II, C-R).

Training costs, which include costs for preparation, materials, employee's time, and instructor's time, have been estimated at $110 per facility plus $14 per employee (Vol. II C-II).

In calculating recordkeeping cost, OSHA has utilized the ADL cost factor (Vol. II C-12) for record maintenance but adjusted this cost to reflect the decrease in the records required by the final standard. In its testimony, AISI expressed the view that recordkeeping would require the addition of one record clerk per facility (Tr. 3165-317 I). However, the final standard significantly reduces the monitoring records and eliminates medical records for employees whose exposure is below the action level. Therefore, application of ADL's cost formula will give a reasonable approximation of the recordkeeping burden.

For most firms, a normal cost for labels has been determined on the basis of ADL's cost of estimate (Vol. II, C-13). No cost has, however, been determined for signs. Detailed data on a plant-by-plant basis were not available to OSHA. Therefore, no costs for this compliance activity are included within the cost estimates shown herein. OSHA recognizes that these costs will occur but is of the opinion that they will be relatively small for all firms covered.

Compliance costs for the following industries have been calculated by OSHA using the compliance factors described above.

PETROLEUM REFINERIES

It is estimated that petroleum refineries will incur greater compliance costs than any other industry sector. This is primarily due to the large number of refinery workers who are exposed to petroleum products, almost all of which contain benzene. There are 48 refineries engaged in the production of benzene and 275 refineries which do not produce benzene but maintain process streams with benzene concentrations.

At the 48 refineries which produce benzene, there are approximately 1,440 exposed workers. On the basis of the exposure data made available by ADL and industry witnesses, it appears that approximately 20% of these employees are exposed above the permissible exposure limit. OSHA has, therefore, applied the ADL sampling protocol and calculated costs on the assumption that initial measurement will show approximately 300 workers exposed above the permissible exposure limit. The remaining exposed workers were divided for cost calculation purposes. Capital investment to reduce exposure levels has been estimated at approximately $24 million extrapolating from a model plant analysis. This cost would be for replacement of seals and controls of leaks (ADL, E-1) and similar indicated improved maintenance and repair. No estimate is made for shut down time since it is anticipated that any controls including modifications of pumps and compressors, would be installed during normal down time (Tr. 536). OSHA does not concur in the API view (Scarborough statement p. 36-7) that additional down time is needed since normal maintenance procedures of highly flammable material require purging of lines, and replacement of seals, pumps, etc. First year operating costs for these refineries are estimated at approximately $600,000.

Cost estimates for the 275 refineries which do not produce benzene were based on API's and ADL's data as to 98,000 exposed employees. (ADL, Vol I, p. 419; add API reference for that number). ADL relied in part on the results of an API questionnaire which indicates a wide range of exposures from negligible to as high as 25 ppm. (ADL pp. 4-16 to 4-17). API's witness indicated that exposures at refineries were very low with most below the proposed permissible exposures limit and many below the action level (Tr. 1467-1468). Other testimony indicates similar exposure patterns (Grospiron statement, pp. 4-6). The API post hearing submissions suggest that refinery workers move frequently from one post to another so that they are not exposed to hydrocarbon vapors for extended periods of time (API, PH5). Initial monitoring results showing 8 hour TWA measurements of 0.13 ppm and lower were cited by API. (API PH 8-9.) In the face of conflicting data of this sort, it was assumed that approximately 5,000 workers initially could be found to be exposed above the permissible exposure limit, that approximately 5,000 would be exposed at levels between the permissible exposure limit and the action level, and that the remaining 88,000 would be below the action level. This distribution indicates somewhat higher exposure levels that API's testimony suggests, but lower levels than the results of the API questionnaire survey might be construed to indicate and is, therefore, intended to be a reasonable estimate. Using this distribution, OSHA estimates first year operating cost for compliance to be approximately $12.8 million. It may be noted that approximately $2 million of the indicated first year cost is for exposure measurements. Since many refineries have apparently made such measurements, actual first year compliance costs may be substantially lower than the estimates listed herein.

The cost of capital investment in engineering controls is estimated to be $110 million, for control of leaks from seals and valves, control of sumps and other disposal areas, and control of other emissions associated with storage, transfer, and gauging activities. If it is assumed that application of engineering controls is highly effective in bringing all worker exposures to levels below the action level, recurring annual costs would decline to approximately $3.5 million. If, alternatively, it is assumed that as many as 5,000 refinery workers may continue to be exposed at levels between the action level and the permissible exposure limit, recurring annual costs would be approximately $4.4 million. OSHA believes the latter assumption is indicative of the upper bound of these annual costs.

COKE PLANTS

Compliance costs related to the production of benzene as a byproduct of coking are associated with activities at the coke batteries and at light oil processing facilities.

There are some 65 affected coke producing facilities. ADL had estimated that there were approximately 17,000 employees potentially exposed to benzene at coke oven batteries. The American Iron and Steel Institute (AISI) challenged this number as being an underestimation of exposed employees. The larger figure of 24,000 workers exposed at the coke oven batteries has been used by OSHA as urged by AISI. (TR 4635, 472-4) This was the number estimated to be exposed to coke oven emissions in the Economic Impact Statement for the Coke Oven Emissions Standard. It is probable that this constitutes an overestimate of workers exposed to benzene at the coke oven batteries since benzene emissions are expected to occur only during portions of the production cycle and not at all locations. (ADL, PC, 15f, 15l).

The AISI number is used, however, to avoid all possible underestimation of compliance costs for coke oven batteries and to indicate an upper bound to the probable costs. On the basis of data supplied (Ex. 135, p. 4 and attachment), it is assumed that initial measurement of exposures will show all workers exposed to benzene concentrations below the action level. Even if this were not the case, OSHA believes that engineering controls required under the coke oven emissions standard would provide adequate control of benzene emissions. On this basis, the estimated compliance costs are approximately $1.2 million for first year operating costs and $860,000 for recurring annual costs. Since employee exposure does not exceed the permissible exposure limit, no capital cost is indicated for compliance with the benzene standard at coke oven batteries.

There are 65 coke oven light oil facilities associated with the coke oven batteries. ADL estimated that 2,370 employees were potentially exposed at the derivative light oil facilities. AISI argued that this number should be 4,000 employees. Consistent with the reasoning followed to estimate the number of workers exposed at coke oven batteries, the number of workers exposed in the light oil byproduct areas has been estimated to be 4,000, the number used in the coke oven emissions study. Data provided by ADL and AISI indicate that a substantial percentage of these workers may be exposed to benzene above 1 ppm. AISI specifically argued that at least 50% of by-products workers are exposed above the permissible exposure limit (AISI brief, p. 95). Consequently, to avoid any possible underestimation of compliance costs and indicate the upper bound of such costs, it has been assumed that all exposed workers will be exposed above the PEL at the time of initial measurement. Using these estimates of exposure, estimated compliance costs are $5.6 million for first year operating costs. OSHA has assumed that, after the installation of engineering controls, all exposures will be below the action level and the permissible exposure level. Since a significant number of light oil plants are old, controls may not result in reduction of exposure to below the action level. AISI expressed the view that engineering controls for older plants might not be feasible (P.I. statement p. 19); however, it appears that this concern was for economic rather than technological reasons. Thus, recurring annual operating costs are estimated at $887,000 for training, labels and recordkeeping. Capital costs are estimated at $19 million for investment in engineering controls.

PETROCHEMICAL INDUSTRY

The petrochemical industry includes firms producing a large variety of products from petroleum feedstocks. Six petrochemicals, accounting for about 90% of domestic consumption (ADL, Ex. 5B, B-96) were analyzed separately by ADL (Ex. 5A, 6-18ff) but since the exposure patterns of each are similar and the ecomonic impacts affect many of the same or competing markets, the costs of petrochemicals have been aggregated by OSHA. The industry sectors included are chlorobenzene, cumene/phenol, cyclohexane, dodecylbenzene, ethylbenzene/styrene, maleic anhydride, nitrobenzene. This follows the analysis provided by ADL and the cost estimates shown are based on ADL's data. Industry participants did not challenge these costs.

There are 92 facilities covered by this standard, with 2,760 exposed workers. (ADL, Ex. 5A, p. 4-4) On the basis of the limited data available on exposure levels (Vol. 1, pp. 4-7; Table 4-8, pp. 4-12 ff), it was assumed that 40 percent of the exposed workforce is exposed to benzene levels below the action level, 40 percent between the action level and the permissible exposure limit, and 20 percent above the permissible exposure limit. On this basis, first year operating costs will be approximately $1 million. Capital investment for engineering controls will be approximately $20.9. Recurring costs will be approximately $118,000. Recurring costs include training, labels, and recordkeeping since it assumed that engineering controls will reduce exposures below the action level.

BULK TERMINALS

Storage and discharge of gasoline, and other petroleum products at bulk terminals is included within the benzene standard.

It is estimated that there are 1,992 terminal facilities with 52,345 exposed workers. Testimony by industry witnesses indicated that by and large, no terminal employees were exposed to benzene levels in excess of 1 ppm (Tr. 1469-70, 1472-73; Ex. 115A,10, pp. 17-19). Using these indicators of possible exposure distributions, OSHA has estimated compliance costs for terminals on the basis of 5,000 workers exposed to benzene between the action level and the permissible exposure limit, exposure with all other workers below the action level. On this basis, first year operating costs would be approximately $5 million and recurring annual costs would be approximately $2 million.

It is estimated that there are 21,106 bulk plants employing 23,471 drivers who are potentially exposed to benzene at terminals (ADL, Ex. 5a). Again using the limited data supplied by API, it was estimated that as many as 2,300 of these workers would be exposed at levels above the action level but below the permissible exposure limit. Conversion to bottom loading on all trucks was projected to reduce all exposures to below the action level. On this basis, first year costs were estimated at approximately $17.9 million, recurring annual costs at $3 million, and capital investment at $51.5 million

OIL AND GAS PRODUCTION

There are approximately 700,000 oil and gas wells in the U.S. and, since any crude oil or natural gas may be expected to contain benzene, all of these are covered by the standard. API indicated that the number of exposed workers is 74,000 and that exposure levels are below the action level (Ex. 115A, 4, Ex. 116). As noted in the section on technological feasibility, workers are normally exposed only intermittently and occasionally, and emissions are limited to points where leaks or spills occur (Galloway, API, Ex. p. 116, pp. 4-5). Proper work practices and good maintenance can minimize these exposures. It may reasonably be inferred, therefore, that all workers are exposed to benzene levels below the action level. Compliance costs, estimated on the basis of the reasoning used above, are approximately $39.6 million for first year operations and $2.6 million for recurring annual operations.

In determining these compliance costs, OSHA has assumed 10 wells per facility in contrast to API's reference to each well as a separate facility. It is probable that even this computation overestimates first year and recurring annual operating costs, since the costing formula for initial measurments allocates substantial charges on a per facility basis, amounting for example to approximately 90 percent of first year costs. The estimated first year costs would be reduced by the amount appropriate for measurements already performed pursuant to the benzene guidelines or the ETS. Monitoring costs would further be reduced where exposures at one workplace are representative of exposures at another.

Transportation

Companies engaged in the transportation of benzene and benzene contaminated products are covered by the benzene standard. OSHA's costs analysis, therefore, covers the transportation by pipeline, marine tanker, barge, tank car and tank truck of benzene and benzene products.

The cost analysis for pipelines is based on approximately 3,000 facilities. This figure was furnished by an API witness (Scarborough) who testified that there were 3,000 pipeline facilities employing 12,000 exposed workers. A witness for Williams Pipe Line Company estimated that there are 17,000 exposed pipeline workers (Bailey). The Williams figures, however, may represent total employment, including administrative and clerical workers. OSHA has utilized the figure of 3,000 facilities in combination with the two estimates of exposed workers to yield a range of estimated compliance costs. On this basis first year operating costs are estimated to be approximately $2.7 million and recurring annual costs for personal protective equipment, training and recordkeeping approximately $800,000. Since the standard does not require labelling of pipelines, no labelling cost is included. Neither have engineering controls been assessed since industry measurements indicate that all workers will be below the action levels (Bailey).

The number of affected tank car facilities (loading and unloading) was estimated by ADL to be approximately 100, with an average of one exposed worker per facility. (Table 5-1) The exposures of these workers will be dependent on the quantity of the various benzene-containing products shipped through the facility, the methods of loading and unloading cars, and the effectiveness of efforts to avoid leaks and spills. Conditions are somewhat similar to other facilities at which benzene and benzene-containing liquids are stored and transferred. Following the sampling protocol used throughout this analysis, costs were assigned on the basis that 20 percent of the exposures are above the permissible exposure limit, 40 percent between the permissible exposure limit and the action level, and 40 percent below the action level. It should be noted that, since it was estimated that only one worker per facility would be exposed, monitoring cost was estimated on the basis of 100 employees monitored. Engineering controls are to be installed at these facilities that will effectively control benzene emissions, bringing the exposure levels of all exposed workers below the action level. On the basis of this reasoning, it was estimated that compliance costs will be approximately $215,000 for first year operations, $16,000 for recurring annual costs, and $200,000 for capital investment.

Analysis of tank truck facilities shows a pattern very similar to that for tank cars. The rationale just described yield the following estimates: 200 affected facilities, 200 exposed workers, first year operating costs approximately $428,000, recurring annual costs approximately $37,000 and capital investment $100,000.

Barges are used for the transportation of benzene from refineries to points of utilization. Barges also transport various refined petroleum products primarily for discharge at bulk terminals. It is estimated that there are 480 employees and 240 barge facilities involved in the transportation of benzene. The record indicates that employee exposure during loading and unloading exceeds the permissible exposure limit (Ex. 5A, 4-24). For cost calculation purposes, OSHA has estimated that 20% of the employees are above the permissible exposure limit, forty percent are between the action level and the permissible exposure limit, and forty percent are below the action level. On this basis first year operating costs are calculated at $526,000, recurring annual costs are calculated at $95,000 and capital investment is estimated to be $420,000.

Barges and marine tankers are also used for the transportation of gasoline. The distribution of these exposures, however, appear to resemble those encountered during gasoline production and distribution. In that case, the average cost of compliance per employee in his transportation, loading and unloading operations would range from $100 to $750 per employee.

Laboratories

Benzene exposures occur in chemical laboratories where benzene and other petroleum-based solvents are used. Where careful laboratory procedures are followed, such as the use of properly functioning hoods and appropriate work practices, exposures should be very low. In order to estimate the range of possible costs, OSHA calculated the cost on the basis of two scenarios. One scenario assumes that 10 percent of the exposed workers are initially exposed above the action level but below the permissible exposure level. The second scenario assumes 5 percent of the exposed workers are initially exposed above the permissible exposure level, 10 percent are exposed between the action level and the permissible exposure level, and all others are below the action level. On this basis, first year operating costs range from approximately $4.5 million to $10.9 million. Recurring annual costs would be approximately $2.4 million, if all workers exposures drop below the action level after the first year, which is expected as a result of appropriate controls and work practices.

Rubber Products

The ADL study examined the potential for benzene exposure in plants manufacturing rubber tires and other products, such as belts, hoses, fittings, and coated fabrics. ADL identified 206 affected tire producing facilities employing 11,400 exposed workers. The study cited data indicating that exposure levels were generally in the range of 1-3 ppm and may be considerably higher (ADL, Ex. 5A, pp. 14-19; Tr. 4-20). Similar testimony was provided by Drs. Tyroler and Harris of the University of North Carolina based on their studies of benzene exposure in the rubber industry. (Tr. 3083-3095) Both studies indicated that control of exposure could be achieved, reducing exposure levels to below the action level, by substitution of materials and enforcement of appropriate work practices, and that engineering controls would not be required. (Tyroler, Tr. 3086) Compliance costs were estimated, therefore, on the basis of all exposed workers being exposed initially above the permissible exposure limit, and exposures reduced to below the action level after controls have been instituted in tire manufacturing plants.

First year operating costs were estimated to be approximately $15.8 million and recurring annual costs approximately $407,000.

The analysis of firms producing other rubber products was based on the same considerations of workers exposed and compliance activities. For these firms first year operating costs were calculated at approximately $18.1 million, and recurring annual costs at approximately $484,000 based on 197 facilities with 13,050 exposed workers.

Other Industries

There are a number of other industries in which benzene exposures will occur due to the use of solvents, containing benzene in the processes or products involved. OSHA has estimated the approximate level of anticipated compliance costs for 16 industries thought to be representative of the range of cost impact involved: Adhesives manufacture and application, paint manufacture and application, paint stripper manufacture and application, printing ink manufacture, and gravure printing, metal can production, manufacture of photographic equipment and supplies, production of motors and generators, commercial printing manufacture of folding paperboard boxes, paper mills, and manufacture of wood household furniture. Firms in these industries are relatively small, use similar materials, will have generally low exposure levels, and can effect control of all worker exposures to below the action level by substitution of materials and enforcement of good work practices. It is OSHA's judgment that most of the employees in these industries are exposed to benzene levels below the action level. However, some data in the record on employee exposures related to the use of these solvents indicate that some workers in particular industries may be exposed at higher levels. OSHA feels that the probable range of costs can be indicated by calculating the costs under two hypothetical exposure patterns: (1) If 10 percent of the exposed workers are exposed above the action level, but below the permissible exposure limit, and all other employees are exposed below the action level; and, (2) if 5 percent of the workers are exposed above the permissible exposure limit, 10 percent between the action level and the permissible exposure limit, and all others below the action level. Costs calculated in this way indicate that, in these industries, the first year operating costs will lie in the range of $2,000 to $3,270 per facility, or $240 to $390 per exposed worker.

OSHA is aware that there are other industries where employee exposure to benzene may occur. However, although an opportunity to present data and views as to the application of the benzene standard to all industry sectors was afforded, no submissions were made on behalf of these industries. In the absence of exposure and cost data, OSHA has been unable to calculate costs for those industries. In summary, OSHA finds that the economic impact of the final benzene standard will not be such as to threaten the financial welfare of the affected firms or the general economy.

Other Economic Impacts

In addition to assessing the compliance cost and economic feasibility of the proposed standard, OSHA evaluated the economic impact of the proposed standard on affected industries and the general economy utilizing the criteria of Executive Order 11821 (39 FR 41501) as amended by Executive Order 11949 (42 FR 1017), and related implementing instructions particularly Secretary's Order 15-75 (40 FR 54484). The evaluation of such impact was made a part of the economic analysis of ADL. The methods of evaluating these impacts and the conclusions reached were extensively discussed during the benzene hearing.

The portion of the ADL study was designed to show the economic effects of compliance with the proposed final benzene standard upon affected industries. OSHA believes that this analysis, as modified to reflect the changes made in specific provisions in drafting the final standard, adequately defines the economic impact of the final standard.

ADL's analysis was, as discussed further below, challenged by witnesses for AISI and API on the ground that it was not sufficiently rigorous. Neither AISI nor API questioning of the ADL costing methodology indicated serious disagreement with the data used. Nor did these witnesses offer any evidence tending to disapprove the conclusions reached by ADL, or suggest specific analytical procedures which, if followed, would clearly improve the study and increase the accuracy of the findings. After careful consideration of each of the questions raised by these and other witnesses, OSHA has made minor modifications in some elements of the cost analysis, and has concluded that the evidence indicates that the economic impact of the final standard may be more widespread than that shown in the ADL study but less severe for the industries analyzed by ADL.

Impact on Energy, Critical Materials, and Other Macroeconomic Variables

The ADL analysis indicated that there would be no significant change in either the supply or demand sides of the markets for energy or critical materials.

Two economists testifying for API argued that these were important areas of potential impact improperly ignored by ADL. Scarbrough, p. 4-5 and Henderson, p. 6-7). Under questioning at the hearings, however, neither witness offered any data indicating that these effects would be significant or put forth any reasons to suggest that they would be. (Tr. 2225-2226, Scarbrough, and Tr. 2212-2214, Henderson.)

Impacts on Prices

The compliance costs identified in the cost analysis provided the basis for estimating the impact on prices, output and labor productivity. The indirect costs resulting from increases in the prices of benzene and products made from it were identified and included in the estimation of impacts on producers and consumers. Capital costs were amortized over 7 years at 10 percent to reflect a normal return on investment. ADL assumed that the price would increase in the long run by the amount of any increase in long run average cost, including the normal return to investment. This provides a reasonable approximation of maximum potential price impact, and OSHA considers it to be acceptable where exact costs cannot be known and precise measurement of price changes is not feasible.

Impact of Prices on Output and Market Structure

Economic impact was judged on the basis of expected overall influence of price increases, product substitution, price elasticity of demand, market growth, volume of imports and exports, market concentration, and compliance cost differentials.

ADL reasoned that buyers of benzene will continue to purchase the same quantities after a small price increase, provided there are no economical substitutes for benzene and provided that they can pass the added cost along in the price of their products. If users of the end products reduce their purchases in response to the cost pass-through, then the industrial purchasers will change the mix of products they manufacture and buy less benzene. The elasticity of consumer demand in turn depends on whether consumers can substitute other goods of the same type for those whose prices increase as a result of the cost pass-through, and whether consumers reallocate their incomes to buy other kinds of goods and services while buying less of the ones whose prices increased.

The magnitude of the expected costs is such that the percentage change in consumer product prices would be very small. No income effect (reallocation of income) of any significance would be expected. Therefore, elasticity would reflect almost entirely the availability of substitutes, as ADL indicated, although this conclusion would be questionable if a large price change was anticipated. ADL's product-by-product discussion of the availability of substitutes for benzene and for the petrochemical products made from benzene showed clearly that in all major product lines substitutes were judged to be technically inferior and more costly, at best, and unavailable with present technology in may instances. (ADL, Ex. 5A, pp. 6-18 to 6-22.) Consumer substitution was also judged to be sharply limited on the basis of the quality and cost of available substitutes. OSHA believes that the reasoning followed by ADL in this qualitative analysis is reasonable.

Using an estimate of relatively inelastic demand for benzene (elasticity = -0.5) (ADL pp. 6-11) and an estimated price increase of approximately 1 percent, ADL estimated that demand for benzene would fall by approximately 0.2 percent (16 million pounds) in the short run and by approximately 0.5 percent (54 million pounds) in the long run for refineries and steel companies producing benzene. Since the demand for benzene is growing moderately, these decreases would represent a slight decrease in the amount of growth, rather than an absolute decline from current levels of production. Foreign trade accounts for only 5 percent of U.S. supply and is not expected to be a significant factor affecting domestic producers, considering the small projected change in price.

The price of benzene is determined by petroleum refiners, who supply 95 percent of the market. Since coke oven producers will incur higher compliance costs than the benzene producing refineries, they will be able to recover only part of their costs through price increases and will be adversely affected by the standard. It appears probable that some producers of benzene from coking operations will choose to sell their light oil to other firms rather than continuing to produce benzene (AISI). Similarly, some refineries now producing benzene may elect to change their product mix to eliminate benzene production. These decisions will be made by benzene producers on the basis of many factors affecting the profitability of benzene production and not on the basis of the impact of this standard alone.

No data were submitted in the hearings that provided a basis for estimating the number of firms that may eliminate benzene production. It was not contended, however, that any producing firm will be forced to close or be forced into an unprofitable position as a result of the compliance costs associated with this standard.

IMPACT OF PRICE ON EMPLOYMENT AND PRODUCTIVITY

Since ADL projected no absolute decrease in total production of benzene, it did not project employment losses for benzene producers. API witnesses questioned this on the basis of their challenge of the analysis of demand elasticity, price change, and reduction of demand. (Scarbrough, pp. 24-25; Henderson) OSHA reasons, however, that if these elements of the analysis are acceptable, the conclusion that production employment will not decline is valid, although some reallocation due to structural changes in the industries may occur.

API also argued that increased labor cost resulting from compliance programs could lead to some substitution of capital for labor. (Henderson, p. 10) In its testimony, ADL pointed out that the productivity of both labor and capital would decline in the same order of magnitude and that this would indicate that substitution would not occur. (TR p. 584) ADL also submitted that the industry sectors in question are quite capital intensive and that for this reason, "it is not anticipated that the proposed regulation would have any observed effect upon the utilization rate for labor in these industries." (ADL post hearing. Ex. 15N)

OSHA recognizes that some decline in overall labor productivity may result from adding compliance personnel such as technicians and hygienists to the work force of benzene producers, although the direct productivity of production workers is not affected. This decline was estimated by ADL to be approximately 1.4 percent for coke producers and 0.6 percent for refineries (ADL, Ex. 5A, Table 6.5, p. 6-16).

BENEFITS

The legislative history and language of the Occupational Safety and Health Act, as distinguished from some other environmental and safety legislation, clearly indicate that Congress has already arrived at a judgment concerning the balancing of cost and benefit, with the result that worker safety and health are to be heavily favored over the economic burdens of compliance. Specifically, Section 6(b)(5) of the Act provides that

the Secretary, in promulgating standards dealing with toxic materials or harmful physical agents under this subsection. shall set the standard which most adequately assures, to the extent feasible, on the basis of the best available evidence, that no employee will suffer material impairment of health or functional capacity even if such employee has regular exposure to the hazard dealt with by such standard for the period of his working life. Development of standards under this subsection shall be based upon research, demonstrations, experiments and such other information as may be appropriate. In addition to the attainment of the highest degree of health and safety protection for the employee, other considerations shall be the latest available scientific data in the field, the feasibility of the standards, and experience gained under this and other health and safety laws.

Thus, while feasibility is an appropriate consideration, the Secretary is directed to set standards which attain the "highest degree of health and safety protection for the employee * * *."

This does not mean, however, that a systematic evaluation of costs and benefits is not to be encouraged within the limits of the estimation techniques. In considering the issue of feasibility in this rulemaking, as in others, OSHA has carefully evaluated the cost of compliance which may be incurred by the directly affected employers and their ability to comply. Additionally, OSHA believes that a standard for a substance which has been found to pose a cancer risk to workers, in this case benzene, must assure maximum benefit (i.e., prevention of serious illness or death), constrained only by the limits of feasibility.

There is general agreement that benzene exposure cases leukemia as well as other fatal diseases of the bloodforming organs. In spite of the certainty of this conclusion, there does not exist an adequate scientific basis for establishing the quantitative dose response relationship between exposure to benzene and the induction of leukemia and other blood diseases. The uncertainty in both the actual magnitude of expected deaths and in the theory of extrapolation from existing data to the OSHA exposure levels places the estimation of benefits on "the frontiers of scientific knowledge." While the actual estimation of the number of cancers to be prevented is highly uncertain, the evidence indicates that the number may be appreciable. There is general agreement that even in the absence of the ability to establish a "threshold" or "safe" level for benzene and other carcinogens, a dose response relationship is likely to exist; that is, exposure to higher doses carries with it a higher risk of cancer, and conversely, exposure to lower levels is accompanied by a reduced risk, even though a precise quantitative relationship cannot be established. In light of the uncertainties in this area of scientific knowledge, OSHA believes that it is required by prudence and by the statutory mandate to adopt a highly protective posture in considering the evidence for health benefits.

Various witnesses argue that cost benefit or cost effectiveness should be the primary criterion for regulatory decision. OSHA was criticized by industry participants for failure to consider what benefits would be derived from reducing the current permissible exposure level for benzene (API brief, page 92, 96; AISI brief, page 89, 97). The Council on Wage and Price Stability (CWPS) also suggested that OSHA estimate the incremental reduction in health risks which are associated with the reduction in employee exposure to benzene (Bosworth letter, 9/12/77, to Bingham).

Both CWPS and industry suggested ways in which this could be achieved. CWPS suggested that OSHA apply the expected lifetime incidence rates of leukemia and the increased incidence rates revealed by the scientific studies to the benzene exposed population to derive the number of expected excess cases attributable to that exposure. An MCA witness, Dr. Wilson, assumed a dose response relationship and calculated that at the current exposure levels only one leukemia and one other cancer would be observed among the benzene exposed workers covered by the proposal every 6 years (Ex. 149-B). This witness further testified that the proposed standard would cost $300 million for every hypothetical life saved (Tr 2750) and concluded as a result that the risk from benzene related deaths is comparable to many other socially acceptable risks.

An API witness contended that rational allocation of OSHA's resources required policies that utilize cost effectiveness as the primary basis for regulatory decisions. He further argued that OSHA should make an assessment of what risks exist from exposure to benzene at 1-10 ppm., which should then be compared with estimates of risks posed by other health hazards such as from cotton dust (Tr. 2157) as well as with the risk associated with other activities of society which he termed generally acceptable risks in order to determine whether the cost of the benzene standard is warranted.

In the face of the record evidence of numerous actual deaths attributable to benzene-induced leukemia and other fatal blood diseases, OSHA is unwilling to rely on the hypothesis that at most two cancers every six years would be prevented by the proposed standard. By way of example, the Infante study disclosed seven excess leukemia deaths in a population of about 600 people over a 25-year period. While the Infante study involved higher exposures than those currently encountered, the incidence rates found by Infante, together with the numerous other cases reported in the literature of benzene leukemia and other fatal blood diseases, make it difficult for OSHA to rely on the Wilson hypothesis to assure the statutorily mandated protection of employees. In any event, due to the fact that there is no safe level of exposure to benzene and that it is impossible to precisely quantify the anticipated benefits, OSHA must select the level of exposure which is most protective of exposed employees.

We recognize that in view of the latency period usually associated with the induction of cancer, significant reductions in mortality may not be seen for many years. However, unless exposures are reduced now, OSHA believes that the mortality rate will not decline and employees exposed to benzene will continue to suffer excess mortality. Finally, it should be noted that this standard for employee exposure to benzene cannot be simply described as a reduction in permissible exposure from 10 ppm. to 1 ppm. as a daily average. The previous standard for worker exposure to benzene permitted daily exposures as high as 50 ppm. for a period up to 10 minutes and repeated exposures of 25 ppm., levels which are known to produce serious hazards of blood diseases.

In making judgments about specific hazards, OSHA is given discretion which is essentially legislative in nature. However, in setting an exposure limit for a substance like benzene, OSHA may not substitute cost benefit criteria for the legislatively determined directive of protecting all exposed employees against material impairment of health or bodily function. Where the health effectiveness of the alternative approaches are extremely uncertain and likely to vary from situation to situation, OSHA must adopt the compliance strategy which provides the greatest certainty of worker protection even if that approach carries with it greater economic burdens for the affected employers. In the case of the benzene standard, the evidence in the record indicates that the costs of compliance are not overly burdensome to industry. Having determined that the benefits of the proposed standard are likely to be appreciable, OSHA is not obligated to carry out further exercises toward more precise calculations of benefit which would not significantly clarify the ultimate decision. Previous attempts to quantify benefits as an aid to decision making in setting health standards have not proved fruitful (41 FR 46742).

Based upon the foregoing and the record as a whole, OSHA finds that compliance with the standard (even if the higher cost estimates suggested by some participants are used) is well within the financial capability of the covered industries. Moreover, although the benefits of the standard cannot rationally be quantified in dollars, OSHA has given careful consideration to the question of whether these substantial costs are justified in light of the hazards of exposure to benzene. OSHA concludes that these costs are necessary in order to effectuate the statutory purpose of the Act and to adequately protect employees from the hazards of exposure to benzene.

VI. SUMMARY AND EXPLANATION OF THE STANDARD

The following sections discuss the various issues raised during the benzene rulemaking proceedings, the individual requirements of the permanent benzene standard and the rationale and policy considerations underlying the provisions of the standard. In developing these requirements, OSHA has considered all the evidence in the benzene record. After consideration of all of this evidence, OSHA has revised and clarified, as described in detail below, certain provisions of its proposed benzene standard.[1]

Scope and Application: paragraph (a) This standard applies to occupational exposure to benzene in all workplaces in all industries where benzene is produced, reacted, released, packaged, transported, handled or otherwise occupationally used, except for the agriculture industry. Thus, the standard applies to "general industry" as well as to the construction and maritime industries. By specific exclusion discussed in detail below, the standard does not apply to the sale, discharge, storage, transportation, distribution or use as a fuel of gasoline and other fuels, subsequent to discharge from bulk terminals. The application of the standard is further limited in that not all of the requirements of the standard will apply to every employer regardless of the conditions in his workplace. The permanent standard, unlike the emergency temporary standard and the proposed standard, incorporates an action level so that the specific provisions of the permanent standard that apply to any particular workplace depend upon the extent of employee exposure to benzene. Additionally, only the labeling requirements of paragraph (k) (2), (3), (4), and (5) and the training requirements of paragraph (j) apply to workplaces where the only benzene is contained in sealed intact containers.

The scope of the final standard has been significantly changed from the scope of the proposal and the emergency temporary standard. OSHA was faced with several alternatives in deciding on the extent of coverage of the benzene permanent standard and in achieving its objective of maximizing the protection afforded employees occupationally exposed to benzene while limiting the burden placed on their employers. Benzene, as pointed out earlier, is used in a vast number of operations in quantities varying from trace amounts in solvents to large amounts of pure benzene. Coverage of all workplaces where benzene is present would, of course, extend the protection of the standard to the greatest number of employees. It would mean, however, that workplaces where benzene is present only in trace amounts and potential employee exposure is minimal would be subject to the same exposure monitoring and medical surveillance requirements as workplaces where the potential for high employee exposures exists. While this would satisfy that part of OSHA's objective of maximizing the protection afforded employees, it would fail to reasonably limit the burden placed on employers.

---

[1] For the convenience of the public, OSHA has included, as part of the title of each section of the summary and explanation of the standard, the particular paragraph of the standard to which the discussion refers.

Proposed Percentage Exclusion

The proposal and the emergency temporary standard both would have exempted from coverage those operations utilizing liquid mixtures containing 1 percent or less benzene by volume. The proposal would have limited this 1 percent exemption to the first year of the standard after which period only liquids containing 0.1 percent or less benzene by volume would be exempted. The proposed percentage exemption was based on some indication that exposures resulting from the use of mixtures containing less than 1 percent benzene would generally be less than 1 ppm (42 FR 27453-4). It was also anticipated that the percentage exclusion would lead to a reduction of the benzene content in solvents or to substitution of other substances for benzene with the result that employee exposure to benzene would be reduced or eliminated. As a result of evidence developed during the benzene rulemaking, however, OSHA has determined that the percent exclusion cannot be supported on this basis. The benzene record indicates that there is no consistent predictable relationship between the percent of benzene in a liquid mixture and the resultant airborne exposure to benzene (Tr. 3120). Studies conducted by the University of North Carolina demonstrated that exposure levels varied considerably during various work operations utilizing liquid mixtures containing the same percent of benzene (Tr. 3090-3091). Support for the conclusion that a less than one percent benzene solution can result in exposures above 1 ppm is found in testimony suggesting that other factors than the percent of benzene in the liquid may be determinative of exposure levels. Thus, NIOSH testified that employees working with No. 6 fuel which contains only 0.1 percent benzene were exposed to levels as high as 60 ppm under certain conditions of confined space, poor ventilation or elevated temperatures (Tr. 754). The University of North Carolina data also revealed that exposure levels in heavy duty tire spraying utilizing rubber solvents were 1.3 ppm when sprayed by the regular full time operator but reached 5.2 ppm and 7.3 ppm when sprayed by a substitute operator (Tr. 3093). This suggests that work practices are also determinative of exposure levels. To maximize protection of employees, it is necessary to reduce their exposure to benzene, to the lowest feasible limit. Since the percentage of benzene in a liquid is not in itself necessarily controlling of the employee's exposure level, the proposed percentage exemption, would not achieve this objective. Accordingly, the final standard does not contain a percentage exemption and thus applies to all liquid mixtures containing benzene regardless of the percent of benzene.

It is OSHA's view that the absence of a percentage exemption does not eliminate all incentive to reduce the amount of benzene in liquid mixtures. Where another substance is totally substituted for benzene, this standard will obviously not apply. Additionally, it should be noted that the inclusion of an action level will effectively limit the burden of this standard where employee exposure is found to be low because of its limited presence in a mixture.

Benzene Substitutes

In lieu of the percentage exception, OSHA considered the option of exempting from the standard certain benzene substitutes, such as toluene and xylene (although they may contain small amounts of benzene). An exemption of this type would encourage employers to discontinue the use of benzene and utilize a substitute in order to avoid the requirements imposed by the benzene standard. These substances themselves, however, do contain varying amounts of benzene and, as stated above, there is no evidence of a consistent direct predictable correlation between the amount of benzene and exposure levels. Thus, it was concluded that toluene and xylene, to the extent they contain benzene and result in benzene exposure, would be covered by the standard.

Action Level

OSHA recognizes that the lack of a percentage exemption substantially expands the coverage of the benzene standard to operations which may otherwise have been exempt. In many of these operations, exposure levels below the permissible exposure limit have already been achieved. To minimize the impact of the standard on those employers who have attained these low exposure levels, the final standard provides for an action level. This action level is a benzene exposure equal to one-half of the permissible exposure limit above which certain precautionary measures, such as periodic monitoring and medical surveillance programs must be conducted and below which only a very limited number of the standard's requirements will apply. Thus employees who have an initial exposure measurement below the 0.5 ppm TWA action level will not have to be monitored periodically. If the initial monitoring exposure measurement is below the action level, no further monitoring is necessary until such time as a redetermination may be required as a result of a process, control or personnel change. Furthermore, medical surveillance of individual employees whose initial exposure measurements are below the action level would not be required. Nor would regulated areas need to be established or engineering or work practices instituted or respirators provided. All other provisions of the standard, however, would apply and the employer would need to train his employees in the hazards of benzene exposure, to properly label his products and to maintain the record of the initial exposure measurement and, if any, the record of the redetermination measurement.

The adoption of an action level had been recommended by several participants to the rulemaking. NIOSH, recommended that an action level for periodic monitoring and routine periodic medical exams be incorporated for the purpose of assuring that no employee is exposed above the permissible exposure limit (Tr. 749-750). Industry participants advocated a level which would trigger the periodic monitoring of employee exposure (Com. 24; Com. 72; P.C. 36 AISI brief p. 74, FT. no. 168). A need for an action level is also suggested by the record evidence that some minimal exposure to benzene occurs naturally from animal and plant matter (Tr. 749-750; 759-760). Naturally occurring benzene concentrations, it appears, may range from 0.02 to 15 parts per billion (Ex. 117, p. 1). Additionally, it was suggested by certain employers that their operations be exempted from the requirements of the standard because those operations involve only intermittent and low level exposures to benzene. The use of the action level concept should accommodate these concerns in all cases where exposures are indeed extremely low since it substantially reduces the monitoring of employees who are below the action level and removes for these employees the requirement for medical surveillance. At the same time, employees with significant overexposure are afforded the full protection of the standard.

In developing this regulatory approach to occupational exposure from benzene, OSHA considered the request of those participants to the rulemaking who pressed for exemption from the benzene standard of their particular operations.

Bulk Terminals

The final standards apply to bulk terminal operators but exclude exposure from gasoline, motor fuels and other fuels subsequent to the discharge from the bulk terminal.

During the rulemaking, bulk terminal employers requested that their gasoline operations be exempted from the benzene standard. Bulk terminals are the primary distribution facilities in the gasoline marketing network. A bulk terminal is the first distribution point of gasoline after the gasoline is processed by the refinery. Gasoline regulatory action to be taken on non-gasoline motor fuels after review of the data and recommendations of the joint Federal task force and evaluation of the information furnished in response to OSHA's notice on gasoline. For these reasons, OSHA has exempted from the permanent benzene standard the storage, transportation, distribution, dispensing, sale or use as a fuel of all fuels (including gasoline, jet fuel, and diesel fuel) subsequent to their discharge from bulk terminals.

CLOSED SYSTEMS

Some participants requested exemption from the standard for workplaces where benzene is present solely in closed systems. The record evidence, however, indicates that closed systems frequently develop leaks, and that it is the amount of the solvent exposed to air (Tr. 3111-3) which, among other variables, is a major determinant in the degree of exposure. Accordingly, OSHA has retained coverage of closed systems in the final benzene standard. However, in the absence of leaks, exposures from closed systems should be below the action level and the impact of the standard in such operations should be minimal.

LABORATORIES

Still another segment of industry requested an exemption from the benzene standard. Many comments requested that research facilities in both industrial and academic laboratories be exempted from the benzene standard (Com. Nos. 14, 20, 57, 62, L.C. 12, L.C. 13) particularly where the permissible exposure limit is not exceeded (Com. 56, 58). Some comments suggested that separate requirements be developed for laboratory use of benzene (L.C. 12, L.C. 13). These requirements could include mandatory use of hoods (L.C. 9, Com. 21) of a specified velocity (L.C. 8, Com. 18) and periodic medical surveillance (Com. 18, 28, 29, 64, 70); annual inspection in lieu of monitoring (Com. 28); written safety rules (Com. 18) and container labeling (Com. 39) were also suggested. In the alternative, it was suggested, OSHA should promulgate separate regulations for all chemicals used in laboratories (L.C. 12, L.C. 13), since laboratories use a variety of chemicals, and compliance with separate regulations for each chemical, it was argued, would be burdensome (Com. 58).

The requests for exemption from the benzene standard or for special treatment of laboratories were largely based on the view that there exists a lesser, or slight, risk of exposure to benzene for laboratory technicians. In support of this contention, laboratories pointed to the manner in which they use benzene. Laboratories, it was stated, generally use small quantities of benzene at a time (Com. 10, 28, 38).

One laboratory reported that sometimes a sample is dissolved in only one ml of benzene (L.C. 2). That laboratory estimated that an individual laboratory technician, who analyzed 25 samples per day, 7 days a week, would use less than four gallons of benzene a year (L.C. 2). When not in use, benzene is stored in closed containers (L.C. 1).

The argument that laboratory technicians are trained in the hazards of chemicals was also made to support the view of lesser risk. One laboratory reported that it conducted blood tests every four months for twenty years of its technicians, and found no problem (Com. 4). Another reported that in thirty-five years, none of its employees developed leukemia even though exposures in the laboratory were between 1 and 20 ppm (Com. 31).

Another reason advanced for exemption or separate treatment of laboratories was the burden and cost of implementing exposure monitoring, engineering controls and medical surveillance (Com. 4, 5, 10). One industry comment argued that responsible research and quality control laboratories normally provide and encourage annual physical examinations that include the essentials of the proposed medical surveillance program (Com. 38).

OSHA has determined that applying the provisions of this benzene standard to laboratories is consistent with OSHA's responsibility to protect the health and safety of workers. The record indicates that exposure levels in laboratories vary greatly. While monitoring samples of one laboratory indicated levels of only .01 ppm (Com. 28), others indicated levels as high as 20 ppm (Com. 31; Ex. 39-3). Furthermore, it appears that protective measures vary from lab to lab (Tr. 3157). One chemist, who worked frequently and for several hours a day with benzene during his six years in a research lab that he described as "one of the most prestigious research institutions in the country," testified that employee exposures were not measured, hoods were not used, and there was no medical surveillance program (TR 3492-6). Moreover, there does not appear to be any firm basis to believe that training alone in handling of chemicals adequately protects all laboratory employees from the hazards of exposure to benzene.

Based on the above evidence, OSHA believes that laboratory employees are exposed to the leukemogenic hazards of benzene and, therefore, must be covered by this standard. OSHA does not subscribe to the view that laboratories should be exempted from the benzene standard and that protection for employees exposed to benzene should await promulgation of separate regulations for all chemicals used in laboratories. OSHA is not currently developing a laboratory standard, and the result of exemption of laboratory workers from the benzene standard would, consequently, deprive them of the necessary protection against the hazards of benzene exposure for some time to come. Furthermore, the provisions of the standard relating to methods of compliance are performance oriented and, therefore, will encompass the suggested special requirements for laboratories. For example, the requirement that the employer institute engineering controls and work practices could be met by laboratories in many instances through the use of properly designed hoods. Also, the inclusion of an action level will minimize the monitoring and medical surveillance requirements in laboratories where employee exposure measurements indeed are low.

Although OSHA has not included special provisions just for laboratories, the provisions of the standard accommodate several of the suggestions of the participants, as noted above. Those laboratory operations where benzene exposure levels are at or below the action level will need to be monitored only initially, and medical surveillance of employees engaged in those operations will not be required. When benzene is sealed in containers, the laboratory employer will be required only to train his employees in the hazards of exposure to benzene, and assure that the container is properly labelled, and provide protective clothing where necessary.

COKE OVEN BATTERIES

Steel industry participants requested an exemption from the benzene standard for coke oven workplaces which are covered by the coke oven emissions standard. Although coke oven emissions at coke oven batteries are regulated by § 1910.1029, it is OSHA's view that there is a need to apply to these workplaces the requirements of the benzene standard, as well. The coke oven standard regulates the benzene soluble fraction of total particulate matter present during the carbonization of coal for the production of coke to protect workers from the risk of developing cancer of the lung and urinary tract. The coke oven standard, however, was not designed to protect coke oven employees from the hazards to the hematopoietic system which can result from exposure to benzene. Benzene exposure can occur from leaks in the ovens or in the collection system or from residual vapors left in the ovens when they are opened and the hot coke is pushed out (Little study, B-21). Accordingly, in order to protect workers from the hematopoietic hazards presented by exposure to benzene, OSHA has concluded that inclusion of all coke oven workplaces in the benzene standard is necessary.

The coverage of coke oven workplaces by both the benzene and the coke oven emissions standard does not create a conflict. OSHA believes that the engineering controls required by the coke oven emissions standard will, in all likelihood, reduce exposures to benzene below the action level. In such a case, the benzene standard will impose on coke oven workers only the burden of conducting initial measurement, retaining a record of that measurement, and the training of employees in the hazards of benzene.

SEALED CONTAINERS

The final standard applies only to the labeling requirements and the training requirements to the storage, transportation, distribution and sale of sealed, intact containers. This is a change from the proposed standard, and the emergency temporary standard which would have applied all the requirements of the benzene standard although benzene were "present" solely in sealed containers. The change in the final standard was made in response to comments to the effect that closed containers of benzene pose no threat of exposure to employees (Com. 10). Participants in the rulemaking pointed out that coverage of closed containers would result in employers being required to monitor exposure and make medical surveillance available to employees, such as maritime employees occasionally handling small quantities of benzene in breakbulk form (i.e., drums or packages) (Com. 35), employees of receiving docks and warehouses, and employees involved in inter- departmental transfers of the containers (Com. 28). Additionally, such coverage would also apply the requirements of the benzene standard to employees of retail stores that sell products containing benzene.

Most participants requested exemption of closed containers (Com. 10, 28, 35) without distinguishing between closed and sealed containers. A state agency comment, however, suggested that only "sealed" containers be exempted (Com. 67). OSHA concurs in the view that properly closed containers do not present a health hazard to employees which warrants application of the monitoring and medical surveillance requirements. OSHA, however, is concerned that containers may, after being opened, be closed in such a fashion as to expose employees to a hazard. For this reason, OSHA has adopted the suggestion that the exemption apply only to *sealed* containers, intending by the use of that term to exempt only containers of benzene which are closed in such a manner as to contain the benzene vapors. However, since sealed containers may, during handling, develop leaks (Tr. 3051-2) and thus expose employees to the hazards of benzene, the exemption is further limited to intact containers. Therefore, where the breakage of the container occurs, employers will be required to monitor and to comply with any other applicable provisions of the standard. To assure that all employees are aware of the hazardous nature of the contents of the containers they are handling and because at some point in downstream occupational activities the containers will be opened, the standard applies the training and the labeling requirements to sealed containers. Finally, it should be noted that the regulation of sealed containers in this standard is consistent with that in other caracinogen standards; §§ 1910.1003-1910.1016 exempt sealed containers from their requirements, except for labeling and transshipment.

Definitions: paragraph (b). The standard contains some appropriate definitions. The purpose of providing definitions for key terms is to clarify the intent of those terms as used in the substantive provisions of the standard. All of the definitions of the proposal have been retained in the final standard.

A new definition has been added to those contained in the proposed standard. This is the definition of "Action level." Action level is defined as a concentration of benzene of 0.5 ppm over an 8 hour workday. Industry participants recommended that monitoring be conducted only where exposure exceeds the permissible exposure limits (PC 36, AISI brief, p. 74). OSHA does not agree. It is generally acknowledged that worker exposure to airborne emissions can vary from day to day in a random fashion. This variation in levels is unavoidable; it is only minimally connected to the precision and accuracy of the method of measurement and does not include variations due to wind, temperature and other environmental factors.

Therefore, even though individual measurements of exposure levels may fall below the permissible limit, some possibility exists that on unmeasured days the employee's exposure may exceed the permissible limit. Thus, the concept of an action level provides, statistically, a means by which the employer may assure himself that his employees will not be exposed to benzene over the permissible exposure level (PC No. 14g, p. 5). In view of these considerations and in order to provide a greater degree of employee protection, OSHA has incorporated in the permanent benzene standard an action level of 0.5 ppm.

For those employees whose exposure level is determined by initial monitoring to be below 0.5 ppm, the employer need not do any further monitoring unless a redetermination of exposure is necessitated because of a process, control or personnel change. For these employees, moreover, the employer is not required to provide medical surveillance. The employer, however, must train these employees, label his benzene products, provide protective clothing where necessary, and retain a record of the initial determination and, if any, record of the redetermination measurement.

Employees whose exposure measurements are above 0.5 ppm, but below 1 ppm must be monitored quarterly and provided with medical surveillance. Where exposures are above the action level but below the permissible exposure limit the requirements for regulated areas, methods of compliance and respiratory protection do not apply but the employer must comply with all other requirements of the standard.

Those employees whose measurements are above 1 ppm must be monitored monthly. All the provisions of the standard, moreover, apply where employees are exposed above the permissible exposure limits.

OSHA has, as suggested by some participants (Com. 46), clarified the definition of benzene to include solids which contain benzene. Some comments suggested that benzene be defined as commercial or high grade benzene (Com. 48, 59, 76). OSHA has not adopted this suggestion because of record evidence to the effect that low concentrations of benzene do not necessarily result in low exposures. Other comments suggested that a definition of benzene exposure be included (Com. 36, 57, 59). The thrust of these suggestions was to limit the application of the proposal's monitoring requirements which applied to workplaces "where benzene is present" and the proposal's medical surveillance requirements which applied to "all employees who are or will be exposed to benzene." For the reasons stated below in the appropriate parts of this preamble relating to these requirements, OSHA has not adopted this suggestion.

A new definition of emergencies was added to the standard to clarify the type of situation requiring the use of a respirator. The definition of emergencies also clarifies the situations in which employees shall be provided with a biological screen.

PERMISSIBLE AIRBORNE EXPOSURE LEVEL. PARAGRAPH (c)(1)

Based upon a thorough review and evaluation of evidence in the record, OSHA has, as stated above, concluded that benzene:

(1) Is a human leukemogen;
(2) Causes bone marrow depression resulting in alterations in peripheral blood; and
(3) Causes chromosomal damage in blood cells.

These conclusions are based on studies ranging from single case reports to retrospective mortality studies and also experimental evidence. These studies represent the best available evidence upon which OSHA must make a decision. This decision is not predicated on any particular study, but is based on an assessment of the entire set of evidence taken as a whole.

OSHA recognizes that only recently has there been a well-designed study which may indicate a leukemogenic response among benzene-exposed animals. (Nelson, Ex. 178) In any event, the best evidence to date is that based on direct human experience. Based upon such evidence, OSHA concludes that benzene is and must be regulated as an occupational carcinogen.

Comment and testimony from industry for the most part, does not dispute the leukemogenic potential of benzene (e.g. Eckhardt, Ex. 115.B. 1, pg. 3; Tabershaw Tr. 2545). However, industry argues that this relationship is valid only at high exposure levels. In support of this view, they cite as evidence that: (a) the documented cases of benzene-induced leukemia were among employees who were, most probably, exposed to high concentrations, and (b) that several studies of undefined numbers of workers who may have been exposed to low levels of benzene did not demonstrate a leukemia excess (The benzene studies are described under Health Effects: Leukemia). Industry also believes that observable blood disorders precede the development of benzene-leukemia and that such hematological abnormalities are reversible (ORC, PC 34, p. 2; Tabershaw Tr. 2543 Jandl. PC 26B, Allied P.C. 22, p. 7). They conclude that the present exposure limits of 10 ppm TWA and 25 ppm ceiling are sufficiently protective to guard against the development of non-malignant blood disorders and, consequently, that limiting exposure to such levels will also protect against leukemia. (ORC, PC 34; Jandl, PC 26B; Tabershaw, Tr. 2546.)

The issue of the levels at which cancer is induced by chemical agents and whether or not there is a "threshold" has been a major issue in every OSHA rulemaking concerning the regulation of occupational carcinogens (See preambles to Carcinogen standard (39 FR 3758); Vinyl Chloride (39 FR 35892); Coke Oven emissions (41 FR 46742). The benzene hearing was no exception. As in the case of arsenic, the lack of an unequivocal animal model (Kraybill, Tr. 758) requires that the Secretary's decision rely primarily on data obtained from human evidence. However, the epidemiologic method is by its very nature, a retrospective view of the evidence, i.e. findings of a recent excess of mortality among workers may relate to initial exposures occurring as much as 20 years or more previously and which most certainly correlate to exposures at higher concentrations than present levels. Because of these variables, it is extremely difficult to derive definite conclusions as to the quantitative health risk to the workers at exposures near 1 ppm. Studies which report negative findings, such as those studies conducted by Stallones, Tabershaw-Cooper and Thorpe suffer from the disadvantage of not being able to clearly define an exposed cohort, i.e. the indentification of a group of workers actually exposed to benzene and to what levels they were, in fact, exposed. (Stallones, Ex. 115.C; Tabershaw-Cooper, Ex. 149A; Thorpe, Ex. 2-34). OSHA recognizes that it is extremely difficult to reconstruct and define employee exposures retrospectively. Therefore, given the uncertainty of the definition of exposure and the potential for dilution of mortality excess among those actually exposed to benzene and other methodological deficiencies, OSHA is reluctant to place substantial reliance upon these negative reports. Thus, there is little definitive information available pertaining to the leukemogenic risk of an adequate size cohort of workers exposed to benzene less than 10 ppm and who have been followed for an adequate amount of time. Furthermore, it is OSHA's view following a careful review of the record that, at the present time, it is impossible to derive any conclusions regarding dose-response relationships for benzene (Goldstein, Ex. 75, p. 2 NRC, Ex. 2-4 p. 11) beyond the general observation that higher exposure levels carry a greater risk than do lower exposure levels. What is apparent however, is that a decrease in exposure level and/or duration will result in a decreased risk of leukemia.

OSHA also recognizes that, in some published cases with adequate documentation, leukemia attributable to benzene exposure developed after observable changes in the peripheral blood, i.e., various cytopenias, pancytopenia and/or aplastic anemia. (Goldstein, Ex. 43B, p. 166.) However, many cases of benzene-induced leukemias were recognized as such only after a patient with overt physical symptoms presented himself to a physician. Similarly, Snyder stated that, in the cases of pancytopenia and aplastic anemia, individuals "exposed to benzene usually do not approach their physicians until late stages in the disease." (Ex. 156.2, p. 2.) As a result, the hematological picture before the onset of disease is often not known. Moreover, because of the well-known reserve proliferative capacity of the bone marrow it is possible that marrow damage may occur without being reflected in the peripheral blood. (Goldstein, Ex. 43.B, p. 175; Wintrobe, Ex. 2-107, p. 676.) Olson acknowledged that alterations in bone marrow activity can occur despite a normal blood count (TR. 2894). Moreover, Goldstein noted that some individuals may have significantly depressed blood values, and yet remain within normal ranges. For this reason, a pancytopenic response to hematotoxic agents may go unrecognized. Therefore, OSHA believes that the scientific evidence is insufficient for adoption of the hypothesis that a preceding prodromal blood syndrome is a necessary prerequisite for the development of benzene leukemia. In this regard, Goldstein in his testimony cautioned that benzene may act as a direct-initiator of a neoplastic response.

OSHA is also aware that in many instances, non-malignant blood disorders resulting from chronic exposure to benzene may be reversed by removal from benzene exposure. However, the apparent high degree of reversibility, as set forth in Jandl's review, is open to question as pointed out in the Health Effects section.

Furthermore, a most characteristic finding which pervades the published literature, is that physiologic factors which cannot be identified a priori, such as nutritional state, genetic constitution, variations in metabolism of benzene and exposure (both occupational and non-occupational) to other marrow depressants, and the variables of age and sex may act to modify the individual's response. (Browning, Ex. 31, pp. 26 ff; Shaw, Tr. 421; Snyder, Ex. 156, p. 2.) For example, Browning stated, "* * * [t]here is no doubt that both men and women differ markedly in their response to similar conditions of exposure." (Ex. 31, p. 2.) Therefore, unqualified identification of high-risk subgroups cannot be made at present. Because of these uncertainties, this standard is designed to protect the most sensitive as well as the more resistant members of benzene-exposed worker populations by limiting exposure to the maximum extent feasible.

It is, therefore, clear from an examination of the record that a determination of a precise level of benzene exposure which presents no hazard cannot be made and that the corollary question of whether a "safe" level of exposure to benzene exists cannot be answered. The agency is aware of and has examined scientific opinion and data submitted by industry that thresholds for carcinogens may exist. However, prudent public health policy requires a conservative course of action until such evidence is of a definitive nature.

In its conclusionary document the International Workshop also stated: "[t]he workshop discussed but could not agree on whether there was a concentration below which there would be no leukemogenic effect clearly attributable to benzene" (Ex. 17, p. 7). Kraybill testified that because there was not enough data at the lower part of the dose-effect curve, he was unable to estimate a "safe" level. As early as 1939, in presenting data on the effects of chronic benzene exposure including leukemia, Hunter stated: "It is doubtful whether any concentration of benzene greater than zero is safe over a long period of time." (Ex. 2-74, p. 354.) In 1974, the MAK-Werte Working Group stated, * * * [O]n the proven carcinogenic (leukaemogenic) effects of benzene and the lack of quantitative measuring data in the low concentration ranges, it is not possible at this time to establish an MAC (maximum allowable concentration) which might be regarded to be without danger." (Ex. 2-58, p. 52.)

Goldwater testified that zero is the level at which there is no risk of getting cancer from benzene (Goldwater Tr. 2500).

Goldstein stated; "As we do not know what the safe level is, the standard would appear to me to be based, must be based on prudent medical approach, which is to keep it at the lowest feasible level" (Tr. 352).

Aksoy stated; "[F]or me, the permissible exposure limit for benzene should be zero. Where it is technically impossible to achieve this, the permissible exposure limit should be lowest possible" (Tr. 149). In OSHA's view, the demonstration of cancer induction in humans at a particular level is not in the regulatory context, a prerequisite to a determination that a substance represents a cancer hazard for humans at that level. Relying upon a substantial body of scientific opinion, OSHA has concluded that, when dealing with a carcinogen, no safe level exists for any given population. For example, the National Cancer Institute's Ad Hoc committee on the Evaluation of Low Levels of Environmental Carcinogens (1970) states.

no level exposure to a chemical carcinogen should be considered toxicologically insignificant for man. For carcinogenic agents, a "safe level for man cannot be established by application of our present knowledge." (Ex. 272, p. 1.)

Thus, OSHA takes the position that in promulgating a health standard for a life threatening hazard such as benzene-induced leukemia, the exposure limits be established on a conservative basis so that worker health is properly safeguarded.

From the point of view of choosing a safe level of exposure, therefore, the permissible exposure limit should be set at zero. However, based on the evidence in the record, it is OSHA's judgment that a zero standard for exposure to benzene is not technologically feasible. In fact, it is clear that certain quantities of benzene are present in the ambient environment as a result of natural phenomena and as artifacts of human activity. While a permissible exposure limit equal to zero plus background would represent the lowest level theoretically possible, OSHA believes that the record shows such an approach is not feasible. Even if such a number could be determined, achieving a standard of zero plus background would require that exposure to benzene be effectively zero so as not to increase employee exposure above background levels. There has been no evidence presented that convinces OSHA that such a complete elimination of benzene in all industries can be achieved by existing or future technology. In determining the apropriate permissible level of employee exposure to benzene, OSHA relies in part on the record of this proceeding and in part on policy considerations which lead the Agency to conclude that, in dealing with a carcinogen or other toxic substance for which no safe level of exposure has been demonstrated, the permissible exposure limit must be set at the lowest level feasible. Such a determination involves a measure of subjective judgment which OSHA believes is justified by the nature of the hazard being dealt with, and the intent of the Act. Section 6(b)5 provides that the standards for toxic substances shall be feasible. That section specifically provides that:

In addition to the attainment of the highest degree of health and safety protection for the employee, other considerations shall be the latest available scientific data in the field, the feasibility of the standards, and experience gained under this and other health and safety laws.

OSHA has determined that 1 ppm TWA, 5 ppm ceiling limit for 15 minutes is the level which most adequately assures, to the extent feasible, the protection of workers exposed to benzene. In making this determination, the Agency recognizes that many industries affected by this standard have already achieved this permissible exposure limit, in a good many of their operations. As shown below for worksites currently above these values, currently available engineering controls and/or modifications of work practices can effectively reduce employee exposure to below the 1 ppm TWA, 5 ppm ceiling. This might suggest that the lower permissible exposure limits could be established, if not for all, at least for some of the industries covered by the benzene standard. OSHA has considered the appropriateness of establishing lower permissible limits for those industry sectors which can acheive these lower limits. It is, however, difficult on the basis of the available evidence to make this determination. The benzene standard applies to a multiplicity of industries and exposures occur from such varying sources as leaks in closed systems to batch operations using solvents. Identification of some operations with lower exposure levels is possible in some industries, but even for these industries exposure information is not available for all operations. Moreover, even where information has been furnished by particular employers as to exposure levels for operations in their plants, it is not possible with any degree of confidence to apply this information to operations of another facility in the same industry. There is a great deal of variability in the configuration of plants, and the physical location of operations within a plant varies from facility to facility. Finally, it is OSHA's view that different levels for different industries would result in serious administrative difficulties. Because of these widespread differences and diversity, OSHA has decided to apply to all affected industries a permissible exposure limit of 1 ppm. Several industrial situations have been examined in making this determination and are discussed below.

Oil and gas production work is conducted in outdoor open air environments with relatively low-benzene content (Tr. 1463) streams maintained in closed systems. (Tr. 1463). Employee exposures in producing operations are intermittent and extremely low (PC 33API Brief p. 141; Ex. 116). Evidence in the record indicates employee exposures ranging between 0 ppm and 2 ppm, with over 80% of the measurements below 0.25 ppm. (Tr. 1467, Ex. 116, Ex. 115.A.4, p. 8). Since the production of crude oil and natural gas occurs in completely enclosed systems, the only employee exposures result from leaking valves (P.C. No. 30) and, as already indicated, these exposures are minimal. Visual inspections can detect these malfunctions and correction can be accomplished largely through replacement of seals. Engineering controls in the form of "double sealed" or "canned" pumps could also be instituted to avoid the leaks.

Pipeline transportation of crude oil and refined products is primarily a closed system operation (Tr. 1460). Potential employee exposure may occur where the products are transferred by truck or rail. (Tr. 1471.) Other operations involving employee exposure are gauging, and sampling operations, all of which are of relatively short duration with minimal employee exposures (Tr. 1472). Measurements taken by industry indicate that most such exposures on an 8 hour TWA basis are below 0.5 ppm. (Ex. 115, A.1.) The only area in which exposures were shown to exceed the permissible exposure limit was in the additive handling operations in which benzene was being utilized as a diluent. Since benzene was not necessary to the process, it was being eliminated from that operation (Tr. 1473).

The unloading and loading of gasoline, crude oil, and other refined products from marine vessels results in limited exposure to shoreside personnel ranging between 0.01 ppm. to 0.82 ppm. with an 8 hour TWA of .2 ppm. (Tr. 2271.) The shoreside personnel's potential for exposure occurs during the connecting and disconnecting of the hoses and loading arms, an activity of limited duration. (Fuller Ex. 115, A.3, p. 4.) The unloading and loading of benzene from marine vessels, however, may result in higher employee exposures. The transportation by barge accounts for the majority of the total benzene transported. (A.D. Little, Ex. 5A, 4.24, 4.42.) The connecting and disconnecting, opening and closing of valves and the monitoring of fill levels are the typical operations in which employees will actively be involved. These operations are all of limited duration and, consequently, employees would spend little of their time involved in them. The remainder of the employees' time would be spent elsewhere removed from exposure. Therefore, work practice controls would be extremely effective in reducing employees' exposures to benzene during the unloading and loading of marine vessels. (A.D. Little Vol. 1, May 1977 p. 4.27.)

One methodology discussed to reduce exposure to vapors is a marine vapor recovery system (Fuller, Ex. 115-A.3). However, at the present time, questions concerning fire and explosion risk, and potential structural damage to barges have, not as yet, been adequately resolved.

The exposures measured during the unloading of gasoline at bulk terminals are low. Sexton, an industry spokesman, reported that the highest long-term samples on individual truck operators was 0.41 ppm., with the vast majority being below 0.2 ppm. (Ex. 115-A.10.) These terminals from which the exposure data was obtained represent examples of the various types of truck loading techniques and different physical characteristics of the terminals, such as loading rates, covered versus uncovered loading racks, and floating roof tops.

Data presented by Williams Pipeline also showed exposure levels, for the most part, below 0.5 ppm. on an 8 hr. TWA basis. It, therefore, appears that one or a combination of these available engineering controls will be sufficient to reduce employee exposure below the permissible exposure limit.

Approximately 94% of the benzene produced is derived from petroleum through the process of catalytic reforming, recovery from pyrolysis of gasoline and hydroalkylation of toluene. Many of the processes used to produce benzene involve completely enclosed units such as reactors. However, these processes do contain vents and other sources of leaks, which can result in significant exposures (A. D. Little, Vol. II-B11).

Control of emissions can be accomplished by such activities as removal and reinstalling equipment such as pumps, compressors, machining of parts, cutting and welding of piplines, fabrication of equipment, replacement of gaskets. Also, the use of automatic gauging devices, closed loop sampling systems, vapor recovery units, floating roof installations and rupture discs has been suggested (Ex. 5A, 4-53).

The majority of employees exposed at petrochemical plants are already within the permissible exposure limit. (Ex. 111-A, pp. 4-7.) At only four of twelve plants surveyed, exposures were above 1 ppm. (ADL. May 1977 p. 10, Table 2), and of those four plants only two were above 5 ppm. The fact that eight of those twelve plants were already within the permissible exposure limit suggests that other plants can be rehabilitated with existing technology to achieve levels within 1 ppm.

The record indicated that exposures to benzene vary between 0.1 ppm and 20 ppm. (Com. 28, 31). The differences in exposure levels appear to be due to the particular protective measures employed in the laboratory. (Tr. 3157, 3492-6). On the basis of this evidence, OSHA has determined that the use of a properly operating laboratory hood is one available engineering control which is currenty reducing employee exposures (See ADL 4-41) well below the permissible exposure limit. The type of control selected will be dependent upon the individual laboratory. (ADL May 1977 pp. 4-40 and 4-41.)

The two segments of the steel industry wherein benzene exposure occurs are the coke oven battery and the light oil distillation plants. Exposures at the coke oven batteries range between 0.02 ppm and 0.31 ppm. (Ex. 135 p. 4 and Attachment 2). It is thus apparent that the steel industry's compliance with the Coke Oven Emission Standard will maintain benzene exposures well below the action level.

Within the distillation plants, equipment is used to distill the benzene out of the light oil. It is in this operation that the steel industries claim that in certain situations it will not be feasible to reduce exposures through engineering controls. It appears, however, that those situations are limited to the older "coke driven by-products plants" (between 30, 40, and 50 years old). (PC 36 AISI Brief, p. 87-88) and does not exist for the newer plants which currently have low exposure levels.

As the Courts of Appeals have emphasized, OSHA is not restricted to the status quo. Standards may be set which require improvement in existing technologies or which require the development of new technology and OSHA is not limited to setting standards based solely on devices currently available, at least where new technology appears on the horizons to limit exposures below the permissible exposure limit. (See e.g. *Society of Plastics Industry v. U.S. Department of Labor*, 509 F. 2d 301 (C.A. 2. 1975) cert. denied.) Certainly here, the evidence indicates that technological controls to reduce exposures, if they are not universally used, do exist (Ex. 5A, 4-37).

Benzene exposure may occur in a variety of industries when solvents containing small quantities of benzene are utilized. Some examples are the rubber industry, the manufacture and use of adhesives, paint manufacturing and application, metal can production and commercial printing. Evidence in the record indicates that, with the exception of paint remover firms for the most part have restricted the use of benzene as a raw materials. It appears, therefore that the percent of benzene coupled with presently available engineering controls, such as local exhaust systems, results in minimal employee exposures, (May 1977 ADL p. 4-27, 4032).

Based on the record of the benzene rulemaking, OSHA has concluded, as indicated above, that the permanent benzene standard is technologically feasible. Furthermore, OSHA has determined that this standard better effectuates the purposes of the act than the national consensus standard for benzene.

Dermal and Eye Exposure Limits: Paragraph (c)(2). The final standard, like the emergency temporary standard and the proposed standard, prohibits all eye contact and skin contact with liquid benzene. This requirement is based on OSHA's policy that, in dealing with a carcinogen, all potential routes of exposure (i.e. inhalation, ingestion, and skin absorption) be limited to the extent feasible.

Although the record evidence does not conclusively establish what the effects of contact with benzene are on the eyes or the skin, participants generally did not oppose a requirement for restricting such contact. Indeed some participants indicated that they already provide protective clothing and equipment to their employees in order to protect them from possible hazards of contact with liquid benzene.

The record evidence on the effect of liquid benzene on the eyes or the skin is extremely limited. No scientific data as to the effects of benzene on the eyes was presented during the rulemaking proceeding. The few studies of skin effects on animals and humans (Ex. 2-46, 2-47, 2-48, TR 2459-90) are not definitive as to the extent of benzene that is absorbed through the intact skin or as to the comparative rate of absorption through damaged skin (TR. 2469-70). Testimony as to skin absorption of benzene was given by NIOSH and by an industry expert witness. NIOSH's view is that benzene is readily absorbed through the skin although the extent of absorption in any particular exposure is unknown. NIOSH however, believes that in instances of an inflamed skin or when benzene is contained in a mixture a "considerable amount" can be absorbed (TR-860). Industry's witness stated that he had no personal view as to whether benzene is absorbed through the skin (TR. 2469) nor whether the rate of absorption of benzene was or was not increased where the benzene is a component of a solvent (TR. 2471-2). The witness also pointed out that he did not know the effect of multiple applications of benzene on the broken skin (TR 2471). He suggested that additional studies be conducted to resolve these issues. This witness nevertheless testified that liquid benzene can damage the outermost protective layer of the skin (TR 2470) and that, although there is no verifying data, "it is ordinarily assumed that penetration of any molecule will be greater through damaged skin that undamaged skin" (TR 2469).

In order to assure that employees are adequately protected against all benzene hazards, it is OSHA's belief that it is appropriate to provide precautions against eye and dermal contact with benzene. In reaching this conclusion, OSHA has taken note that the threshold limit value for benzene adopted by the ACGIH for 1976, and the intended change of that TLV, bear a "skin" notation which refers "to the potential contribution to the overall exposure by the cutaneous route including mucous membranes and eye, either by airborne, or more particularly by direct contact with the substance. Vehicles can alter skin absorption. This attention-calling designation is intended to suggest appropriate measures for the prevention of cutaneous absorption so that the threshold limit is not invalidated" (TLVs for chemical substance in Workroom Environment with Intended Changes for 1976, p. 5). OSHA further notes that maximum allowable concentrations of benzene in other countries also carry a "skin" notation.

OSHA has reviewed the various suggestions for changes in some of the requirements of the provision on eye and dermal limits. Some participants suggested that the provision apply to "liquid benzene" to make clear that the limitation did not apply to eye and skin exposure to benzene vapors (For example, see Com. 53). OSHA has adopted this suggestion since it was never intended that the provision apply to other than liquid benzene. Some comments suggested that the limitation apply only to liquid benzene of a commercial grade or higher (Com. 48; Com. 76) so as to exclude solvents with trace amounts of benzene (Com. 59). There is no evidence, however, to suggest that the absorption rate depends on the amount of benzene present in the liquid (TR 2472, 2482).

Regulated Areas: paragraph (d). The final standard contains requirements for regulated areas. The purpose of establishing regulated areas is to limit the exposure at above 1 ppm to as few employees as possible. The burden of the regulated area provison on the employer is expected to be minimal since the provisions require the employer merely to identify and control access to regulated areas and to notify the applicable OSHA area office of their existence and condition.

The standard requires that regulated areas be established where airborne exposures are above the permissible exposure limit and that access thereto be limited to authorized persons. Regulated areas must be established at all worksites where the permissible exposure limit is exceeded. Although transportation was not specifically listed in the proposed regulated area paragraph, the regulated area requirements of the permanent standard apply to the transportation of benzene, as well as to workplaces where benzene is produced, reacted, released, packaged, stored, handled or used. The scope of the regulated area subparagraph is, therefore, identical to the scope of the standard. This identity of scope has the result that, while the loading of gasoline at the bulk terminals may require the designation of a regulated area, the transportation of the gasoline from the terminals does not, in as much as that transportation activity is exempted from the standard.

Some participants misunderstood the regulated area requirement in the proposal and thought that regulated areas must be established even where exposures are below the permissible exposure limits. (Tr. 3136.) The standard, however, requires establishment of regulated areas only where exposures are above 1 ppm as an 8 hour time weighted average or 5 ppm averaged over a 15 minute period.

Some participants recommended that regulated area requirements not apply to laboratories (Ex. 6, No. 37, 70) or to coke oven batteries (Ex. 135; p. 2 PH; 36 AISI brief, p. 17). It is OSHA's view that existence of a hazard, rather than the type of operation, should be the basis for establishment of regulated areas. Therefore, if the exposures in laboratories or coke oven batteries were to exceed the permissible exposure limits, regulated areas would have to be established at those facilities. Moreover, OSHA feels that the burden is minimal compared to the protection afforded employees.

The standard requires that the OSHA area office be notified within 30 days following the establishment of a regulated area. Each employer must notify OSHA of the existence of any regulated areas within his establishment and of the conditions existing within such areas. Since regulated areas are required to be established only where exposures to benzene exceed the permissible exposure level, employers whose operations are all below that level need not file the report. This requirement is designed for compliance purposes. One comment suggested that OSHA should require, as it did in the Emergency Temporary Standard on Benzene, notification of the use of benzene in every workplace where it is present (Com. 62). It is clear from the record evidence that benzene is present in some amount or other in a vast number of workplaces and that, consequently, a notification of use provision would be administratively burdensome to OSHA and employers without necessarily resulting in more effective enforcement of the standard. For this reason, the final standard, as the proposal, requires merely the notification of regulated areas. Other comments suggested that laboratories and other small users of benzene be exempted from this notification provision on the ground that their use of benzene is infrequent (Com. 70) and that the filing of this notice by "big users" would be sufficient to set priorities for facility inspections. OSHA does not agree. The amount of benzene used by research and testing laboratories will vary from time to time depending on the projects being worked on. Finally, the suggestion presupposes that facility inspections would be conducted solely on the basis of the frequency of use of benzene. However, even where the use of benzene is infrequent, impairment of employee health may occur particularly where the exposure is above the permissible exposure limit. Other industry comments requested that employers be relieved of notification of regulated areas where temporary regulatory areas of short duration are established for maintenance work or for special projects or when leaks in closed systems occur (Com. 46, 50). In view of the hazardous nature of benzene exposure, OSHA deems it essential to require notification of regulated areas in all cases where employees are exposed above the permissible exposure limit. Moreover, in a facility where there are frequent leaks, the employer will be alerted by the necessity to report these leaks to the possible need to take additional measures to protect employees against such exposures. If the possibility exists that the permissible exposure limit will be exceeded from time to time, the maintenance of a regulated area at the location of the leaks will, of course, minimize exposure and minimize the reporting requirement. Some industry participants recommended that the standard include a method for "deregulating" the regulated areas once exposures fall below the permissible exposure limits. (Com. 45, Com. 53, Com. 57) OSHA does not feel that a formal method for deregulation is necessary. Since the standard does not require maintenance of a regulated area except where exposures are above the permissible exposure limit, the regulated area requirement does not apply where the employer has reduced the exposure levels below the permissible exposure limit. Nor would a deregulation procedure contribute to the protection of employees. Moreover, if formal deregulation were required, it would be necessary for the employer to file an additional notification with the OSHA area office should conditions change thereafter and the exposure again exceed the permissible exposure limits. In view of these considerations, a formal deregulation procedure would appear to be unnecessarily burdensome.

Monitoring: paragraph (e). Significant changes have been made in the monitoring section of the final standard in response to comments by participants. Briefly, the final standard requires all covered employers to make measurements to determine whether any employee may be exposed to airborne concentrations of benzene, and imposes different measuring requirements depending on whether exposure measurements are above or below certain levels.

The monitoring requirements are imposed pursuant to Section 6(b)(7) of the Act (29 U.S.C. § 655) which mandates that any standard promulgated under section 6(b) shall, where appropriate, "provide for monitoring or measuring of employee exposure at such locations and intervals, and in such manner as may be necessary for the protection of employees." The purposes of monitoring are to determine the extent of exposure, to identify the source of exposure to the hazard and to enable the employer to select proper control methods and evaluate the effectiveness of the selected methods. Thus, monitoring enables employers to meet the legal obligation of the standard to assure that their employees are not exposed to benzene in excess of prescribed levels. Additionally, monitoring enables employers to notify the employees of their exposure level, as required by Section 8(c)(3) of the Act, and provides information necessary to the examining physician.

The need to conduct exposure monitoring was generally accepted by participants in the rulemaking process (P.C. 35MCA brief, p. 55-b; P.C. 36;

ORC brief, p. 7). Many comments, however, objected to the proposal's requirement to measure airborne exposures in all workplaces "where benzene is present" (P.C. 33; API brief, p. 127a; P.C. 34; ORC brief, pp. 7-8). Participants argued that the use of the term "present" is so broad as to encompass each and every employee at a facility (Ex. 6, No. 53). It was also argued that benzene is present in the ambient atmosphere (Ex. 6, No. 43; Ex. 84B. 18, Ex. 84A, p. 7) with the result that every work operation in every city would probably have to be monitored (Ex. 6, No. 43, p. 12). OSHA's objective is to minimize all occupational exposures to benzene. The term "presence" was used in the proposal to convey the intent that, whenever the exposure of any employee to benzene concentrations resulted from workplace operations at the place of employment, the employer was required to measure that exposure. OSHA does not intend that employees, who might be exposed solely from other sources, such as ambient levels of benzene, be covered by the monitoring requirements. In view of the confusion as to the meaning of the term "presence," the final standard does not use that term but instead specifies the general classes of occupational activities which can result in employee exposure to benzene, and requires monitoring where any of these activities are conducted.

In conducting the monitoring of exposures, the standard does not require that each individual employee's exposure level be measured. Although individual measurement is the ultimate indicator of employee exposure, OSHA believes that a requirement for individual measurements may be too burdensome. Accordingly, the standard requires that the measurements be made by monitoring which is representative of each employee's exposure to benzene over an eight-hour period without regard to the use of respirators. It should be noted that the requirement for representative monitoring does not preclude an employer from taking individual exposure measurements of each of his employees; individual measurements are certainly considered to be representative; and the representative monitoring requirement is the minimum that the employer must meet.

In establishments having more than one work operation involving the use of benzene, the monitoring to be representative must be performed for each type of employee exposure within each operation. One participant requested that area sampling be permitted in order to lessen the burden and cost of monitoring each employee with a different job function (Ex. 6, No. 49, p. 4). Although the final standard does not specifically require personal sampling, monitoring under

the standard must determine breathing zone exposures. Appendix B, IV, therefore, recommends that air sampling be taken in the employee's breathing zone. Area samples are generally not as direct a measure of employee exposure, and consequently, may not meet the requirement for representative monitoring, although where area sampling can be correlated with breathing zone exposures, area sampling may be used. OSHA, however, notes that while there are techniques to correlate area sampling with breathing zone exposures, these are generally more burdensome than personal sampling and involve much more sophisticated data collection and analysis, including the performance of personal sampling to assure the correlation.

Some participants suggested that employers with several places of employment, in which there are workplaces with identical processes or operations, be permitted to monitor only a representative number of such locations (Ex. 6; Ex. 50, p. 3; Ex. 60, p. 7). The fact that work operations are in different geographical locations would not in itself preclude representative monitoring. However, it is the employer's responsibility to assure that identification of conditions, characteristics, activities, climate, etc. exist so that monitoring at particular sites would actually be representative of employee exposure at other sites.

The employer is also required to use a method of monitoring and measurement with an accuracy (at a confidence level of 95%) of not less than plus or minus 25% for concentrations of benzene of 1 ppm or more. Methods of measurement are presently available to detect benzene to this accuracy level (Tr. 342-3) and one such method is described in Appendix A, II, E. Industry participants expressed concern that consistent compliance with these accuracy requirements will be impossible. (PC 36; AISI brief pp. 106-108.) The record, however, indicates that, even where benzene is present with other organic solvents, the OSHA Analytical Method will still enable employers to measure low levels of benzene well within the accuracy requirement of the standard. (Tr. p. 343.) Indeed, it was suggested at the hearing that there is an alternate method for sampling, passive dosimeters, which may comply with the accuracy reqirements contained in the proposal. (Tr. 2381.)

As earlier indicated, the standard requires all covered employers to initially measure the airborne exposure of all of their employees. Some participants objected to this requirement and recommended that initial monitoring be limited to those workplaces where there is a probability of exposure to benzene in excess of the permissible exposure limit (P.C. 36; AISI brief, pp. 73-74; P.C.34; ORC brief; p. 8) or in excess of 10 ppm (MCA brief, p. 50). It is OSHA's view that a "probability" is too speculative where a carcinogen is involved, and that a more definitive basis is needed for determining employee exposure. It is, therefore, OSHA's decision to require initial monitoring at any level of exposure to a carcinogen. Initial measuring of exposure is necessary for the protection of employees particularly where, as here, the permissible exposure level is not a "safe" or "no effect" level but is predicated upon feasibility. Moreover, initial monitoring does not place an undue burden on the employer since he need not continue the monitoring for those employees whose exposures are below the action level.

The standard requires that the initial monitoring be conducted and the results thereof obtained within 30 days of the effective date of the standard. Several comments requested that a longer period, ranging between two months (Ex. 176) to six months (P.C. 36; AISI brief, p. 75; TR 3137), be allowed for employers to meet this requirement. OSHA has retained the thirty day requirement. Since the final standard has a delayed effective date of thirty days after date of publication in the FEDERAL REGISTER, employers will have sixty days to comply with this requirement. OSHA feels that this period is sufficient to enable employers to secure sampling equipment, take samples, and obtain the results. Moreover, the standard permits employers, who have monitored within the last year, to utilize these measurements for purposes of compliance with the initial monitoring requirements, provided that the sampling and analytical method used meets the accuracy test of this standard and provided that the employer maintains a record of these measurements and notifies employees of their exposure levels. Employers who have already monitored their employees' exposures within this period will, therefore, not have to conduct initial monitoring unless, because of a process, control or personnel change, they are required to redetermine exposures. In addition, to the extent that the initial monitoring requirement cannot be complied with because of the unavailability of professional or technical personnel or of materials or equipment, the temporary variance procedures of the section 6(b)(6)(A) of the Act may be appropriately utilized. As noted, however, it is not expected that this will be generally necessary.

The frequency of monitoring employee exposure has been modified in the final standard. The proposal would have required monthly monitoring where exposures were above the permissible exposure limit and quarterly monitoring of those employees whose exposure was below that level. The final standard does not require employers to repeat the monitoring where the initial measurements are below the action level, unless there is some change in operations which might alter the exposures. Where the exposure measurements are above the action level but not in excess of the permissible exposure limit, the employer is required to monitor quarterly. Where initial or subsequent monitoring indicates that exposures are above the permissible exposure limit, the employer must monitor monthly. In developing these requirements, OSHA has given serious consideration to the numerous suggested modifications of the proposal's requirements. These ranged from suggestions to lessen the frequency of repeated monitoring to suggestions that monitoring frequencies should depend on whether the process is stable or highly variable (Ex. 6, No. 40), or should be based on performance variables, such as frequency of use, quantity used (Ex. 6, No. 70) and engineering controls (Ex. 6, No. 12, No. 43), or should be at the discretion of a professional (Ex. 6, No. 43). Some participants objected to periodic monitoring where initial monitoring reveals exposures below the permissible exposure limit (P.C. 36; AISI brief, p. 76; Com. 36, 10, 37). OSHA agrees that periodic monitoring is not necessary where exposures are shown to be below the action level. Accordingly, OSHA has eliminated the repeat monitoring requirement for operations below the action level. Where initial measurements are above the action level but do not exceed the permissible exposure limit, periodic monitoring is necessary to assure that exposures remain within that range and do not rise to above the permissible exposure limit. OSHA believes that quarterly monitoring is sufficient to detect substantial changes in exposures and to assure that employee exposures remain below the permissible exposure limit. Some participants objected to monthly monitoring where initial measurements are above the permissible exposure limits (P.C. 36 AISI brief, p. 76, Ex. 6, No. 50). Their rationale is that repeat monitoring serves no purpose since the hazard has already been identified (Ex. 170). They argued that the requirement to establish regulated areas where exposures are above the permissible exposure limit is a sufficient incentive for employers to remonitor a workplace as soon as they believe that the exposure has been reduced (P.C. 36 AISI brief, p. 76-7). As previously pointed out, monitoring serves other purposes than initital determination of employee exposure. Monitoring is a check on the efficiency of engineering controls: increased levels alert the employer to the need to check and possibly alter these controls. Increases in benzene levels may also necessitate changes in types of respirators. Exposure measurements remind employees and employers of the continued need to protect against hazards, and are also useful to the examining physician in determining whether the exposed employee is at an increased risk. OSHA, therefore, is of the view that repeated monitoring, where initial measurements are above the permissible exposure limit, is necessary to protect employees. Other participants expressed the view that this monitoring should be less frequent than monthly—that is, should be quarterly (P.C. 31 ORC brief, p. 9; L.C. 3); semi-annually (P.C. 36 AISI brief, p. 78; Ex. 6, No. 57), annually (Ex. 6, No. 59, No. 76), or as frequently as significant changes are believed to have occurred (L.C. 19). In view of the leukemogenic hazard of exposure to benzene, it is OSHA's view that monthly monitoring, where exposures are above the permissible exposure limit, is necesary to protect the health of exposed employees.

The final standard also provides that employers may discontinue monitoring for those employees for whom two consecutive measurements taken seven days apart show exposures to be below the action level. Where employee exposure measurements fall below the permissible exposure limit but are at or above the action level, the employer may alter the monitoring schedule for those employees from monthly to quarterly after two consecutive measurements taken seven days apart indicate the reduction in levels.

The standard further requires that, whenever there has been a production, process, control, or personnel change which may result in new or additional employee exposure, or whenever the employer suspects that a change of employee exposure may occur, the employer must repeat the required monitoring. Redetermination of exposures in such instances is necessary so that the employer may take the necessary action to protect his employees, such as providing appropriate respiratory equipment, or instituting engineering controls. Redetermination is required only where there is a new exposure or an increase in exposure levels is suspected (Ex. 6, No. 76, No. 50). The required redetermination where there is a change of personnel was added to the final standard in view of evidence that exposure levels can change when different employees perform the work (Tr. 3093). This requirement does not mean, however, that the employer must remonitor every time there is a personnel change but he must redetermine exposures when the work practices of the substitute employee are such that an increase in exposure may result. Redetermination of exposure must also take place after the cleanup of spills and the repair of leaks, ruptures or other breakdowns. One comment suggested that redetermination not be required in these instances (L.C. 19). OSHA has not adopted this suggestion. Spills, leaks, etc., can result in very high exposure levels (Tr. 1299; P.C. 30, p. 48), and the requirement to redetermine exposures after cleanup or repair provides one method of ascertaining that proper corrective methods have been instituted and employee exposures are not significantly altered.

The final standard further requires that employers notify each of their employees of the exposure measurement which represents that employee's exposure. This requirement is discussed in detail under Recordkeeping.

It should also be noted that paragraph (m) of the standard requires the employer to allow employees or their designated representatives an opportunity to observe the monitoring. The specific provisions of paragraph (m) are discussed below.

Methods of compliance: paragraph (f). The final standard, as the proposed standard, requires employers to institute engineering and work practice controls to reduce employee exposure to benzene to or below the permissible exposure limits, except to the extent that such controls are not feasible. This requirement is in accord with OSHA's policy that feasible engineering and work practice controls must be used as the primary methods of reducing employee exposures. This policy is based on the view that the most effective means of controlling employee exposures is to contain concentrations at their source through use of mechanical means combined with work practices rather than reliance on the variability of human behavior so critical to the successful use of respirators. Thus, the standard also provides that, in situations where feasible engineering controls and work practices are insufficient to reduce exposure to the permissible limits, the controls must nonetheless be used to reduce exposures to the lowest achievable level, and then be supplemented by the use of respiratory protection.

In reaching the decision to require engineering and work practice controls as the primary methods of reducing benzene exposures, OSHA has carefully considered the objections of various participants. While recognizing that in many situations engineering and work practice controls are the preferred methods of reducing exposures to or below the permissible levels, several participants recommended that the hierarchy of control measures be eliminated and employers be allowed the freedom of selecting the control measure to be instituted. (Tr. 3140).

These participants particularly objected to the requirement that engineering controls and work practices be used to reduce exposure to the lowest achievable level even if that level is above the permissible exposure limit. Various arguments were made in support of this position. One participant pointed out that, where the controls will not reduce exposures sufficiently, respirators will have to be used anyway. This employer suggested that OSHA mandate the continued search for controls which are sufficient to reduce employee exposure, but allow employers to select any appropriate method of reducing exposures (Com. 43). As was pointed out in the benzene hearing (Ex. 59, p. 12), as well as in other OSHA rulemaking proceedings, respirators are the least satisfactory means of control because of difficulties inherent in their design and use. Respirators are capable of providing good protection only if they are properly selected for the types and concentrations of airborne concentrations present, properly fitted and refitted to the employee, worn by the employee, and replaced when they have ceased to provide protection. While it is theoretically possible for all of these conditions to be met, it is more often the case that they are not. Consequently, the protection of employees by respirators is not always effective and is, therefore, permitted only in certain specified circumstances. For example, proper facial fit, is essential but, due to variations in fit, individual facial dimensions and the limited range of the facepiece configurations, such fit is difficult to achieve. (Tr. 332-335) Often the work involved is strenuous and the increased breathing resistance of the respirator reduces their acceptability to employees. Safety problems presented by respirators must also be considered. Respirators limit vision. Speech is also limited. Voice transmission through a respirator can be difficult, annoying and fatiguing. Movement of the jaw in speaking also causes leakage. Communication may make the difference between a safe efficient operation, on the one hand, and confusion and panic, especially in difficult and dangerous jobs, on the other hand. Moreover, skin irritation can result from wearing a respirator in hot, humid conditions and such irritation can cause considerable distress and disrupt work schedules. It is clear, therefore, that respirators cannot be considered as the primary means of employee health protection. Nevertheless, respirators do provide some protection and OSHA has concluded that under certain limited circumstances, where no alternatives are available, respirators may be used to reduce employee exposures.

Industry participants further argued against a system of priorities because of alleged unfeasibility of engineering controls. Most of these participants' objections to methods of compliance, however, lie in the area of economic rather than technological feasibility (Com. 49, p. 5; Com. 54, p. 2; Com. 73, p. 3). As discussed below, OSHA has determined that this standard is feasible and that the implementation of engineering and work practice controls is for the most part technologically feasible. However, OSHA realizes that, under some particular circumstances, engineering and work practice controls may not be technologically feasible in a particular work operation. Therefore, the standard explicitly recognizes that an employer may demonstrate the infeasibility of engineering and work practice controls as to one or more operations in a particular process, and in these circumstances use respirators to provide the required protection. The question of whether an employer has met its burden of establishing that engineering and work practice controls are infeasible in a particular work operation involves the consideration of many complex factors and a rational balance process. Factors such as levels of exposure, useful remaining life of the equipment and the effort made by the employer to implement such controls are relevant.

In addition to the obligation to institute engineering and work practice controls, except to the extent that such controls are not feasible, the final standard also requires that each employer establish and implement a plan, including schedules for reducing exposures to within the permissible exposure limit or to the greatest extent feasible, solely by engineering and work practice controls. These written plans must be furnished upon request for examination and copying to representatives of the Assistant Secretary and the Director. These plans must be reviewed and updated periodically to reflect the current status of exposure control. Some participants felt that the requirement for written plans was burdensome and unnecessary (Com. 45, 59). OSHA, however, views the requirement for written plans as an essential part of the compliance program since it will encourage employees to actually achieve the controls and also provide the necessary documentation to OSHA, employers and employees of the compliance methods chosen, the extent to which controls have been instituted and plans to institute further controls to achieve safe and healthful workplaces.

A compliance issue relating to the requirement for engineering controls and work practices in this final benzene standard involves the relationship of that requirement to the covered employers' legal obligation as to engineering controls under the prior benzene standard (29 CFR 1910.1000, Table Z-2). The new standard, at paragraph (f)(1), requires the implementation of engineering controls and work practices. OSHA views the two standards as a continuum of enforceable obligations which have been crystallized in the standards promulgation. More specifically, in evaluating whether the employer has instituted controls, OSHA compliance personnel would consider not only the employer's performance under the new standard but also his prior obligation under its predecessor.

Respirator protection: paragraph (g). The final standard contains requirements for respiratory protection. The standard requires employers to provide and assure the use of respirators whenever the permissible exposure level is exceeded. Paragraph (g) makes clear, however, that respirators are only to be used when other means of control are not feasible, and that the use of respirators is not a substitute for engineering and work practice controls. The standard allows respirators to be used to achieve compliance with the permissible exposure levels only in certain specific situations. The reason for the limitations is that respirators are, as stated above, the least satisfactory means of exposure control.

The standard contains a respirator selection table to enable the employer to provide the type of respirator which affords the proper degree of protection. While the employer must select the appropriate respirator from the table for each work operation on the basis of the airborne concentrations of benzene, he may always select a respirator providing greater protection—that is, one prescribed for higher concentrations of benzene than present in his workplace. It should also be noted that OSHA does not anticipate, as previously suggested in the proposal, that a conflict between the respirator selection table of the benzene standard and that of the coke oven standard will arise. Industry date (Ex. 136, attach No. 2) indicates that benzene exposures at coke oven batteries are generally below 1 ppm and respiratory protection for benzene exposure will generally not be required.

The standard further requires that the employer select respirators from among those approved by NIOSH under 30 CFR Part 11. Participants pointed out (Com. 41; Tr. 752; TR 139) that, although the standard's respiratory selection table permits the use of chemical cartridge respirators for concentrations of 50 ppm or less, NIOSH does not approve chemical cartridge respirators for benzene. OSHA is well aware of the fact that NIOSH does not approve organic vapor cartridge or canister respirators for substances, such as benzene, with poor warning qualities and that such respirators do not generally have end-of-service life indicators (Tr. 140). The standard's requirement for NIOSH approval, however, is designed to assure that only devices approved and certified by NIOSH for use against organic vapors be used, notwithstanding the fact that such certification carries a stated or implied prohibition against the use of these devices for substances with poor warning properties. The alternative would be to chance employee protection by allowing use of non-approved devices, or to require supplied air respirators or self-contained breathing devices which in many cases would give rise to extreme operational difficulties (Tr. 142). Since organic vapor cartridge and canister respirators currently approved by NIOSH have been shown to be effective in absorbing benzene vapors, neither of these alternatives is necessary.

To compensate for the lack of adequate warning properties of these respirators, the standard requires frequent replacement of the cartridge or canisters (Tr. 140). The standard requires that the cartridges or canisters be changed at the end of their service life or at the end of the shift in which they are first used, whichever comes first. Some participants objected to the requirement to replace air purifying canisters and cartridges at the end of the shift. They felt that it would lead to waste (Com. 40), would do little to encourage research and development of an end of service life indicator (Com. 41), and would be too frequent where benzene concentrations are low and cartridge life is long (Com. 65). Recommendations for longer periods varied (Com. 53; L.C. 3; L.C. 19; Com. 60). OSHA has considered these recommendations but has concluded that, in order to assure that employees receive the maximum protection from benzene vapors, cartridges or canisters must be changed at least at the end of each shift. Should NIOSH, at some later date, certify a benzene end of service life indicator, which demonstrates that the canister or cartridge is useful past the end of the shift, OSHA would allow the cartridge or canister to be used for the extended period. Accordingly, the final standard has been changed to make this clear and hence encourage research and development of an end of service life indicator for benzene.

The standard further requires that the employer institute a respiratory protection program in accordance with 29 CFR 1910.134. This section contains basic requirements for proper selection, use, cleaning and maintenance of respirators. The employer is also required to assure that the respirator is properly fitted and to allow employees to wash their faces and respirator face pieces. In addition, paragraph (j)(i) of the standard requires that employees be properly trained in the use of respirators.

The standard requires that the employer shall provide respirators and other clothing and equipment required for protection from exposure to benzene at no cost to the employee. OSHA has allocated the costs of respirators and clothing and equipment required for protection from benzene exposure to the employer in order to effectuate the purposes of the Act. The employers' costs in providing respirators and protective equipment and clothing were considered in the economic assessment. This language clarifies OSHA's position which has long been implicit in health standard proceedings under section 6(b) of the Act.

Protective Clothing and Equipment: paragraph (h). The final standard retains the requirement that, where the employee is exposed to eye contact or repeated skin contact with benzene, the employer must provide and assure that employees wear the appropriate protective clothing and equipment. This requirement compels employers to furnish, where necessary, such items as goggles, face shields or gloves, or footwear.

The available scientific evidence on the intact and unbroken skin has been discussed above under Dermal and Eye Exposure Limits. Although the evidence on skin absorption of benzene is inconclusive, it is an "undisputed assumption" that the penetration of any benzene molecule will be greater through damaged skin (Tr. 2469-70). The record evidence also indicates that there are situations where employees are immersed or drenched with benzene liquids (Tr. 289, 292, 299). Liquid benzene on the skin may cause erythemia and blistering, and dry scaly dermatitis may develop on prolonged or repeated exposure (Ex. 2-3). Burns may occur from benzene spills (Tr. 292). One purpose of the protective clothing and equipment requirement is to protect employees from dermatitis and burns. Additionally, since benzene is a carcinogen and since breaks in skin surfaces commonly occur, OSHA has concluded that it is necessary to require personal protective clothing and equipment although the full effect of liquid benzene on the eye and the skin is as yet unknown. To do otherwise would be to leave employees unprotected and expose them to what might ultimately be shown as a most serious hazard.

Few comments were received on the protective clothing and equipment requirements. Some participants indicated that they already provide such protection to their employees (L.C. 4). One participant suggested that the extent of body exposure, which would trigger the need for the protective clothing, be delineated in the standard. (Com. 55.) OSHA, however, believes that the likelihood of eye or repeated skin contact and the resultant need for protective clothing must be assessed on the basis of individual workplace operations. One comment objected to the need for protective clothing or equipment where 1 percent or less benzene is present in the liquid benzene. There is, however, nothing to indicate that such percentages would create a sufficiently reduced hazard to warrant total elimination of the requirement.

The final standard requires the use of "impermeable" protective clothing and equipment. This requirement was in the emergency temporary standard. The proposal, however, would have simply required "appropriate protective clothing and equipment." Where eye and skin protection are required, the protective clothing and equipment to be "appropriate" must not allow liquid benzene to reach the eyes or skin. To make this clear, OSHA has adopted the language of the emergency temporary standard. The impermeable clothing must prevent all penetration of benzene. Thus, where impermeable gloves are used, liquid benzene should not seep through the cuff or any other opening in the glove. This could very well increase the hazard to the employee. It appears, on the basis of testimony from the industry expert witness, that, when a chemical substance on the skin is covered with a totally inclusive material, the rate of permeability of the substance through the skin is greatly increased (TR 2464; 2473).

Hygiene Facilities

Although normally required in carcinogen standards (See Coke Oven and Cancer Policy), the final benzene standard does not contain any requirement for hygiene facilities, such as waste disposal, housekeeping, showers, change rooms, or laundering of clothing.

The question of what, if any, hygiene facilities should be included in the final standard was at issue in the benzene rulemaking proceeding. The most common view of participants in that proceeding was that hygiene facilities were not necessary because of the volatility of benzene (L.C. 19). Almost all participants felt that no hazard would be created by the absence of hygiene facilities. One witness, however, testified that, in his opinion, the wife of a benzene exposed employee, who laundered her husband's work clothes daily died from benzene induced leukemia (Tr. 314). Unfortunately, no further information concerning this case was furnished. OSHA, consequently, finds that the record does not establish the need for a requirement for hygiene facilities.

Medical surveillance: paragraph (i). This standard requires that employers make available a medical surveillance program for those workers who are exposed to benzene at or above the action level of 0.5 ppm TWA. In addition, the regulation contains an emergency provision for the biological monitoring of employees exposed to a massive release of benzene.

Evidence contained in the record clearly indicates that a medical surveillance program is an appropriate measure. Hematologists, company physicians and other participants have recommended the inclusion of a medical surveillance provision in the final standard (Ex. 106; P.C. 30; Ex. 211; Ex. 17, Ex. 179, Bommarito, Tr. 3382). Also, a review of the record discloses that it is a current common practice of many industries to routinely examine their employees for evidence of benzene toxicity. (e.g. Wodka, Tr. 1263; Com. 26; Joyner, Tr. 2286; Dow, Tr. 2967, 3021; Ex. 77H.)

Section 6(b)(7) of the Act requires that employers make available medical examinations to ascertain whether the health of workers is adversely affected by exposure to toxic substances, where appropriate. The requirements contained in this regulation are designed to detect changes in the hematopoietic system resulting from chronic exposures to benzene which may be manifested in a range of blood disorders, including leukemia.

It is widely recognized that the major target organ in chronic benzene toxicity is the hematopoietic system, especially the bone marrow. This system has the attribute that a sample of the peripheral blood offers a unique biological "window" on the functioning and health of this system, providing much direct visual information not readily available for other organs or systems.

It is OSHA's view that a medical surveillance program directed toward the early detection of hematopoietic dysfunction by examination of peripheral blood samples is an appropriate measure. This determination receives support from: (1) Studies showing that early blood dyscrasias have been detected in workforces screened by laboratory tests (and often in workers displaying no obvious physical symptomatology) (Greenburg, Ex. 2-8, p. 419; Savilahti, Ex. 2-95, p. 1; Goldstein, Ex. 43B, p. 138); and (2) that even in cases of acute myelogenous leukemia, a period of remission, unfortunately brief, may be effected in a substantial proportion of those afflicted with this fatal disease by appropriate treatment. (ORC/Jandl, P.C. 34, p. 73-74). OSHA's decision is also supported by evidence that some cases of benzene-induced blood disorders may be reversed upon cessation of exposure and by appropriate treatment. (ORC/Jandl, P.C. 34, p. 39).

The medical surveillance provision of the proposed standard was the object of much comment and testimony. For example, over 45 separate comments addressed this issue in prehearing submissions alone. Additionally, substantial testimony was adduced during the hearing, and additional views were submitted to OSHA in the form of post-hearing comments. These suggestions addressed both the content and frequency of the examinations as well as what information should be elicited from the medical/work history portion.

OSHA, in formulating the provisions of the medical surveillance section, was aware of several problems. While there is no disagreement that the blood tissues are the most important organ relative to chronic benzene toxicity, there are as yet no consistent patterns of abnormalities, especially at low exposure levels, and no specific pathognomonic tests of benzene toxicity (Rosen, PC 23A, p. 2). Compounding this problem is the fact that it is difficult to separate cases of early benzene hematoxicity from more common minor blood disorders (Goldstein, Tr. 354). Associated with the problems of exam content is one of testing frequency. There are simply no precise estimates of the time course of the various abnormalities to allow determination of optimal testing frequency. Opinions based upon the lifespan of the blood cells are equally inapplicable in determining the frequency of such exams (Battle, P.C. 26C). While realizing that the more frequent the exams the greater the probability of detecting blood disorders induced by benzene, OSHA is also aware of the practical considerations of a medical surveillance program. Unlike the coke oven standard which applies to a defined number of workers employed in well characterized work sites, benzene is a component (usually as a contaminant) of a wide variety of refined petroleum solvents and, therefore, some benzene exposure occurs in a myriad of work sites and work operations. OSHA also recognizes that to be effective a medical surveillance program must be acceptable to a majority of workers subjected to the examination, the tests must display a high degree of accuracy and reproducibility, and the program should be able to be performed in a routine manner without unduly taxing medical resources. Given the above facts, the prescribed medical surveillance program is designed to accommodate the purpose of detecting early blood disorders resulting from benzene exposure without being overly burdensome and unduly intrusive.

The medical surveillance program consists of three main elements: (1) An initial and periodic examination for all those employees exposed to airborne benzene concentrations at or above the action level, (2) evaluation of abnormal findings by a hematologist and, (3) a biological screen for workers who may have received excessive exposure as a result of an emergency.

Several commentators and hearing participants recommended to OSHA that the medical screening program should be, in some way, related to employee exposure, and several specifically suggested that an action level be utilized as the determining boundary. (Com. 10; Com. 21; Com. 26; Com. 36; Com. 46; Com. 47; Com. 53; Com. 70; Com. 72; Com. 76; L.C. 3; PC 36; AISI brief). Many others felt that routine medical surveillance applied to all workers exposed to benzene without regard to degree of exposure, as was specified in the proposed standard, was an excessive requirement (e.g. Ex. 6-36). OSHA has examined these analyses and, as a result, has significantly altered the content of the surveillance program in the final standard. Furthermore, it is OSHA's view that the requirements of the medical surveillance program cannot be left to discretion of the examining physician because some physicians may not be fully aware of the wide range of effects resulting from benzene exposure.

The final standard requires that employers must make available an initial and periodic examination to any worker exposed at or above the action level. While it is the Agency's position that there is no "safe" level for benzene exposure, OSHA believes that the vast majority of workers exposed below the action level will not manifest blood dyscrasias, while those at or above this value are at greater risk, the precise extent of which is unknown. The inclusion of an action level concept will therefore limit medical surveillance requirements for those employees as to whom the probability of identifying benzene-induced blood disorders is in all likelihood minimal, and will focus such efforts on a population at increased risk.

INITIAL EXAMINATION

The initial examination has two components: a medical/work history and a series of laboratory blood tests. This differs from the proposed standard which drew no distinctions between initial histories and examinations and subsequent ones. The final standard does, however, make such a distinction. The purpose of the history is to alert the physician to problems which may indicate an individual's increased sensitivity to benzene. The permanent standard requires the taking of a history with the data elements specified in the proposal.

Eli Lilly (Com. 28) and Dupont (Com. 46) generally agreed with the proposed history requirements, while Dow Chemical agreed with the intent, i.e., that a complete medical history with an emphasis on blood problems is an essential part of a benzene medical surveillance program. They felt, however, that the wording should be simplified so that the worker could supply the information him/herself with minimal professional assistance. (Ex. 154.) OSHA, recognizing the merits of this suggestion, has not chosen to mandate the precise wording of the history taken, but only to require that specific areas should be queried. The employer, therefore, has considerable latitude with respect to the actual format. Whether the employee answers the medical history or is questioned by a health professional is also left to the discretion of the employer. The standard does require, however, that all aspects of the medical surveillance program be under the supervision of a licensed physician.

One hematologist objected to the inclusion of questions pertaining to a family history of hematologic neoplasms, noting that there is little evidence associating these disorders with a predisposing increased sensitivity to benzene (ORC/Jandl P.C. 34, Add. 5, p. vii). Another hematologist also criticized this provision on the basis that the information obtained was both unrealiable and of unknown relevance to benzene toxicity (P.C. 26C). Jandl also stated that genetically related hemoglobin alterations (hemoglobinopathies) are not highly relevant and are therefore of no predictive value (ORC/Jandl, P.C. 34). OSHA agrees that there is no known or even suspected association of benzene-induced blood dyscrasias with some hemoglobinopathies such as sickle cell anemia. However, Aksoy has reported findings that individuals with thalessemia (another hemoglobin disorder) may be more sensitive to the hematotoxic effects of benzene (Ex. 44). (It is not clear from Aksoy's testimony whether or not thalassemias are at greater risk of developing leukemias resulting from benzene exposure.) Moreover, he also observed a familial link with leukemia resulting from benzene exposure. (Ex. 24). In view of Aksoy's findings, OSHA believes that it is appropriate that the questions in the medical history relating hemoglobinopathies and a family history of malignant blood diseases remain in the standard.

Battle recommended that the health history also include questions concerning possible exposure to marrow toxins, such as the use of insecticides and volatile cleaning agents in the home (P.C. 26C). Since exposure to these types of agents may have significant impact on the blood picture of examined workers, queries regarding exposure to other marrow toxins both in and outside of the work environment have been added to the history requirements.

The initial laboratory examination requires that a series of classical blood tests and also measurements of serum bilirubin and reticulocyte count be performed. As noted, there is no consistent pattern of blood abnormalities resulting from benzene exposure. Various investigations claim that the earliest sign of benzene hematotoxicity is anemia or leukopenia, or changes in red cell indices. Some believe that red cell enlargement (macrocytosis) (Goldstein, Ex. 75; Goldwater, Ex. 140) or an absolute lymphopenia (Goldstein) may be suggestive of early marrow damage. (Occasional observations of a paradoxical hypercellularity in response to benzene exposure have been reported, e.g., by Aksoy; however, the basic hematological workup should detect this particular response).

An examination of the record demonstrates that there is general agreement that the following laboratory hematological tests should be part of an initial or baseline examination:

(a) Red Cell count
(b) White Cell count
(c) Hematocrit
(d) Hemoglobin
(e) MCV—(mean corpuscular volume)
(f) MCH—(mean corpuscular hemoglobin)
(g) MCHC—(mean cell hemoglobin concentration)

These tests are routinely performed in most blood examinations; they provide basic information about all three blood cell lines. The normal values are well known, and the tests are usually automated with the attendant benefits of being highly reproducible and accurate.

Differential counts of white blood cells, also required by the standard, are not amenable to automation and are therefore performed manually under microscopic examination by trained technicians. Battle commented that white blood cell counts are no longer performed on a routine basis in his clinic and that such an assay is usually only performed in response to abnormal cell counts (P.C. 26C). However, other commentators have recommended the inclusion of a differential count as part of a basic hematologic examination (both the initial and interval tests) (ORC/Jandl, P.C. 34, pp. 46-51, Goldstein Ex. 75). In addition to providing information on the relative number of various white cell types, the differential count provides an opportunity to observe aberrant cellular morphology and to note any unusual proportions of immature cells. Jandl feels that scanning of the slide by a trained technician for these types of findings is of the utmost importance (ORC/Jandl, P.C. 34, p. 45). For the above reasons, a differential count is a required part of the laboratory examination. In addition to the basic laboratory tests, serum bilirubin and reticulocyte counts must also be done in the initial laboratory exam. These tests may provide information indicative of hemolytic states and other abnormalities. Because of the non-specificity of these two assays relative to benzene toxicity, the time-consuming nature and instability of the reticulocyte preparations, these particular tests were the object of much comment and criticism. Several commentators stated that although these tests were of little value on a routine basis, they should be made part of a preplacement or baseline exam (Com. 28, 72). Based on a review of this issue, OSHA believes that these two particular blood examinations should be included as a part of the initial laboratory tests only, to provide a more comprehensive picture of a worker's hematological profile, for possible later comparative purposes.

Several hematologists stressed the importance of obtaining pre-exposure data as a baseline against which latter comparisons could be made (ORC/Jandl, P.C. 34, p. 36; Battle, P.C. 26C; Goldstein Ex. 75, p. 5). The initial exam requirement of the standard applies to both employees already exposed to benzene at or above the action level as well as to new or reassigned employees who may be exposed at or above the action level. In the first case, the blood data obtained is not, of course, a true non-exposed baseline, but may be useful at a later date should significant changes in blood values occur. In addition, since the standard requires that an initial laboratory exam be performed for all new or reassigned employees prior to their actual exposure to benzene at or above the action level, pre-exposure baseline data will gradually become increasingly available in the years subsequent to promulgation of this standard.

As an aid to physicians for comparing the results of the most recent blood testing with early data, the recordkeeping paragraph requires that, at the completion of the differential count, the slide of a peripheral blood smear be made permanent and stored for possible future reference. Retention of the slide allows the examining physician to directly compare the most recent findings with earlier ones, especially with respect to possible changes in morphology of the blood cells; counting data alone would not permit this type of comparison.

Doc Chemical suggested that kidney and liver function tests should be performed as a part of the preplacement examination (Tr. 2967). OSHA recognizes that these laboratory examinations may identify conditions indicative of an impaired ability to detoxify benzene. However, given the low permissible exposure limit mandated by this standard, OSHA concluded that requirement of these tests would be excessively burdensome. In this regard, Goldstein testified that he saw no value to incorporate these tests (Tr. 380).

PERIODIC EXAMINATION

The medical surveillance section requires that a brief updated history be taken semiannually at the time of one of the blood examinations. Battle suggested that such a history include queries as to exposure to drugs or chemicals as well as recent illnesses since the last history (P.C. 26C). Because these agents may act to adversely affect the hematopoietic system and may be reflected in the blood picture, OSHA agrees with these particular suggestions and has incorporated all of them into the requirements for the interval history.

The periodic laboratory tests are similar to those specified for the initial exam, except that serum bilirubin and reticulocyte assays are deleted since, as stated above, the two assays are required only in establishing a baseline. Comments received by OSHA concerning the frequency of interval blood examination ranged from quarterly (ORC, P.C. 34, p. 11; Tr. 3382, Tr. 270) to 3 times a year (Com. 28; Tr. 3328; Battle, P.C. 26C), semiannually. (Tr. 3133) to those who felt the frequency should in some manner vary according to the exposure level (e.g., Tr. 2965, 6-10, 6-37, 6-47, 6-64, 6-76, 41-3, 41-19). Dow expressed the view that the frequency should be the decision of the responsible physician and related to the medical condition of the employee and the environmental control of the specific work area (Tr. 2965). As detailed previously, there is insufficient information in the record to allow a precise determination of optimum testing frequency. No completely reliable information exists on the number of months which elapse between the first laboratory signs of a blood disorder and the onset of clinical disease or what effect a particular time delay has on therapeutic outcome. There is no reason for casualness about the testing frequency though, since even if a blood change were to progress steadily downhill, early detection and treatment may provide substantial benefit. It is OSHA's judgement, following examination of the various options available, that a six month interval between routine blood examinations is appropriate.

The standard contains a provision, applicable to both the initial, periodic and emergency exams, requiring that employees' abnormal test results must be referred to a hematologist for further evaluation should certain specified warning signs appear. This provision, which was not a part of the proposal, is in response to recommendations that: (1) A hematologist be included in some manner in the medical surveillance program (ORC/Jandl, P.C. 34); (2) specific quantitative guidelines including ranges (Ex. 6-60) be given physicians and; (3) that additional laboratory tests should be given if an abnormal count becomes manifest or significant changes occur relative to baseline values (NPRA, Tr. 3325-3331; AISI, Tr. 3263-64; ARCO, P.C. 32C). With respect to item (1), an industry participant testified that it was already the practice at his company to refer workers to hematology specialists if blood abnormalities are discovered (Joyner, Tr. 2286-87), and Dow remarked that they discussed abnormal blood findings of workers with the hematologists (Tr. 3021). Also Arco commented that, if abnormal conditions arise among its employees exposed to benzene, follow-up examinations are given and medical treatment performed until blood tests reveal values within the normal range (P.C. 32C). Also, Dow disclosed that, for its employee exposed to benzene, an increased medical surveillance program would result if conditions such as an undiagnosed anemia or leukocytosis were detected (Tr. 2984-85). Sakol has noted that ineffective and inappropriate medical treatment was given to workers apparently suffering hematological disorders from benzene exposure which ultimately evolved into a frank leukemia (Tr. 303). This incident further illustrates the need for a specialist's evaluation, if abnormalities are detected.

Again it should be emphasized that the question of what constitutes the earliest laboratory evidence of chronic benzene toxicity is not known with certainty (ORC/Battle, P.C. 26C, p. 3). It is known that the level of all 3 blood cell lines may be variously affected by exposure. However, examination of the benzene literature reveals that the hematological boundaries of what is considered "normal" have varied. More recent tabulation of normal *ranges* are those published by NIOSH (Ex. 2-2, p. 135); Wintrobe (Ex. 2-107, p. 1791, 1794, 1795) and those submitted by Jandl (ORC/Jandl, PC 34, p. 29). The normal range of hematological values presented generally agree quite well, especially those given by Jandl and Wintrobe. It is OSHA's conclusion that laboratory findings beyond these ranges must require additional evaluation by a blood specialist. In specifying the values beyond which findings are considered abnormal, OSHA utilized the values in the above citations to yield the widest range (often the range limits do not exactly coincide) in order to minimize the burden on the employer in borderline cases.

However, as illustrated by Goldstein, it may be possible for some individuals to experience a significant hematological response to benzene exposure and yet exhibit blood values within the range normal for the population as a whole (Goldstein, Tr. 356). To identify such workers for referral to a hematologist, the standard also includes as a trigger, an allowable deviation figure for certain indicies, relative to an earlier determined value. The advantage of this approach is that the individual acts as his own control. As early as 1926, Greenberg in recommending criteria for exclusion of workers from further benzene exposure, utilized as one of his criteria changes in the blood picture based on results from previous examination of individual employees. He recommended a determined value of a 25% decline in red and white cell levels. The red cell count, hemoglobin and platelet count are generally very stable indicies and usually do not vary in individual cases by more than ±10% from baseline values. OSHA believes that a deviation of ±15% or more in these indicies compared to the laboratory findings obtained in the most recent test is of sufficient concern to require the attention of a specialist. As white cell counts commonly exhibit greater variation, a percentage deviation trigger would be too restrictive and is therefore not included in the standard. In lieu of a percentage deviation, limits of normal ranges are prescribed.

EMERGENCY SITUATIONS

The emergency medical surveillance provisions reflect OSHA's concern for those employees normally subject to low average exposures but who, because of equipment breakdown or other causes, may be exposed to massive doses of benzene. These workers may be at a relatively high risk for developing adverse hematological effects.

If a worker is exposed to a massive release of benzene, the employer must provide for each individual so exposed a urinary phenol assay at the end of the work shift in which the emergency occurred. If the results of such a test corrected to a specific gravity of 1.024 are less than 75 mg/l, no further testing is required. However, if a urinary phenol result is greater than 75 mg/l, indicating an average exposure above 10 ppm (Watts, Ex. 2–109), then a complete blood count including a differential count must be performed as soon as practicable.

If the red blood cell or platelet count or hemoglobin differ more than ±15% from the most recent prior exam's findings, the worker's test results shall be referred to a hematologist for additional evaluation. Also, if the levels of the three formed elements lie outside of the prescribed ranges further evaluation by a hematologist shall be required.

The use of a urinary phenol assay to determine whether additional tests are indicated for workers exposed to high concentrations of benzene is supported by evidence in the record. The Sun Oil Company reported that special urinary phenol screening tests are prescribed for all persons when 10 ppm exposure levels are suspected (Ex. 77F).

For all types of medical examinations, the employer is required to provide the physician with certain information. This information includes a copy of the regulation, a description of the affected employee's duties as they relate to the employee's exposure, the results of the employee's exposure measurement, if any, or the employee's anticipated or estimated exposure level, a description of any personal protective equipment used or to be used, and information from previous medical examinations of the affected employee to the extent that it is not readily available to the physician. The purpose in making this information available to the physician is to aid in the evaluation of the employee's fitness to work in areas in which the exposure is at or greater than the action level. It should be noted that the standard does not require that a copy of the regulation be given to the physician for each employee. One copy would be sufficient, provided the employer informs the physician which employees are covered by this standard. Information that relates to individual employees or categories of employees (such as the description of job duties) need be transmitted to the physician only once, unless, for example, the duties change. Exposure measurements will be cumulative so that the results of each monitoring since the last examination are to be sent to the physician. However, since sampling may be done on a representative basis, the language of the standard requires the physician to be given either the employee's actual exposure measurements, if available, or the estimated level.

Several criticisms and suggested changes, discussed below, were also received by OSHA in regard to paragraph (i)(6) of the proposed medical surveillance requirements. In the final standard, the employer is required to obtain a written opinion from the examining physician containing: the physician's opinion as to whether the employee has any detected medical conditions which would place the employee at an increased risk of material impairment of health from exposure to benzene, the results of the medical tests performed, and any recommended limitations upon the employee's exposure to benzene and upon the use of protective clothing and equipment such as respirators. This written opinion must not reveal specific findings or diagnoses unrelated to occupational exposure, and a copy of the opinion must be provided to the affected employee.

Dow Chemical expressed the opinion that the notification requirements were overly burdensome and that the physician should be required to notify the employer and the employee only for the following reasons: Evidence of benzene toxicity, recommended limitations on worker exposure to benzene for any reason, or a health condition that precludes the use of protective clothing and equipment. (Tr. 2966). The Rubber Manufacturer's Association offered the comment that the physician's written opinion should be supplied to the employee only if requested (Ex. 6–59, p. 5). One commentator cautioned that because of malpractice considerations, a physician might be reluctant to sign such an opinion (National Paint and Coatings Association, Inc. (NPCA, Ex. 41–14, p. 15). Also because of malpractice considerations, the American Iron and Steel Institute recommend that this section be deleted (P.C. 36, p. 109). ORC suggested language changes for this physician's opinion requirement (ORC, P.C. 34, p. 11, Ex. 6–76, p. 11). The NPCA also warned that unless the physician's opinion paragraph was altered, denial of clearance for many prospective employees may result (Ex. 41–14, p. 15). There was also comment concerning with whom the doctor's opinion should be shared. ORC felt that any medical findings or diagnoses should be kept strictly confidential between the worker and the examining physician (P.C. 34, p. 11; Com. 76). Finally, a suggestion was received from industry that paragraph (i)(6)(ii) which limits the physician's opinion to specific findings and diagnoses related only to occupational exposure should be deleted on the basis that the employer has a right to know if there are any reasons related to work or not, that might have an adverse effect upon an employee or may possibly jeopardize fellow workers. (American Coke and Coal Chemicals Institute, Ex. 176, p. 9).

The purpose in requiring that specific findings or diagnoses unrelated to occupational exposure not be included in the written opinion is to encourage employees to submit to medical examination by removing the fear that employers may find out information about their physical condition that has no relation to occupational exposures.

The purpose in requiring the examining physician to supply a written opinion containing the above mentioned analyses is to provide the employer with a medical basis to aid in the determination of initial placement and to provide information on a continuing basis concerning whether or not the worker is at increased risk as a result of his/her benzene exposure. Requiring that the opinion be in written form will serve as an objective check that the employers have actually had the benefit of the information in making these determinations. Likewise, the requirement that the employee be provided with a copy of the physician's written opinion will assure that the employee is informed of the result of the medical exam and may take appropriate action. Comments suggesting that the physician's opinion be communicated only if there are problems or only upon request are not acceptable to OSHA on the basis that the employer has the ultimate responsibility to assure the protection of the worker's health. Where the findings are negative, transmittal of the doctor's opinion also provides necessary information to both parties that the employee's health has not been adversely affected and provides documentary evidence that the prescribed tests were performed and were evaluated.

Other aspects of paragraph (i) of the medical surveillance section are to be discussed under Recordkeeping, paragraph (l).

MANDATORY REMOVAL AND RATE RETENTION

Among the issues in the benzene rulemaking were whether OSHA should include a mandatory removal requirement—that is, a provision prohibiting the exposure of an employee to benzene if the employee would be placed at increased risk of material impairment to health because of such exposure, and whether OSHA should include a rate retention provision—that is, a provision requiring the transfer of such employee to another job or providing that removal for medical reasons should not result in loss of earnings or seniority status to the affected employee. These issues, as OSHA has previously stated (41 FR 46780), are related and must be addressed together. Both employee and industry participants expressed their views as to several aspects of these issues in prehearing comments, in testimony during the hearing and in post hearing arguments. Subsequent to the close of the record in this proceeding, however, OSHA conducted an informal public hearing on mandatory removal and rate retention for workers exposed to lead as part of the rulemaking proceeding on lead. Consideration of the critical issue of medical removal protection is being undertaken for several pending standards together. Once this consideration is completed, OSHA will consider the extent to which the conclusions on medical removal protection are appropriate for benzene and will propose the inclusion of those provisions in the benzene standard. The final standard published today, therefore, does not address the issues of mandatory removal and medical removal protection.

Employee Information and Training: Paragraph (j). The standard requires each employer to provide training to each of his employees who is or may be exposed to benzene. The need to train employees was not disputed by participants in the rulemaking proceeding. Some comments, however, suggested limiting the training to certain employees. One industry comment requested that workers in closed system operations be excluded from training (Com. 47). Testimony at the hearing, however, revealed that leakages in closed systems are not unusual occurrences and that employee exposure during leaks can reach high levels. One comment suggested the exclusion of workers in open system operations where the benzene content is 1 percent or less (Com. 47). However, as stated above, it has been established that a consistent predictable relationship between the amount of benzene in a mixture and exposure levels does not exist and this suggestion has, therefore, been rejected. Another comment suggested the exclusion of laboratory personnel from training requirements on the ground that laboratory personnel have a good understanding of the hazardous nature of benzene (Com. 70). A laboratory technician, however, testified that during his six years working with many toxic chemicals, including benzene, he received no training (Tr. 3493) and indeed neither he nor his coworkers were aware of the fact that there was an OSHA standard limiting exposure to benzene (Tr. 3492). Information and training are essential for the protection of an employee. Each employee can do much to protect himself if he is fully informed of the hazards in his workplace and the protective equipment he should use. Furthermore, each employee, who is fully informed of the obligations which the standard imposes upon the employer, can determine if he is working in a safe and healthful environment. For the reasons stated herein, OSHA believes that training must be provided to all employees in workplaces where benzene is present.

A trade association, while not disputing the need for training, suggested that employee training not be included in the benzene standard but await development of a comprehensive employee information and training standard based on the report of the Advisory Committee on Hazardous Materials Labeling and/or NIOSH's criteria document entitled "A Recommended Standard * * *".

An Identification System for Occupationally Hazardous Materials" (L.C. 14). OSHA deems it necessary for the protection of the health of employees to include the training requirements in the benzene standard at this time.

The standard specifies the contents of the training program. The information which must be imparted to the employee must include the nature of benzene related health problems, the necessity for exposure control, and the purposes of medical surveillance and respiratory protections. No participants objected to any of these items. Two participants suggested that employees should also be trained in "early symptom diagnosis to detect acute leukemia" (Ex. Dow Venable statement, pg. 3; Tr. 3192; Ex. 154; Tr. 2963-3056). These participants, however, testified that the most common manifestations of acute leukemia are fatigue and nosebleeds and that these symptoms are not specific to acute leukemia (Tr. 3193).

The signs and symptoms of benzene-induced diseases are described in considerable detail in Appendix A and Appendix C. Both the proposed standard and the final standard specifically require that the employees be informed, among other things, of the information contained in Appendix A and Appendix C. (In this connection, the employee should be instructed to report promptly the development of any of these symptoms which could be attributed to benzene exposure.) In view of the requirement to include the information contained in the appendices as part of the training program, an additional requirement on early symptom diagnosis would be redundant. Similarly, since Appendix B details the volatility of benzene, OSHA has not adopted NIOSH's suggestion (Baier statement, Ex. 84A p. 11) that the standard further emphasize training on the flammability of benzene.

In addition to informing employees, the standard requires that the employer make available to his employees a copy of the standard and its appendices. This requirement is intended to assure that employees understand their rights and duties under the standard.

A public participant suggested that employers be required to hold classes for the training of each employee. (Ex. 183 p. 5) OSHA has not included any specific requirement to this effect in the final standard, preferring to leave the manner of training up to the individual employers. Some employers will need to train individuals on a one-to-one basis, depending on the number of employees at a particular workplace or involved in particular operations. Employers with a few employees may find it burdensome or disruptive of work schedules to conduct training in class sessions.

The employer is also required to provide, upon request, all materials related to the training program to the Secretary and the Director. This requirement is intended to provide an objective check of compliance with the content requirements of the training program.

The standard requires that training of employees be conducted within ninety days of the effective date. Newly assigned employees must be trained at the time of initial assignment. The standard further requires that training be repeated annually. These requirements were not explicitly in the proposed standard; however, they are consistent with OSHA's policy on training (see, Coke Ovens Emissions Standard) and have been included in the final standard to assure that employees are trained promptly and receive continuing education essential to the protection of their health.

Signs and Labels: paragraph k. The requirements of the standard regarding the posting of warning signs and affixing caution labels remain primarily the same as those in the proposal with some difference in the legend of these signs and labels. These requirements are consistent with section 6(b)(7) of the Act which prescribes the use of labels or other appropriate forms of warning to apprise employees of the hazards to which they are exposed.

Signs. The standard requires the posting of warning signs in regulated areas. Some industry comments objected to this requirement on the ground that the training requirements of the standard are sufficient for informing employees of the hazards of benzene (Com. 28). In light of the serious nature of the hazard of exposure to benzene, OSHA does not agree. Not every employee will have completed his training before entering an area of operation where the exposure to benzene is over the permissible exposure level. Moreover, even trained employees will need to be reminded of the locations and the dangers of entering these areas. Additionally, other workers, such as employees of independent maintenance contractors who are authorized to enter particular regulated areas, need to be warned of the hazard and reminded to use protective equipment. OSHA, therefore, believes that both signs and training are necessary to adequately apprise employees of the hazards of benzene.

Other comments requested that laboratories be exempted from the sign requirements (Com. 70) on the ground that laboratory personnel are familiar with the hazards (Com. 20) and because there are already too many other hazardous substances in these work places (Com. 28) for which signs are posted (Com. 37). One comment expressed the view that there is no hazard from exposure to benzene in laboratories or other physical areas where the content of benzene in the mixture used is small (Com. 63, 59) or where benzene is used infrequently (Com. 70). As stated above, there is no established consistent predictable relationship between the amount of benzene and the exposure level, nor does infrequency of use eliminate the hazard. Signs are required by the standard only in regulated areas where, by definition, the exposure is above the permissible exposure limit. Furthermore, as stated above, OSHA believes that training in hazards alone is not sufficent to inform employees of the potential health hazards of entering or working in a regulated area and the need for the use of protective equipment.

Since warning the employees is necessary, the standard explicitly provides that the employer must assure that nothing which detracts from the required information appears near or on the sign. Laboratory and other employers who have other signs posted in the regulated area, therefore, have the responsibility for meeting this requirement.

Another comment pointed out that signs would be required even where benzene was in a closed system regardless of lack of employee exposure (Com. 65). This comment appears to overlook the fact that signs would be required only in those areas where employee exposure is over the permissible exposure limit and that, if there is no benzene exposure from the closed system, no sign is required. This same comment suggested that signs would be required where benzene was in a sealed container. Since there is no exposure to employees when benzene is in sealed containers, the standard exempts sealed intact containers from all requirements but the labelling and training requirements; hence signs would not be required in areas where benzene is present solely in sealed intact containers. Another comment suggested elimination of sign posting requirements for gasoline products (Com. 55). Since benzene is present in gasoline there is, of course, a benzene hazard attendant to exposure to gasoline. However, sampling data indicates that exposure to benzene is usually below 1 ppm, which would, in such cases, obviate the need for signs.

The standard specifies the wording of the signs. The purpose of this provision is to assure that the proper warning is given to employees. Some participants suggested that the word "Cancer" be replaced by "Leukemia Suspect Agent" (AISI, Princ. Wit, Ex. 156. Tab I, p. 34; PC 36; AISI brief, p. 108; PC 35; MCA brief pp. 103-5, Com. 39). In view of the fact that it has been established that benzene actually induces leukemia, OSHA does not agree that "Leukemia Suspect Agent" adequately informs employees of the hazard. For the same reason, OSHA has not adopted the suggestion of another comment that the sign merely warn of the presence of breathing zone vapors which "may" be hazardous, the need to wear respirators and that unauthorized persons should stay out (Com. 69). One comment pointed out that the sign requirements do not warn of other serious health hazards that may result from exposure to benzene. Addition of the various other health hazards would increase the wording on the sign with the likely result of detracting from the cancer warning. The reference to the most serious hazard is, in OSHA's view, a sufficient alert to employees that entering the regulated area may be hazardous. Some comments objected to the use of the term "Cancer Hazard" (Com. 26) on the ground that it would only serve to unduly alarm employees (Com. 28, 50). One comment suggested "Cancer Suspect Agent" be used to conform with 29 CFR 1910.1017 which regulates occupational exposures to vinyl chloride. Since exposure to benzene creates an increased risk of leukemia, and since leukemia is cancer of the blood, OSHA does not believe that the hazard is overstated by the use of the term "Cancer Hazard." Nor has OSHA adopted the suggestion that the word "Leukemia" replace the word "Cancer" (Com. 76). While the term "Leukemia" is more specific of the nature of the hazard, "Cancer" is the more familiar term and thus more readily alerts the employee.

Another comment (Com. 69) suggested that the word "Caution" be substituted for "Danger" to conform to 29 CFR 1910.145 (c)(1) and (2) which provide that danger signs should be used only where an immediate hazard exists, and caution signs used only to warn against potential hazards or to caution against unsafe practices. The suggestion was based on the view that OSHA failed to show that there exists a cause-effect relationship between employee exposure to benzene concentrations below 10 ppm. In view of the epidemiological evidence of the human carcinogenicity of benzene, OSHA feels that the use of the word "Danger" is appropriate. Furthermore, the word "Danger" is used to attract the attention of workers, to alert workers to the fact that they are in a hazardous area, i.e., an area where the permissible exposure limit is exceeded, and to emphasize the importance of the message to follow. The use of the word danger is consistent with the Coke Ovens Emissions Standard and other recent health standards.

Two comments objected to the legend "Respirator Required" pointing out that respirators are not always required in regulated areas. The reason given was that, although the concentration within a regulated area will be above 1 ppm, the employee would not have to use a respirator if his 8-hour TWA were below 1 ppm and his 15 minute ceiling is below 5 ppm (Com. 46, 59). OSHA recognizes that some employees entering regulated areas may not be exposed above the permissible exposure limit. It is likely, however, that there will be many employees who work in the regulated areas for such periods of time or at such levels of exposure that they would without respirators be exposed to well above the permissible exposure limit. For this latter group, respirators would be required to reduce their exposure levels. To assure that these employees are adequately protected, it is necessary that the sign alert them to the need to wear respirators. Moreover, since even persons who are in a regulated area occasionally and are not exposed over the permissible exposure limits may face an increased risk of leukemia, OSHA does not believe that the respirator warning is such an overstatement as to necessitate elimination of that warning from the sign.

One comment pointed out that the warning sign ignores the flammability of benzene (Com. 50). NIOSH also recommended that the potential danger for fire be displayed on signs with equal prominence with the toxicological hazards. (Tr. 755) OSHA agrees. Benzene, as has been stated, is a highly flammable liquid. Accordingly, in the final standard, OSHA has added a requirement that the signs also contain the legend "Flammable—No Smoking" with equal prominence to the legend "Cancer Hazard."

Labels. The standard requires the use of caution labels on all containers of products containing any amount of benzene. This requirement imposes upon the employer the obligation to assure that all such containers within his workplace are at all times properly labeled. Some industry comments requested that OSHA define "container" (Com. 46, 55, 65) in terms of sizes and types of containers (Com. 53; Com. 72). Participants also appeared to be confused as to whether storage tanks, pipelines, tank cars and tank trucks are such containers as would be require labeling (Com. 60). The purpose of the labeling requirement is to assure that, wherever benzene is present in any quantity, employees are apprised of the hazard. The size or type of the container or its use as a storage vessel is, therefore, not material to this objective. In imposing the labeling requirement, however, OSHA does not intend to include pipelines, or to include trucks or other vessels transporting benzene products in sealed containers. At many points of pipelines, there is no employee exposure nor can the point at which benzene enters a product stream often be determined. Benzene in liquid mixtures, other than in gasoline, is often transported in closed containers, in which case labeling of the container is sufficient without a need for labeling the transport vehicle. Where benzene products not in containers are transported in barges or other tank vessels, such as tankers and tank trucks, it is, of course, necessary that the vessel be labeled. With regard to gasoline, the standard does not cover activities past the bulk terminal and, therefore, gasoline trucks would not require labelling. In view of the above, the standard specifically excludes from the labeling requirements pipelines and any transport vessels or vehicles where the benzene product is transported in sealed containers.

Some participants suggested that the labeling requirements apply only to containers which have liquids with greater than 1% benzene (Com. 53, 72). It was also suggested that labels be affixed only to containers from which exposure to benzene levels above the permissible exposure limit might be reasonably expected to occur so as to exclude a requirement to label finished products which might contain trace amounts of benzene (Com. 59). Since there is no known safe level of exposure to benzene, OSHA feels that, for employees to be adequately apprised of the hazard, all containers must be labeled in the same manner.

The standard requires that the caution labels remain affixed when benzene products leave the employer's workplace. Some comments questioned OSHA's jurisdiction to impose such a requirement (Com. 50, 55, 59). The purpose of this requirement is to assure that all employees, not only those of a particular employer, are apprised of the hazardous nature of benzene exposure. It is OSHA's view that informing employees of the hazards to which they are exposed is an important element in reducing occupational disease and injury and one of the significant purposes of the Occupational Safety and Health Act. Section 6(b)(7) of the Act, which explicitly provides for regulation requiring the use of labels or other appropriate forms of warning to apprise employees of the hazards to which they are exposed, is broadly drawn. This section does not limit the employer's obligation of informing employees of hazardous conditions to the employer's own employee. When an employer manufactures, formulates or sells a product containing a toxic substance, that employer is exposing not only his own employees but also the employees of other employers involved in handling, transporting or using the product. The extent of the obligation to inform should be commensurate with the extent of the exposure. This is especially true where the manufacturer, formulator or seller will in many cases be the only employer capable, through his unique knowledge, of providing the information needed for protection of employees. A narrower reading of the statutory authority would defeat the protective purposes of the Act by withholding from employees down the line who come into any contact with the product, adequate information as to the hazard. Furthermore, the use of the labels will alert other employers, who utilize or handle the product and who would not otherwise know of the presence of benzene in their workplace, of their obligation to comply with the standard. OSHA, therefore, feels that this requirement is necessary and appropriate to effectuate the purposes of the Act.

The standard prescribes the legend that must be included on the label. This is to assure that employees are alerted to the fact that they are handling benzene and to the hazard involved. The record evidence establishes that a variety of code names have been used for benzene (Ex 61, Sakol) and that employers (Tr. 3383; 3516), workers (Tr. 3402) and physicians (Tr. 305-6) often do not know that the product contains benzene. Some participants suggested that labels state the percentage of benzene in the product (Com. 58, 21, 59). In order to keep the label information to a minimum and since there is no consistent predictable relationship between the amount of benzene in the product and the percentage level, the final standard does not require any such statement on the label.

Recordkeeping and reporting. The provisions for recordkeeping, reporting, availability of records and the transfer of records are similar to those in the proposal. Such changes as have been made in the final standard are in response to the request of public participants that OSHA reduce the burden imposed by these types of requirements. These provisions implement the requirements of section 8(c)(1) and (3) of the Act, and are consistent with OSHA's general policy concerning records and reports.

Records: paragraph (1) The standard requires a limited amount of recordkeeping. Employers must maintain exposure measurement records and medical records.

The need for keeping these records was generally accepted by the participants in the benzene rulemaking. The common purposes of these recordkeeping requirements is to assure the employer's compliance with the control measures designed for the protection of his employees, and to collect data vital to epidemiological and diagnostic investigations in order to resolve such questions as dose-response relationships in blood diseases caused by exposure to benzene. Furthermore, these records are useful for the employer by enabling him to identify areas of his operations where there is or may be a problem. Exposure measurement records assist in pinpointing those processes or operations which require additional efforts to reduce exposure to benzene. Medical records provide an awareness of the incidence of industrial illness in the employer's establishment and assist the employer to focus on any aspect of his operation which may materially impair the health of his employees. The required records are also useful to the employee: Exposure records assure employees that their exposure to benzene is being monitored and also inform them of the hazards to which they are exposed; medical records assist their examining physician to accurately diagnose and treat any health problem and assure proper evaluation of their health.

The standard specifies the information which must be included in the record. The exposure measurement record must include information which accurately reflects the exposure of each individual employee. The medical record must include the work history of each individual employee and information accurately reflecting his health. A few participants disputed the need for the content of some of these records. One comment suggested deletion of the requirement in paragraph (l)(1)(i)(b) to include as part of the exposure record a description of sampling on the ground that it is the same requirement as in paragraph (l)(1)(i)(a) for a description of the sampling procedure used to determine representative employee exposure. (Com. 76) This participant has misunderstood these requirements. The "description of sampling" in subparagraph (b) refers to the type of equipment used for collecting the concentrations of benzene, the rate at which these samples were collected and the minimum period during which the collection took place. The "sampling procedure" referred to in subparagraph (a) refers to the criteria or method used by the employer to determine that the samples he has taken are representative of the particular employee's exposure. Although these paragraphs are directed to different requirements, OSHA has eliminated from subparagraph (a) of this provision the word "sampling" in order to avoid such confusion on the part of other employers.

Another comment expressed the view that requiring employers to include in the record all of the information provided to the examining physician is an undue administrative burden (Com. 59). Where the information is communicated in writing to the doctor who will examine the employee, the employer can meet his requirement by simply retaining in his files a copy of this communication. Where, however, the information is not furnished in writing, it is essential to record it to assure that the examining physician has been advised of all the information needed for evaluation of the employee's health. Moreover, since this information includes exposure conditions and symptoms, the information is most useful not only to the current examining physician but also to a physician who years later evaluates the health of the employee. OSHA, therefore, has retained in the standard the requirement to record the information provided to the examining physician.

An industry comment pointed out that it is not necessary for the company physicians, who maintain the medical records on behalf of the employer, to retain a copy of the regulation and its appendices as part of the record of each individual employee, but that retention of one master copy in the files would be sufficient if there is a reference to it in each employee's file (Com. 46). OSHA agrees that retention of the regulation and appendices in each employee's file unnecessarily increases the space needed for the records and, accordingly, has given employers the option in the final standard of either including the regulation and appendices in each employee's file or of retaining a master copy of these documents and referencing them in each employee's medical file.

Some industry participants testified that their facilities would need additional record clerks to maintain the information in the required records (TR 3165-3171). In view of the reduction in the final standard of the number of employees for whom exposure and medical monitoring must be conducted and reduction at the frequency rate suggested by the proposal, the need for additional recordkeeping personnel is significantly reduced. Furthermore, since the standard does not require the use of any particular form for maintaining the prescribed information, employers may incorporate these records into already existing recordkeeping systems, some of which are computerized, and thus further minimize the need for additional personnel.

The standard requires that the exposure measurement records and medical surveillance records be maintained by the employer for at least 40 years or for the duration of employment plus twenty years, whichever is longer. Several participants objected to the length of this retention period. One comment expressed the view that the retention period was unnecessarily burdensome where the only exposure to benzene is from liquid products containing small amounts of benzene (Com. 55). Both because there is no known safe level for exposure to benzene and because there is no direct linear correlation between the percentage of benzene in a liquid mixture and exposure levels, OSHA feels that there is no basis for different retention requirements based on the amount of benzene in liquids. Moreover, in view of the provision for an action level in the final standard, the number of exposure measurements and medical examinations which must be recorded will be greatly minimized where the small amounts of benzene actually do result in exposures below the action level.

Another comment suggested that the retention period for the medical surveillance records is unreasonably long (Com. 59). OSHA is aware of the burden of keeping both the medical and the exposure records for this length of time. However, OSHA feels that it is essential that the records be retained for this period. The record evidence establishes that the latency period for benzene induced leukemia is unknown (Ex. 2B-283, p. 4) and that, as in the case of other chemical carcinogens, the symptoms of carcinogenesis—in this case, benzene induced leukemia—may not appear for many years after the initial exposure (Tr. 385, 391, 410-11, 424). The retention period of forty years has, therefore, been selected to assure that the entire latency period would be covered. Furthermore, as stated in OSHA's standard for Coke Ovens Emissions (41 FR 46782), OSHA believes that it is essential to scientific investigations that exposure and medical records be maintained for the same length of time, and that the retention period for these records be at least forty years.

One comment suggested that independent contractors, who customarily hire employees only for temporary work at a particular facility, be allowed to send the medical records to the Director of NIOSH upon expiration of the employee's employment (Com. 63). As is noted below, it is OSHA's policy for all employers that the records be maintained at the place of employment so long as it is in existence.

The standard does not require that the employer, on any regular basis, send copies of his records to OSHA. The employer is, however, required upon request to make the records available for examination and copying to designated representatives of OSHA and NIOSH. This requirement is necessary both for compliance purposes and for scientific investigations. Records of required exposure monitoring are, in addition, to be made available, for examination and copying, to current employees and their representatives; and exposure monitoring records indicating their own exposure must be made available to former employees or their designated representatives. One participant suggested deletion of the reference to the term "or their designated representative" (Com. 48) thus restricting access to appropriate exposure records only to the current or former employee himself. OSHA is of the view that denying access to the records of individual exposure by designated representatives could result in a denial of access to the information by the employee where he is incapacitated and unable to inspect the records or simply not able to understand them. This would defeat the purpose of the availability provision which is to assure current employees that their exposure is properly monitored and assure former employees of access to information necessary for the continued protection of their health. Furthermore, the Act recognizes (sections 2(b)(13); 8(c)(3); 8(f)(1)) the legitimate role of employee representatives in occupational safety and health. One industry comment suggested that employee representatives be formally designated in writing and that access to records also be requested in writing (Com. 15). OSHA does not deem it necessary to include such a requirement in the standard, but employers may establish procedures for access to records so long as the procedures do not restrict the employees' and former employees' to access.

The standard also requires that medical records also be made available, upon request, for examination and copying to the designated physicians or representative of both current and former employees. Some comments questioned the fact that the standard enables not only the employer but also the employee and his representative and OSHA and NIOSH to have access to medical records without specifying confidentiality or otherwise limiting circulation of the information (Com. 26, 48). OSHA recognizes that a physician's records may contain a wide range of personal and medical information deemed to be confidential or private. For this reason, the standard limits the contents of the medical record to such information as is related to benzene exposure. Indeed, the standard requires in paragraph (i) (6) (ii) that the employer advise the physician that the physician's opinion, which becomes a part of the medical record, should not reveal findings or diagnoses unrelated to occupational exposure. The need of the employer, the employee and OSHA and NIOSH to have access to this information has already been thoroughly discussed. Disclosure of the information to other persons is, of course, subject to protective requirements of any applicable laws or regulations.

To assure that the records will be preserved for the required retention period of forty years, the standard requires an employer, who ceases to do business, to transfer his records to his successor and, in the event that there is no successor, to transfer the records to the Director of NIOSH. One industry comment suggested that this section be deleted. (Com. 59). Allowing employers to dispose, at will, of the records of exposure measurements and medical surveillance would result, more often than not, in destruction of these records thus depriving employees of information necessary for evaluation of their health, as well as depriving NIOSH of information valuable to its scientific investigations.

Another participant suggested elimination of the requirement that the records be transferred to NIOSH by registered mail. (Com. 68). OSHA has adopted this suggestion and the permanent standard eliminates the requirement to use registered mail.

Reports. Paragraphs (d)(3), (e)(5), (f)(2), (i)(4)(i). The standard imposes notification requirements upon employers. This requirement is discussed above under *Regulated Areas.*

The standard also obligates the employer to give certain information to employees. The employer must notify each employee of the exposure measurement representative of his exposure. This notification must be made regardless of what the exposure level of the employee is. Two comments suggested that availability of exposure records was sufficient to inform employees of the results of their monitoring (Com. 46 Com. 53). Several comments suggested that the notification be required only where the employees' exposure is above the permissible exposure level (Com. 10, 15, 21, 41, 46, 53, 59) and that exposures below the permissible level be available to the employee upon request (Com. 10). OSHA believes that, consistent with section 8(c)(3) of the Act, every employee has the right to know what his exposure level is and whether it is above or below the permissible exposure level. Moreover, since the permissible exposure level is a feasibility level and not a "safe" level, the employee must know, for proper evaluation of his health by a physician in the present and future, the level of benzene to which he is exposed. Accordingly, the suggestions to modify the employee notification requirements have not been adopted.

The standard further requires that, where the employee's exposure is over the permissible exposure level, the employer must also state in the notification what corrective action the employer is going to take to reduce the exposure level. This is necessary to assure employees that the employer is making every effort to furnish them with a safe and healthful work environment, and implements section 8(c)(3) of the Act.

Notifications to employees of their exposure levels must be made in writing. Several comments objected to this requirement. Some comments suggested that the employer be permitted to notify employees by posting of a notice on the workplace bulletin board (Com. 15, 49, 59), thus relieving the employer of the burden of notifying each and every employee. It is OSHA's view that direct notification to each employee is necessary in order to assure that the employee, as required by section 8(c)(3) of the Act, is apprised of all hazards to which the employee is exposed.

The standard further requires that the notification be within five working days after receipt of the employer's measurement results. Several industry comments requested a change in the five day requirement so as to allow additional time for employers to comply. (Com. 47, 49, 54, 59). OSHA feels that, under ordinary circumstances, the five day limit fulfills the statutory requirement for promptness (section 8(c)(3)) yet allows employers sufficient time for the written notification.

The standard further obligates the employer to provide the employee with a copy of the examining physician's written opinion. One industry comment pointed out the large volume of letters per year which this notification would entail (Com. 53). However, in view of the necessity for each employee to have an evaluation of his health, OSHA is of the view that furnishing each employee with a copy of the physician's letter is not unreasonably burdensome.

Written compliance programs are also required by the standard. These are discussed above under *Methods of Compliance.* The employer is required to submit his compliance program, on request, to OSHA and NIOSH. Additionally, he must make this program available at the worksite for examination by OSHA and NIOSH and by his employees or their authorized representatives. One participant suggested that employers be required to give to each employee a notice of availability of the compliance program rather than merely making it available to the employee or his representative. OSHA feels that this would place an unnecessary burden upon the employer, and has not incorporated the suggestion.

Observation of monitoring: paragraph (m). The final standard requires that the employer provide affected employees or their designated representatives with an opportunity to observe the measuring of employee exposures. This opportunity is specifically required by section 8(c)(3) of the Act. The standard also sets forth certain procedures which must be complied with in connection with the observation of the monitoring. These procedures are designed to assure that the right to observe is meaningful and that the health and safety of the observer is protected during the observation.

A few industry participants requested that the observation of monitoring provision be deleted from the standard (Com. 65). Union participants objected to any limitations on this requirement (Grospiron statement, Ex. 11.A, p. 12). The arguments by industry were several: (1) The requirement fails to limit the number of employees who can observe the monitoring; (2) it increases the employer's costs of monitoring; (3) employees lose time from their jobs; (4) it exposes employees to additional hazards; (5) it increases the liability potential for non-employees who are injured while observing; and (6) will result in labor/management confrontations (Com. 65, 70: L.C. 4). In view of the statutory mandate that "regulations shall provide employees or their representatives with an opportunity to observe such monitoring or measuring," OSHA has not adopted the suggestions to delete this provision.

One comment suggested that a one-time limit on observation of monitoring be established for each employee (Com. 60). OSHA has not adopted this suggestion since the statute does not impose any such limit and it does not appear to be reasonable. For example, a requirement to this effect would prohibit an employee or his representative who has observed the original monitoring to observe the redetermination of exposure which is required by the standard after a process or control change.

EFFECTIVE DATE

In order to assure that affected employers and employees will be informed of the existence of the provisions and terms of this standard, and that employers and employees are given an opportunity to familiarize themselves with any new requirements, the effective date of this standard will be March 13, 1978. Until such date, 29 CFR 1910.1000, Table Z-2 will continue to apply to those operations covered by the new 29 CFR 1910.1028. Subsequent to that date, 29 CFR 1910.1000, Table Z-2 will continue to apply to all operations exempted under the new 29 CFR 1910.1028.

APPENDICES

Three appendices have been included in this permanent standard. These appendices have been included primarily for information purposes. None of the statements contained therein should be construed as establishing a mandatory requirement not otherwise imposed by the standard or as detracting from an obligation which the standard does impose.

The information contained in Appendices A and B is designed to aid the employer is complying with requirements of the standard. The information in Appendix C primarily provides information needed by the physician to evaluate the results of the medical examination. Appendix C also lists other types of examinations, not required by the standard, which may help the physician in making an accurate assessment as to whether continued exposure to benzene will place the employee at an increased risk. It should be noted that paragraph (j)(1) (i) and (iv) specifically require that the information contained in Appendix A and C be provided to employees as part of their information and training program.

Some suggestions were made by participants regarding the contents of these appendices. One participant suggested that Appendix A, Part III-A include the requirement as to frequency of change of respirator cartridges and canisters (Com. 8). OSHA has not adopted this suggestion since the requirement is clearly set forth in the standard, and the employer is required as part of the training program to give each employee a copy of the standard. Another participant (Com. 59) suggested that Appendix A, I, B2 state that eye contact and repeated skin contact "be avoided" instead of "prohibited." OASH does not agree since the standard prohibits such contact. This same participant suggested that the last sentence of Appendix A, IV, A be changed to indicate that medical assistance should be sought only if irritation of eye or face persists. This limitation has not been added for the reason that OSHA feels that the more conservative advice imparted by the proposed paragraph better alerts the employees to the possible hazards of this type of exposure. The third suggestion of this participant was that Appendix A state that "medical assistance" be sought rather than that "a doctor" be called when large quantities of benzene are breathed. A modification of this suggestion has been adopted and it is now recommended that medical assistance or a doctor be called for as soon as possible.

VII. AUTHORITY

This document was prepared under the direction of Eula Bingham, Assistant Secretary of Labor for Occupational Safety and Health, U.S. Department of Labor, 200 Constitution Avenue NW., Washington, D.C. 20210.

Accordingly, pursuant to sections 4(b)(2), 6(b), 6(c) and 8(c) of the Occupational Safety and Health Act of 1970 (84 Stat. 1592, 1593, 1596, 1596, 29 U.S.C. 653, 655, 657), the specific statutes referred to in section 4(b)(2), Secretary of Labor's Order No. 8-76 (41 FR 25059), and 29 CFR Part 1911, Part 1910 of Title 29, Code of Federal Regulations, is hereby amended by deleting the emergency temporary standard for occupational exposure to benzene at § 1910.1028 and adding a new permanent standard for occupational exposure to benzene as § 1910.1028, and by amending Table Z-2 of § 1910.1000. In addition, pursuant to section 4(b)(2) of the Act (84 Stat. 1592; 29 U.S.C. 653), OSHA has determined that this new standard in § 1910.1028 is more effective than the corresponding standards now in Subpart B of Part 1910, and in Parts 1915, 1916, 1917, 1918, and 1926 of Title 29, Code of Federal Regulations. Therefore, these corresponding standards are superseded by this new § 1910.1028. This determination, and the application of the new standard to the maritime and construction industries, are implemented by adding a new paragraph (c) to § 1910.19 and by revoking § 1910.20.

Signed at Washington, D.C., this 31st day of January, 1978. These amendments are effective on March 13, 1978.

EULA BINGHAM,
Assistant Secretary of Labor.

Part 1910 of Title 29 of the Code of Federal Regulations is hereby amended as follows:

1. A new paragraph (c) is added to § 1910.19, to read as follows:

§ 1910.19 *Special provisions for air contaminants.*

* * * * *

(c) Section 1910.1028 shall apply to the exposure of every employee to benzene in every employment and place of employment covered by §§ 1910.12, 1910.13, 1910.14, 1910.15, or § 1910.16, in lieu of any different standard on exposure to benzene which would otherwise be applicable by virtue of any of those sections.

§ 1910.20 [Revoked]

2. Section 1910.20 is revoked.

§ 1910.1000 [Amended]

3. Table Z-2 of § 1910.1000 is amended by adding a footnote following the words "Benzene (Z37.40-1969)," and by adding the following below Table Z-2:

4. Section 1910.1028 is revised to read as follows:

§ 1910.1028 Benzene.

(a) *Scope and application.* (1) This section applies to each place of employment where benzene is produced, reacted, released, packaged, repackaged, stored, transported, handled, or used.

(2) This section does not apply to:

(i) The storage, transportation, distribution, dispensing, sale or use as fuel of gasoline, motor fuels, or other fuels subsequent to discharge from bulk terminals; or

---

[1] Occupational exposures to benzene are subject to the requirements of § 1910.1028 except as specifically exempted by § 1910.1028(a)(2). Exposures exempted by § 1910.1028(a)(2) are covered by this § 1910.1000

(ii) The storage, transportation, distribution or sale of benzene in intact containers sealed in such a manner as to contain benzene vapors or liquid, except for the requirements of paragraph (k) (2), (3), (4), and (5), and paragraph (j) of this section.

(b) *Definitions.* "Action level" means an airborne concentration of benzene of 0.5 ppm, averaged over an 8-hour work day.

"Assistant Secretary" means the Assistant Secretary of Labor for Occupational Safety and Health, U.S. Department of Labor, or designee.

"Authorized person" means any person required by his duties to enter a regulated area and authorized to do so by his employer, by this section or by the Occupational Safety and Health Act of 1970. "Authorized person" includes a representative of employees who is designated to observe monitoring and measuring procedures under paragraph (m) of this section.

"Benzene" (C_6H_6) (CAS Registry No. 00071432) means solid, liquefied or gaseous benzene. It includes mixtures of liquids containing benzene and the vapors released by these liquids.

"Bulk terminal" means a facility which is used for the storage and distribution of gasoline, motor fuels or other fuels and which receives its petroleum products by pipeline, barge or marine tanker.

"Director" means the Director of the National Institute for Occupational Safety and Health, U.S. Department of Health, Education, and Welfare, or designee.

"Emergency" means any occurrence such as, but not limited to, equipment failure, rupture of containers, or failure of control equipment which may, or does, result in a massive release of benzene.

"OSHA Area Office" means the office of the Occupational Safety and Health Administration having jurisdiction over the geographic area where the affected workplace is located.

(c) *Permissible exposure limits*—(1) *Inhalation*—(i) *Time-weighted average limit (TWA).* The employer shall assure that no employee is exposed to an airborne concentration of benzene in excess of 1 part benzene per million parts of air (1 ppm) as an 8-hour time-weighted average.

(ii) *Ceiling limit.* The employer shall assure that no employee is exposed to an airborne concentration of benzene in excess of 5 ppm as averaged over any 15 minute period.

(2) *Dermal and eye exposure limit.* The employer shall assure that no employee is exposed to eye contact with liquid benzene; or to skin contact with liquid benzene, unless the employer can establish that the skin contact is an isolated instance.

(d) *Regulated areas.* (1) the employer shall establish, within each place of employment, regulated areas where benzene concentrations are in excess of the permissible airborne exposure limit.

(2) The employer shall limit access to regulated areas to authorized persons.

(3) *Notification of regulated areas.* Within 30 days following the establishment of a regulated area, the employer shall report the following information to the OSHA Area Office:

(i) The address of each establishment which has one or more regulated areas;

(ii) The locations, within the establishment, of each regulated area;

(iii) A brief description of each process or operation which results in employee exposure to benzene in regulated areas; and

(iv) The number of employees engaged in each process or operation within each regulated area which results in exposure to benzene, and an estimate of the frequency and degree of exposure within each regulated area.

(e) *Exposure monitoring and measurements.*—(1) *General.* (i) Determinations of airborne exposure levels shall be made from air samples that are representative of each employee's exposure to benzene over an eight (8) hour period.

(ii) For the purposes of this section, employee exposure is that exposure which could occur if the employee were not using a respirator.

(2) *Initial monitoring.* (i) Each employer, who has a place of employment where benzene is produced, reacted, released, packaged, repackaged, stored, transported, handled or used shall monitor each of these workplaces and work operations to accurately determine the airborne concentrations of benzene to which employees may be exposed.

(ii) The initial monitoring required under paragraph (e)(2)(i) of this section shall be conducted and the results obtained within 30 days of the effective date of this section. Where the employer has monitored after January 4, 1977 and the monitoring satisfies the accuracy requirements of paragraph (e)(6) of the section, the employer may rely on such earlier monitoring to satisfy the requirements of paragraph (e)(2)(i) of this section, unless there has been a production, process, personnel or control change which may have resulted in new or additional exposures to benzene or the employer has any other reason to suspect a change which may have resulted in new or additional exposures to benzene, and provided that the employer maintains a record of the monitoring in accordance with paragraph (l)(1) and notifies each employee in accordance with paragraph (e)(5).

(3) *Frequency.*—(i) *Measurements below the action level.* If the measurements conducted under paragraph (e)(2)(i) of this section reveal employee exposure to be below the action level, the measurements need not be repeated, except as otherwise provided in paragraph (e)(4) of this section.

(ii) *Measurements above the action level.* If the measurements reveal employee exposure to be in excess of the action level, but below the permissible exposure limit, the employer shall repeat the monitoring at least quarterly. The employer shall continue these quarterly measurements until at least two consecutive measurements, taken at least seven (7) days apart, are below the action level, and thereafter the employer may discontinue monitoring, except as provided in paragraph (e)(4) of this section.

(iii) *Measurements above the permissible exposure limit.* If the measurements reveal employee exposure to be in excess of the permissible exposure limits, the employer shall repeat the measurements at least monthly. The employer shall continue these monthly measurements until at least two consecutive measurements, taken at least seven (7) days apart, are below the permissible exposure limits, and thereafter the employer shall monitor at least quarterly.

(4) *Additional monitoring.* Whenever there has been a production, process, personnel or control change which may result in new or additional exposure to benzene or whenever the employer has any other reason to suspect a change which may result in new or additional exposures to benzene, such as spills, leaks, ruptures, or breakdowns, the employer shall repeat the monitoring which is required by paragraph (e)(2)(i) of this section.

(5) *Employee notification.* (i) Within 5 working days after the receipt of the measurement results, the employer shall notify each employee in writing of the exposure measurements which represent that employee's exposures.

(ii) Where the results indicate that the employee's exposure exceeds the permissible exposure limits, the notification shall also include the corrective action being taken or to be taken by the employer to reduce exposure to or below the permissible exposure limit.

(6) *Accuracy of measurement.* The employer shall use a method of measurement which has an accuracy, to a confidence level of 95 percent, of not less than plus or minus 25 percent for concentrations of benzene greater than or equal to 1 ppm.

(f) *Methods of compliance.*—(1) *Priority of compliance methods.* The employer shall institute engineering and work practice controls to reduce and maintain employee exposures to benzene at or below the permissible exposure limits, except to the extent that the employer establishes that these controls are not feasible. Where feasible engineering and work practice controls are not sufficient to reduce employee exposure to or below the permissible exposure limits, the employer shall nonetheless use them to reduce exposures to the lowest level achievable by these controls, and shall supplement them by the use of respiratory protection.

(2) *Compliance program.* (i) The employer shall establish and implement a written program to reduce exposures to or below the permissible exposure limits solely by means of engineering and work practice controls required by paragraph (f)(1) of this section.

(ii) The written program shall include a schedule for development and implementation of the engineering and work practice controls. These plans shall be revised at least every six months to reflect the current status of the program.

(iii) Written plans for these compliance programs shall be submitted, upon request, to the Assistant Secretary and the Director, and shall be available at the worksite for examination and copying by the Assistant Secretary, the Director, and the employees or their authorized representatives.

(iv) The employer shall institute and maintain at least the controls described in his most recent written compliance program.

(g) *Respiratory protection.*—(1) *General.* Where respiratory protection is required under this section, the employer shall select, provide and assure the use of respirators. Respirators shall be used in the following circumstances:

(i) During the time period necessary to install or implement feasible engineering and work practice controls;

(ii) During maintenance and repair activities in which engineering and work practice controls are not feasible;

(iii) In work situations where feasible engineering and work practice controls are not yet sufficient to reduce exposure to or below the permissible exposure limits; or

(iv) In emergencies.

(2) *Respirator selection.* (i) Where respiratory protection is required under this section, the employer shall select and provide, at no cost to the employee, the appropriate respirator from Table 1 below and shall assure that the employee uses the respirator provided.

(ii) The employer shall select respirators from among those approved by the National Institute for Occupational Safety and Health under the program of 30 CFR Part 11.

(3) *Respirator program.* The employer shall institute a respiratory protection program in a accordance with § 1910.134(b), (d), (e) and (f).

(4) *Respirator use.* (i) Where air purifying respirators (cartridge, canister, or gas mask) are used, the employer shall, except as provided in paragraph (g)(4)(ii) of this section, replace the air-purifying canisters or cartridges prior to the expiration of their service life or the end of shift in which they are first used, whichever occurs first.

(ii) Where a cartridge or canister of an air purifying respirator has an end of service life indicator certified by NIOSH for benzene, the employer may permit its use until such time as the indicator shows the end of service life.

(iii) The employer shall assure that the respirator issued to the employee exhibits minimum facepiece leakage and that the respirator is properly fitted

(iv) The employer shall allow each employee who wears a respirator to wash his or her face and respirator facepiece to prevent skin irritation association with respirator use.

TABLE I.—*Respiratory protection for benzene*

Airborne concentration of benzene or condition of use	Respirator type
(a) Less than or equal to 10 p/m.	(1) Any chemical cartridge respirator with organic vapor cartridge; or
	(2) Any supplied air respirator.
(b) Less than or equal to 50 p/m.	(1) Any chemical cartridge respirator with organic vapor cartridge and full facepiece;
	(2) Any supplied air respirator with full facepiece;
	(3) Any organic vapor gas mask; or
	(4) Any self-contained breathing apparatus with full facepiece.
(c) Less than or equal to 1,000 p/m.	(1) Supplied air respirator with half mask in positive pressure mode.
(d) Less than or equal to 2,000 p/m.	( ) Supplied air respirator with full facepiece, helmet, or hood, in positive pressure mode.
(e) Less than or equal to 10,000 p/m.	(1) Supplied air respirator and auxiliary self-contained facepiece in positive pressure mode; or
	(2) Open circuit self-contained breathing apparatus with full facepiece in positive pressure mode.
(f) Escape	(1) Any organic vapor gas mask; or
	(2) Any self-contained breathing apparatus with full facepiece.

(h) *Protective clothing and equipment.* Where eye or dermal exposure may occur, the employer shall provide, at no cost to the employee, and assure that the employee wears impermeable protective clothing and equipment to protect the area of the body which may come in contact with liquid benzene. Eye and face protection shall meet the requirements of § 1910.133 of this Part.

(i) *Medical Surveillance*—(1) *General.* (i) The employer shall make available a medical surveillance program for employees who are or may be exposed to benzene at or above the action level and employees who are subjected to an emergency.

(ii) The employer shall assure that all medical examinations and procedures are performed by or under the supervision of a licensed physician, and provided without cost to the employee.

(2) *Initial examinations.* (i) Within thirty days of the effective date of this section, or before the time of initial assignment, the employer shall provide each employee who is or may be exposed to benzene at or above the action level with a medical examination, including at least the following elements:

(a) A history which includes past work exposure to benzene or any other hematologic toxins; a family history of blood dyscrasias including hematological neoplasms; a history of blood dyscrasias including genetically related hemoglobin alterations, bleeding abnormalities, abnormal function of formed blood elements; a history of renal or liver dysfunction; a history of drugs routinely taken, alcoholic intake and systemic infections; a history of exposure to marrow toxins outside of the current work situation, including volatile cleaning agents and insecticides;

(b) Laboratory tests, including a complete blood count with red cell count, white cell count with differential, platelet count, hematocrit, hemoglobin and red cell indices (MCV, MCH, MCHC), serum bilirubin and reticulocyte count; and

(c) Additional tests where, in the opinion of the examining physician, alterations in the components of the blood are related to benzene exposure.

(ii) No medical examination is required to satisfy the requirements of paragraph (i)(2)(i) of this section if adequate records show that the employee has been examined in accordance with the procedures of paragraph (i)(2)(i) of this section within the previous six months.

(3) *Information provided to the physician.* The employer shall provide the following information to the examining physician for each examination under this section:

(i) A copy of this regulation and its appendixes;

(ii) A description of the affected employee's duties as they relate to the employee's exposure;

(iii) The employee's representative exposure level or anticipated exposure level;

(iv) A description of any personal protective equipment used or to be used; and

(v) Information from previous medical examinations of the affected employee which is not readily available to the examining physican.

(4) *Physician's written opinions.* (i) For each examination under this section, the employer shall obtain and provide the employee with a copy of the examining physician's written opinion containing the following:

(a) The results of the medical examination and tests;

(b) The physician's opinion concerning whether the employee has any detected medical conditions which would place the employee's health at increased risk of material impairment from exposure to benzene;

(c) The physician's recommended limitations upon the employee's exposure to benzene or upon the employee's use of protective clothing or equipment and respirators.

(ii) The written opinion obtained by the employer shall not reveal specific findings or diagnoses unrelated to occupational exposures.

(5) *Periodic examinations* (i) The employer shall provide each employee covered under paragraph (i)(1)(i) of this section with a medical examination at least semi-annually following the initial examination. These periodic examinations shall include at least the following elements:

(a) A brief history regarding any new exposure to potential marrow toxins, changes in drug and alcohol intake and the appearance of physical symptoms relating to blood disorders;

(b) A complete blood count with red cell count, white cell count with differential, platelet count, hemoglobin, hematocrit and red cell indices (MCV, MCH, MCHC); and

(c) Additional tests where in the opinion of the examining physician, alterations in the components of the blood are related to benzene exposure.

(ii) Where the employee develops signs and symptoms commonly associated with toxic exposure to benzene, the employer shall provide the employee with a medical examination which shall include those elements considered appropriate by the examining physician.

(6) *Emergency situations.* If the employee is exposed to benzene in an emergency situation, the employer shall provide the employee with a urinary phenol test at the end of the employee's shift. The urine specific gravity shall be corrected to 1.024. If the result of the urinary phenol test is below 75 mg/ml, no further testing is required. If the result of the urinary phenol test is equal to or greater than 75 mg/ml, the employer shall provide the employee with a complete blood count including a red cell count, white cell count with differential, and platelet count as soon as practicable, and shall provide these same counts one month later.

(7) *Special examinations.* (i) Where the results of any tests required by this section reveal that any of the following conditions exist, the employer shall have the test results of the employee evaluated by a hematologist:

(a) The red cell count, hemoglobin or platelet count varies more than 15 percent above or below the employee's most recent values;

(b) The red cell count is below 4.4 million or above 6.3 million per mm^3, (for males), or below 4.2 million or above 5.5 million per mm^3 (for females);

(c) The hemoglobin is below 14 grams percent or above 18 grams percent (for males) or below 12 grams percent or above 16 grams percent (for females);

(d) The white cell count is below 4,200 or above 10,000;

(e) The thrombocyte count is below 140×10^9 cells per mm^3 or above 440×10^9 cells per mm^3.

(ii) In addition to the information required to be provided to the physician under paragraph (i)(3) of this section, the employer shall provide the hematologist with the medical record required to be maintained by paragraph (l)(2) of this section.

(iii) The hematologist's evaluation shall include a determination as to the need for additional tests, and the employer shall assure that these tests are provided.

(j) *Employee information and training*—(1) *Training program.* (i) The employer shall institute a training program for all employees assigned to workplaces where benzene is produced, reacted, released, packaged, repackaged, stored, transported, handled or used and shall assure that each employee assigned to these workplaces is informed of the following:

(a) The information contained in Appendices A and B of this section;

(b) The quantity, location, manner of use, release, or storage of benzene and the specific nature of operations which could result in exposure above the permissible exposure limits as well as necessary protective steps;

(c) The purpose, proper use, and limitations of personal protective equipment and clothing required by paragraph (h) of this section and of respiratory devices required by paragraph (g) of this section and § 1910.134 (b), (d), (e) and (f);

(d) The purpose and a description of the medical surveillance program required by paragraph (i) of this section and the information contained in Appendix C of this section; and

(e) The contents of this standard.

(ii) The training program required under paragraph (j)(1)(i) of this section shall be provided within 90 days of the effective date of this section or at the time of initial assignment to workplaces where benzene is produced, reacted, released, packaged, repackaged, stored, transported, handled or used, and at least annually thereafter.

(2) *Access to training materials.* (i) The employer shall make a copy of this standard and its Appendices readily available to all affected employees.

(ii) The employer shall provide, upon request, all materials relating to the employee information and training program to the Assistant Secretary and the Director.

(k) *Signs and labels.* (1) The employer shall post signs in regulated areas bearing the following legend:

DANGER

BENZENE

CANCER HAZARD

FLAMMABLE—NO SMOKING

AUTHORIZED PERSONNEL ONLY

RESPIRATOR REQUIRED

(2) The employer shall assure that caution labels are affixed to all containers of benzene- and of products containing any amount of benzene, except:

(i) Pipelines, and

(ii) Transport vessels or vehicles carrying benzene or benzene products in sealed intact containers.

(3) The employer shall assure that the caution labels remain affixed when the benzene or products containing benzene are sold, distributed or otherwise leave the employer's workplace.

(4) The caution labels required by paragraph (k)(2) of this section shall be readily visible and legible. The labels shall bear the following legend:

CAUTION

CONTAINS BENZENE

CANCER HAZARD

(5) The employer shall assure that no statement which contradicts or detracts from the information required by paragraphs (k)(1) and (k)(4) of this section appears on or near any required sign or label.

(l) *Recordkeeping.*—(1) *Exposure measurements.* (i) The employer shall establish and maintain an accurate record of all measurements required by paragraph (e) of this section.

(ii) This record shall include:

(a) The dates, number, duration, and results of each of the samples taken, including a description of the procedure used to determine representative employee exposures;

(b) A description of the sampling and analytical methods used;

(c) Type of respiratory protective devices worn, if any; and

(d) Name, social security number, and job classification of the employee monitored and all other employees whose exposure the measurement is intended to represent.

(iii) The employer shall maintain this record for at least 40 years or the duration of employment plus 20 years, whichever is longer.

(2) *Medical surveillance.* (i) The employer shall establish and maintain an accurate record for each employee subject to medical surveillance required by paragraph (i) of this section.

(ii) This record shall include:

(*a*) The name, and social security number of the employee;

(*b*) A copy of the physicians' written opinions, including results of medical examinations and all tests, opinions and recommendations;

(*c*) The peripheral blood smear slides of the initial test, the most recent test, and any test demonstrating hematological abnormalities related to benzene exposure;

(*d*) Any employee medical complaints related to exposure to benzene;

(*e*) A copy of this standard and its appendices, except that the employer may keep one copy of the standard and its appendices for all employees provided that he references the standard and its appendices in the medical surveillance record of each employee;

(*f*) A copy of the information provided to the physician as required by paragraphs (i)(3)(ii) through (i)(3)(v) of this section; and

(*g*) A copy of the employee's medical and work history related to exposure to benzene or any other hematologic toxins.

(iii) The employer shall maintain this record for at least 40 years or for the duration of employment plus 20 years, whichever is longer.

(3) *Availability.* (i) The employer shall assure that all records required to be maintained by this section shall be made available upon request to the Assistant Secretary and the Director for examination and copying.

(ii) The employer shall assure that employee exposure measurement records as required by this section be made available for examination and copying to affected employees or their designated representatives.

(iii) The employer shall assure that former employees and the former employees' designated representatives have access to such records as will indicate the former employee's own exposure to benzene.

(iv) The employer shall assure that employee medical records required to be maintained by this section be made available upon request for examination and copying to a physician or other individual designated by the affected employee or former employee.

(4) *Transfer of records.* (i) When the employer ceases to do business, the successor employer shall receive and retain all records required to be maintained by paragraph (l) of this section for the prescribed period.

(ii) When the employer ceases to do business and there is no successor employer to receive and retain the records for the prescribed period, the employer shall transmit these records by mail to the Director.

(iii) At the expiration of the retention period for the records required to be maintained under paragraph (l) of this section, the employer shall transmit these records by mail to the Director.

(m) *Observation of monitoring.*—(1) *Employee observation.* The employer shall provide affected employees, or their designated representatives, an opportunity to observe any measuring or monitoring of employee exposure to benzene conducted pursuant to paragraph (e) of this section.

(2) *Observation procedures.* (i) When observation of the measuring or monitoring of employee exposure to benzene requires entry into areas where the use of protective clothing and equipment or respirators is required, the employer shall provide the observer with personal protective clothing and equipment or respirators required to be worn by employees working in the area, assure the use of such clothing and equipment or respirators, and require the observer to comply with all other applicable safety and health procedures.

(ii) Without interfering with the measurement, observers shall be entitled to:

(*a*) Receive an explanation of the measurement procedures;

(*b*) Observe all steps related to the measurement of airborne concentrations of benzene performed at the place of exposure; and

(*c*) Record the results obtained.

(n) *Appendices.* The information contained in the appendices is not intended, by itself, to create any additional obligations not otherwise imposed or to detract from any existing obligations.

APPENDIX A—SUBSTANCE SAFETY DATA SHEET, BENZENE

I. SUBSTANCE IDENTIFICATION

A. *Substance.* Benzene.

B. *Permissible Exposure.* Except as to the use as fuels of gasoline, motor fuels and other fuels subsequent to discharge from bulk terminals.

1. *Airborne.* 1 part of benzene vapor per million parts of air (1 ppm); time-weighted average (TWA) for an 8-hour workday for a 40-hour week, with a 15 minute ceiling concentration of 5 ppm.

2. *Dermal.* Eye contact and skin contact with liquid benzene shall be prohibited.

C. *Appearance and odor.* Benzene is a clear, colorless liquid with a pleasant, sweet odor. The odor of benzene does not provide adequate warning of its hazard.

II. HEALTH HAZARD DATA

A. *Ways in which the benzene affects your health.* Benzene can affect your health if you inhale it, or if it comes in contact with your skin or eyes. Benzene is also harmful if you happen to swallow it.

B. *Effects of overexposure.* 1. Short-term (acute) overexposure: If you are overexposed to high concentrations of benzene, well above the levels where its odors are first recognizable, you may feel breathless, irritable, euphoric, or giddy; you may experience irritation in eyes, nose, and respiratory tract. You may develop a headache, feel dizzy, nauseous, or experience unsteadiness in walking. Severe exposures may lead to convulsions.

2. *Long-term (chronic) exposure.* Repeated or prolonged exposure to benzene, even at relatively low concentrations, may result in various blood disorders, ranging from anemia to leukemia, an irreversible, fatal disease. Many blood disorders associated with benzene exposure may occur without physical symptoms.

III. PROTECTIVE CLOTHING AND EQUIPMENT

A. *Respirators.* Respirators are required for those operations in which engineering controls or work practice controls are not feasible to reduce exposure to the permissible level. If respirators are worn, they must have a National Institute for Occupational Safety and Health (NIOSH) seal of approval, and cartridges or canisters must be replaced before the end of their service life, or the end of the shift, whichever occurs first. If you experience difficulty breathing while wearing a respirator, tell your employer.

B. *Protective Clothing.* You must wear impervious protective clothing (such as boots, gloves, sleeves, aprons, etc.) over any parts of your body that could be exposed to liquid benzene.

C. *Eye and Face Protection.* You must wear splash proof safety goggles if it is possible that benzene may get into your eyes. In addition, you must wear a face shield if your face could be splashed with benzene liquid.

IV. EMERGENCY AND FIRST AID PROCEDURES

A. *Eye and face exposure.* If benzene is splashed in your eyes, wash it out immediately with large amounts of water. Call a doctor as soon as possible.

B. *Skin exposure.* If benzene is spilled on your clothing or skin, remove the contaminated clothing and wash the exposed skin with large amounts of water and soap immediately. Wash contaminated clothing before you wear it again.

C. *Breathing.* If you or any other person breathes in large amounts of benzene, get the exposed person to fresh air at once. Apply artificial respiration if breathing has stopped. Call for medical assistance or a doctor as soon as possible.

D. *Swallowing.* If benzene has been swallowed and the patient is conscious, do not induce vomiting. Call for medical assistance or a doctor immediately.

V. MEDICAL REQUIREMENTS

If you are exposed to benzene at a concentration at or above 0.5 ppm on an 8-hour time-weighted average, your employer is required to provide a medical history and laboratory tests within 30 days of the effective date of this standard and semiannually thereafter if you are continually exposed at or above 0.5 ppm. These tests shall be provided without cost to you. In addition, if you are accidentally exposed to benzene (either by ingestion, inhalation, or skin/eye contact) under conditions known or suspected to be toxic exposure to benzene, your employer is required to make special tests available to you.

VI. Observation of Monitoring

Your employer is required to perform measurements that are representative of your exposure to benzene and you or your designated representative are entitled to observe the monitoring procedure. You are entitled to receive an explanation of the measurement procedure, observe the steps taken in the measurement procedure, and to record the results obtained. When the monitoring procedure is taking place in an area where respirators or personal protective clothing and equipment are required to be worn, you or your representative must also be provided with, and must wear the protective clothing and equipment.

VII. Access to Records

You or your representative are entitled to see the records of measurements of your exposure to benzene upon request to your employer. Your medical examination records can be furnished to your physician or designated representative upon request to your employer.

VIII. Precautions for Safe Use, Handling and Storage

Benzene liquid is highly flammable. It should be stored in tightly closed containers in a cool, well ventilated area. Benzene vapor may form explosive mixtures in air. All sources of ignition must be controlled. Use nonsparking tools when opening or closing benzene containers. Ground or bond metal benzene containers. Fire extinguishers, where provided, must be readily available. Know where they are located and how to operate them. Smoking is prohibited in areas where benzene is used or stored. Ask your supervisor where benzene is used in your work area and for additional plant safety rules.

Appendix B—Substance Technical Guidelines, Benzene

I. Physical and Chemical Data

A. *Substance identification.*
1. Synonyms: Benzol, benzole, coal naptha, cyclohexatriene, phene, phenyl hydride, pyrobenzol. (Benzine, petroleum benzine, and benzine do not contain benzene.)
2. Formula: C_6H_6 (CAS Registry Number: 000071432)

B. *Physical data.*
1. Boiling Point (760 mm Hg); 80.1C (176F)
2. Specific Gravity (water = 1): 0.879
3. Vapor Density (air = 1): 2.7
4. Melting Point: 5.5C (42F)
5. Vapor Pressure at 20 C (68F): 75 mm Hg
6. Solubility in Water: .06%
7. Evaporation Rate (ether = 1): 2.8
8. Appearance and Odor: Clear, colorless liquid with a distinctive sweet odor.

II. Fire, Explosion, and Reactivity Hazard Data

A. *Fire.*
1. Flash Point (closed cup): -11 C (12F)
2. Autoignition Temperature: 580 C (1076F)
3. Flammable Limits in Air, % by Volume: Lower 1.3%, Upper: 7.5%
4. Extinguishing Media: Carbon dioxide, dry chemical, or foam.
5. Special Fire-Fighting Procedures: Do not use solid stream of water, since stream will scatter and spread fire. Water spray can be used to keep fire exposed containers cool.
6. Unusual fire and explosion hazards: Benzene is a flammable liquid. Its vapors can form explosive mixtures. All ignition sources must be controlled when benzene is used, handled, or stored. Where liquid or vapor may be released, such areas shall be considered as hazardous locations. Benzene vapors are heavier than air; thus the vapors may travel along the ground and be ignited by open flames or sparks at locations remote from the site at which benzene is handled.
7. Benzene is classified as a *1 B flammable liquid* for the purpose of conforming to the requirements of 29 CFR 1910.106. A concentration exceeding *3250 ppm* is considered a potential fire explosion hazard. Locations where benzene may be present in quantities sufficient to produce explosive or ignitable mixtures are considered *Class I Group D* for the purposes of conforming to the requirement of 29 CFR 1910.309.

B. *Reactivity.*
1. Conditions contributing to instability: Heat.
2. Incompatibility: Heat and oxidizing materials.
3. Hazardous decomposition products: Toxic gases and vapors (such as carbon monoxide).

III. Spill and Leak Procedures

A. *Steps to be taken if the material is released or spilled.* As much benzene as possible should be absorbed with suitable materials, such as dry sand or earth. That remaining must be flushed with large amounts of water. Do not flush benzene into a confined space, such as a sewer, because of explosion danger. Remove all ignition sources. Ventilate enclosed places.

B. *Waste disposal method.* Disposal methods must conform to other jurisdictional regulations. If allowed, benzene may be disposed of: (a) Absorbing it in dry sand or earth and disposing in a sanitary land fill; (b) if small quantities, by removing it to a safe location from buildings or other combustible sources, pouring it in dry sand or earth and cautiously igniting it: (c) if large quantities, by atomizing it in a suitable combustion chamber.

IV. Monitoring and Measurement Procedures

A. *Normal monitoring program:* Measurements taken from the purpose of determining employee exposure are best taken so that the representative average 8-hour exposure may be determined from a single 8-hour sample or two (2) 4-hour samples. Short-time interval samples (or grab samples) may also be used to determine average exposure level if a minimum of five measurements are taken in a random manner over the 8-hour work shift. Random sampling means that any portion of the work shift has the same chance of being sampled as any other. The arithmetic average of all such random samples taken on one work shift is an estimate of an employee's average level of exposure for that work shift. Air samples should be taken in the employee's breathing zone (air that would nearly represent that inhaled by the employee). Sampling must be performed by gas adsorption tubes or alternative methods meeting the requirements of the standard with subsequent chemical analysis, by gas chromatography. Methods meeting the prescribed accuracy and precision and requirements are available in the "NIOSH manual of Analytical Methods."

V. Miscellaneous Precautions

A. High exposures to benzene can occur when transferring the liquid from one container to another. Such operations should be well ventilated and good work practices must be established to avoid spills.
B. Use non-sparking tools to open benzene containers which are effectively grounded and bonded prior to opening and pouring.
C. Employers must advise employees of all plant areas and operations where exposure to benzene could occur. Common operations in which high exposures to benzene may be encountered are: the primary production and utilization of benzene, and transfer of benzene.

Appendix C—Medical Surveillance Guidelines for Benzene

I. Route of Entry

Inhalation; possible skin absorption.

II. Toxicology

Benzene is primarily an inhalation hazard. Systemic absorption may cause depression of the hematopoietic system and leukemia. Inhalation of high concentrations can affect the central nervous system function. Aspiration of small amounts of liquid benzene immediately causes pulmonary edema and hemorrhage of pulmonary tissue. The extent of absorption through the skin is unknown. However, absorption may be accelerated in the case of injured skin, and benzene may be more readily absorbed if it is present in a mixture or as a contaminant in solvents which are readily absorbed. Defatting action of benzene may produce primary irritation upon repeated or prolonged contact with the skin. High concentrations are irritating to the mucuous membranes of the eyes, nose, and respiratory tract.

III. Signs and Symptoms

It is not clear to what extent benzene is absorbed through the skin, however, direct contact may cause erythema or blistering. Repeated or prolonged contact may result in drying, scaling dermatitis, or precipitate development of secondary skin infections. Local effects of benzene vapor or liquid on the eye are slight. Only at very high concentrations is there any smarting sensation in the eye. Inhalation of high concentrations of benzene may have an initial stimulatory effect on the central nervous system characterized by exhiliration, nervous excitation, and/or giddiness, followed by a period of depression, drowsiness, fatigue, or vertigo. There may be sensation of tightness in the chest accompanied by breathlessness and ultimately the victim may lose consciousness. Convulsions and tremors occur frequently, and death may follow from respiratory paralysis or circulatory collapse in a few minutes to several hours following severe exposures.

The insidious effect on the blood-forming system of prolonged exposure to small quantities of benzene vapor is of extreme importance. The hematopoietic system is the chief target for benzene's toxic effects which are manifested by alterations in the levels of formed elements in the peripheral blood. These effects have been noted to occur at concentrations of benzene which may not cause irritation of mucous membranes, or any unpleasant sensory effects. Early signs and symptoms of benzene morbidity are varied and often not overtly apparent and not specific for benzene exposure. Subjective complaints of headache, dizziness, and loss of appetite may precede or follow clinical symptomology. Bleeding from the nose, gums, or mucous membranes and the development of purpuric spots may occur as the condition progresses. Rapid pulse and low blood pressure in addition to a physical appearance of anemia may accompany a subjective complaint of shortness of breath. Clinical evidence of leukopenia, anemia, and thrombocytopenia, singly or in combination, have been frequently reported.

Bone marrow may appear normal, aplastic, or hyperplastic and may not in all situations correlate with peripheral blood forming tissues. There are great variations in the susceptibility to benzene morbidity which prohibits the identification of "typical" blood picture. The onset of effects of prolonged benzene exposure may be significantly delayed after the actual exposure has ceased.

IV. TREATMENT OF ACUTE TOXIC EFFECTS

Remove from exposure immediately, give oxygen or artificial resuscitation if indicated. Flush eyes and wash contaminated skin. Symptoms of non-specific nervous disturbances may persist following severe exposures. Recovery from mild exposures is usually rapid and complete.

V. SURVEILLANCE AND PREVENTIVE CONSIDERATIONS

A. GENERAL

The principal effects of benzene exposure forming the basis for this regulation are alterations of the hematopoietic system as reflected by changes in the peripheral blood and leukemia. Consequently, the medical surveillance protocol is designed to observe on a regular basis, blood indices for early signs of these effects.

Tests must be performed frequently enough to discover individuals who may be unusually sensitive and likely to develop marrow abnormalities, to monitor those who experience accidental overexposure and to provide early detection of delayed evidence of toxicity.

All workers who are or will be exposed to 0.5 parts per million (ppm) or greater benzene as an eight-hour time-weighted average are to be given the opportunity for a medical examination. Initial examinations are to be provided within 30 days of the effective date of this standard or at the time of initial assignment and interval examinations semiannually thereafter. There are special provisions for medical tests in the event of hematolgica abnormalities or for emergency situations.

F. HEMATOLOGY GUIDELINES

The following information excerpted from the analysis of Dr. Jandl, Chief of Hematology, Harvard School of Medicine, may be useful to physicians in conducting the medical surveillance program.

"A minimum battery of tests is to be performed by strictly standardized methods in the circumstances described above.

1. Red cell, white cell, and platelet counts must be performed using an automated (Coulter) counter. The normal range for the red cell count is approximately 4.4 to 6.0 million cells/mm^3, the values for women being about 0.4 million cells lower than for men. A decline from a normal to a subnormal value, or a rise to a supra-normal value, are indicative of potential toxicity, particularly should there be a decline. The normal total white blood count is approximately 6,200 plus or minus 2,000/mm^3. For cigarette smokers and white count will be higher, the upper range of "normal" being approximately 1,000 cells higher than 8,200. Either a decline from normal to subnormal or a rise from normal to supra-normal, should be regarded as a potential indication of benzene toxicity. The normal platelet count is 250,000 with a range of 140,000 to (at most) 400,000/mm^3. A decline to below 140,000 or a rise to above 400,000 should be regarded as possible evidence of benzene toxicity.

The reticulocyte count is performed by technical assistants using a cover-slip smear (see below). In my opinion, the preferred technique for this purpose is the so-called "dry-method" employing brilliant cresyl blue (BCB) for staining the filaments of reticulum within red cell, and counter-staining with Wright's stain. The extreme range of normal for reticulocytes is 0.4 to 1.5 percent of the red cells, the usual range being 0.5 to 1.2 percent of the red cells, but the typical value is in the range of 0.8 to 1.0 percent. There is an advantage of using the BCB reticulocyte staining technique (followed by counter-staining with Wright's stain) in that visible evidence (i.e., the stained, mounted reticulocyte smears) may be stored, and if kept filed in the dark may later be retrieved for reexamination and comparisons. A decline in reticulocytes to levels of less than 0.4 percent is to be regarded as possible evidence (unless another specific cause is found) of benzene toxicity requiring accelerated surveillance. An increase in reticulocyte levels to above 1.5 percent may also be consistent with (but is not as characteristic of) benzene toxicity.

2. The single most important routine surveillance test is an expert technician's careful examination of the peripheral blood smear. As with the reticulocyte count, the smear should be with fresh uncoagulated blood obtained from a needle tip following venipuncture or from a drop of earlobe blood (capillary blood). If necessary, the smear may under certain limited conditions be made from a blood sample anticoagulated with EDTA (but never with oxalate or heparin). When the smear is to be prepared from a specimen of venous blood which has been collected by a commercial Vacutainer^R type tube containing neutral EDTA, the smear should be made as soon as possible after the venesection. A delay of up to 12 hours is permissible between the drawing of the blood specimen into EDTA and the preparation of the smear if the blood is stored at refrigerator (not freezing) temperature. As with the reticulocyte preparations, the smear should be made on cover slips only. Under no circumstances should peripheral blood (or bone marrow aspirate) intended for examination be smeared on microscope slides, a technique which produces artifacts in blood cells and distorts the white cell differential count by severely maldistributing them. Dry blood smears should be stained with Wright's stain which should be filtered at least weekly to remove precipated dye (saturated completely by methylene blue-eosinate derivates).

3. The minimum mandatory observations to be made from the smear and a discussion of their significance now follows. The observations are four:

a. The differential white blood cell count.
b. Description of abnormalities in the appearance of red cells.
c. Description of any abnormalities in the platelets.
d. A careful search must be made by the technician throughout the better areas of every blood smear for immature white cells such as band forms (in more than normal proportion), any number of metamyelocytes, myelocytes. Any nucleated or multinucleated red blood cells should be reported. Very large "giant" platelets of fragments of megakaryocytes must be recognized. Should only a single one of these abnormalities be found, it should be reported.

An increase in the proportion of band forms among the neutrophilic granulocytes is an abnormality deserving special mention for it represents a very early change which should be considered as an early warning of benzene toxicity in the absence of other causative factors (most commonly infection). Likewise, the appearance of metamyelocytes in the absence of other probable cause is to be considered a possible indication of benzene-induced injury.

An upward trend in the number of basophils, which normally do not exceed about 2.0 percent of the total white cells, is to be regarded as possible evidence of benzene toxicity. A rise in the eosinophil count is less specific but also may be suspicious of toxicity if it rises above 6.0 percent of the total white count.

The normal range of monocytes is from 2.0 to 8.0 percent of the total white count an average of about 5.0 percent. About 20 percent of individuals reported to have mild but persisting abnormalities caused by exposure to benzene show a persisting monocytosis which is sometimes striking. The findings of a monocyte count which persists at more than 10 to 12 percent of the normal white cell count (when the total count is normal) or persistance of an absolute monocyte count in excess 800/mm^3 should be regarded as a possible sign of benzene-induced injury.

A less frequent but more serious indication of benzene-induced injury to the bone marrow is the findings in the peripheral blood of the so-called "pseudo" (or acquired) Pelger-Huet anomaly. In this anomaly many, or sometimes the majority, of the neutrophilic granulocytes possess two round nuclear segments—less often one or three round segments—rather than three normally elongated segments. When this anomaly is not hereditary, it is often but not invariably predictive of subsequent leukemia. However, only about two percent of patients who ultimately develop acute myelogenous leukemia show the acquired Pelger-Huet anomaly.

An uncommon but ominous sign, one which cannot be detected from the smear, but can be suspected easily by a "sucrose water test" or peripheral blood, is transient paroxysmal nocturnalhemoglobinuria (PNH), which may first occur insidiously during a period of established aplastic anemia and maybe followed within one to a few years by the appearance of rapidly fatal acute myelogenous leukemia. Clinical detection of PHH, which occurs in perhaps only one or two percent of those destined to have acute myelogenous leukemia, may be difficult; if the presumptive "sucrose water test" for it is positive, the technician may perform the somewhat more definitive Ham test, also known as the acid-serum hemolysis test.

e. Individuals documented to have developed acute myelogenous leukemia years after initial exposure to benzene, have (see above) progressed through preliminary phases of hematologic abnormality. In many instances pancytopenia (i.e., a lowering in the counts of all circulating blood cells of bone marrow orgin—but not to the extent implied by the term "aplastic anemia") preceded leukemia for many years. Seldom does relative scarcity of a single type blood cell (or of platelets) represent a harbinger of imminent acute leukemia. However, the finding of two or more cytopenias, or of pancytopenia, must be regarded as highly suspicious of more advance although still reversible, benzene toxicity. When "pancytopenia" develops and becomes associated with the apearance of immature cells (myelocytes, myeloblasts, erythroblasts, etc.), with abnormal cells (pseudo Pelger-Huet anomaly, atypical nuclear heterochromatin, etc.), or with inappropriate elevations of monocytes, basophils, or eosinsophils, the findings must be regarded as evidence of benzene overexposure unless proved other-wise. These and other aggregates of alterations are frequently term "preleukemia," a term whose meaning is good when used retrospectively, but less good when used prospectively where it has only inferential value. Many severely aplastic patients manifested the ominous finding of 5-10 percent myeloblasts in the marrow, occasional myeloblasts and myelocytes in the blood and 20-30% monocytes; these represented the beginning of recovery rather than the early stage of overt AML. Thus, a considerable proportion of "preleukemias" in benzene poisoning fail to progress to leukemia. Indeed, some have been observed to revert to normal after withdrawal of the afflicted person from toxic exposure. Nonetheless, the chance that "preleukemic" (changes in general) will evolve to leukemia is considerable: at least 20 to 40 percent of persons (only a few of whom were benzene-exposed) with these blood changes develop acute myelogenous leukemia. Certain tests may substantiate the person's prospects for progression or regression. One such test would be an examination of patient's bone marrow. But the decision to perform a bone marrow aspiration or needle biobsy is one that should be made by the hematologist. The findings to be sought there would be: hypoplasia or aplasia; an excess of immature forms; vacuolation in erythroblasts and myelocytes—a phenomenon induced by many toxins apart from benzene, including chloramphenicol and alcohol; and by infections.

The findings of basophilic stippling in circulating red blood cells (usually found in 1 to 5% of red cell during marrow injury), and detection in the bone marrow of what are termed "ringed sideroblasts" must be taken seriously, as they have been noted in recent years to be frequent premonitory signs of subsequent acute leukemia.

In several recent reports dealing with relatively few patients, peroxidase-staining of circulating or marrow neutrophil granules, employing benzidine dihydrochloride, has revealed as a "preleukemic" finding the disappearance of, or diminution in, perosidase in a sizable proportion of the granulocytes. Granulocyte granules are normally strongly peroxidase positive. A steady decline in leukocyte alkaline phosphatase is also suggestive of early acute leukemia. Exposure to benzene commonly causes an early rise in serum iron, often but not always associated with a fall in the reticulocyte count. Thus serial measurements of serum iron levels provide a means of determining whether or not there is a trend representing sustained suppress of erythropoiesis.

Measurement of serum iron, determination of peroxidase and of alkaline phosphatase activity in peripheral granulocytes can be performed by technical assistants.

(Secs. 4, 6, 8, 84 Stat. 1593, (29 U.S.C. 653, 655, 657); Secretary of Labor's Order 8-76 (41 FR 25059); 29 CFR part 1911.)

[FR Doc. 78-3417 Filed 2-3-78; 1:12 pm]

Appendix C

BENZENE: REGULATION AS AN INTENTIONAL INGREDIENT OR AS A CONTAMINANT; PRODUCTS REQUIRING SPECIAL LABELING; AND SUBSTANCES REQUIRING SPECIAL PACKAGING

(Consumer Product Safety Commission; Reprinted from the <u>Federal Register</u>, Friday, May 19, 1978, Part IV)

Appendix C

[6355-01]

CONSUMER PRODUCT SAFETY COMMISSION

[16 CFR Part 1145]

CONSUMER PRODUCTS CONTAINING BENZENE AS AN INTENTIONAL INGREDIENT OR AS A CONTAMINANT UNDER THE CONSUMER PRODUCT SAFETY ACT

Proposed Rule to Regulate

AGENCY: Consumer Product Safety Commission.

ACTION: Proposed rule.

SUMMARY: The Commission proposes to regulate consumer products, except gasoline and solvents or reagents for laboratory use, containing benzene as an intentional ingredient or as a contaminant, under the Consumer Product Safety Act (CPSA) instead of under the Federal Hazardous Substances Act (FHSA), to address the risks of blood disorders, chromosomal abnormalities, and leukemia associated with benzene inhalation. According to the CPSA, the Commission may not regulate under the CPSA a risk of injury that could be eliminated or reduced to a sufficient extent under the FHSA, unless the Commission finds by rule that it is in the public interest to do so. The Commission has preliminarily found that it is in the public interest to regulate these products under the CPSA.

DATE: Comments concerning this proposal must be received by June 30, 1978. It is expected that this rule will become effective when any final CPSA rule concerning benzene is effective.

ADDRESS: Comments should be sent to: Office of the Secretary, Consumer Product Safety Commission, Washington, D.C. 20207.

FOR FURTHER INFORMATION CONTACT:

Francine Shacter, Office of Program Management, Consumer Product Safety Commission, Washington, D.C. 20207, 301-492-6557.

SUPPLEMENTARY INFORMATION: The purpose of this notice is to propose a rule under section 30(d) of the Consumer Product Safety Act (CPSA), 15 U.S.C. 2079(d), to regulate consumer products, except gasoline and solvents or reagents for laboratory use, containing benzene as an intentional ingredient or as a contaminant, under the CPSA. Although the risk of injury from blood disorders, chromosomal abnormalities, and leukemia associated with these products could be eliminated or reduced to a sufficient extent by action under the Federal Hazardous Substances Act (FHSA), 15 U.S.C. 1261-1274, the Commission has preliminarily determined that it is in the public interest to regulate these products under the CPSA.

The Commission believes it is in the public interest to regulate these benzene-containing products under the CPSA rather than the FHSA, since the rulemaking proceedings appropriate to regulate these products under the FHSA are likely to be lengthy and resource-consuming.

The Commission also believes it is in the public interest to regulate these products under the CPSA since it may be more difficult for interested persons to participate in rulemaking proceedings under the FHSA than the CPSA because the proceedings under the FHSA are likely to be more complex and formal.

In order to ban hazardous household substances under section 2(q)(1)(B) of the FHSA, the Commission must follow the provisions of section 701(e) of the Federal Food, Drug and Cosmetic Act (FFDCA), 21 U.S.C. 371(e). Section 701(e) of the FFDCA provides for a two stage rulemaking. After the Commission proposes a regulation, analyzes comments, and issues a final regulation, adversely affected persons may submit objections to the rule and request a public hearing on the objections. If the objections are legally valid, those parts of the regulation objected to are stayed and a hearing is held to receive evidence that is relevant and material to the issues raised by the objections.

After completion of the hearing, the Commission may issue a final regulation. However, this regulation must be based on substantial evidence of record obtained at the hearing, and must set forth detailed findings of fact on which the order is based. The Commission's past experience with formal rulemaking under section 701(e) has been that these proceedings are lengthy, resource-consuming, and often inefficient in eliciting information on which the Commission may make a decision.

In contrast, section 9(a)(2) of the CPSA requires bans to be issued in accordance with the informal rulemaking provisions of the Administrative Procedure Act (5 U.S.C. 553). Although the CPSA requires the Commission to provide an opportunity for oral presentation of comments before issuing a ban, the oral presentation required by section 9(a)(2) is informal and nonadversarial. A ban under section 8 of the CPSA, while providing an adequate opportunity for public participation, is less likely to involve extensive delay, since the Commission is not required to utilize formal rulemaking. In addition, since the proceedings under section 8 of the CPSA are less complex and formal than those under the FHSA, the Commission is more likely to obtain participation from interested persons including consumers, than if the rulemaking is conducted under the FHSA.

The Commission has also preliminarily determined that it is in the public interest to regulate consumer products containing benzene as an intentional ingredient or as a contaminant under the CPSA rather than the FHSA, since the CPSA, unlike the FHSA, provides that persons who knowingly violate the act are subject to civil penalties. The civil penalty provision may provide additional incentive for compliance under the CPSA.

However, if the Commission does decide to regulate these products under section 8 of the CPSA, the Commission recognizes that automatic repurchase of banned hazardous products would not be available, as it is in the case of a ban under the FHSA. Under section 15 of the FHSA, 15 U.S.C. 1274, and the accompanying regulations at 16 CFR 1500.202 and .203, hazardous substances that have been banned must be repurchased by the manufacturers, distributors, and retailers of the hazardous substance, regardless of whether the substance was banned at the time of its sale. In contrast, a rule under section 8 of the CPSA that bans products distributed in commerce or manufactured after the effective date of the ban would not provide for repurchase of the banned hazardous products.

Elsewhere in this FEDERAL REGISTER the Commission has proposed a ban of benzene-containing consumer products under section 8 of the CPSA.

PROPOSAL

The Commission, therefore, proposes below to regulate consumer products, except gasoline and solvents or reagents for laboratory use, containing benzene as an intentional ingredient or as a contaminant under the CPSA instead of the FHSA.

Accordingly, pursuant to provisions of the Consumer Product Safety Act (section 30(d), Pub. L. 92-573, 86 Stat. 1231, as amended 90 Stat 510; 15 U.S.C. 2079(d)), the Commission proposes that title 16, chapter II, subchapter B, part 1145 be amended by adding the following new § 1145.6.

§ 1145.6 Consumer products, except gasoline and solvents and reagents for laboratory use, containing benzene as an intentional ingredient or as a contaminant; risk of blood disorders, chromosomal abnormalities, and leukemia associated with benzene inhalation.

The Commission finds that it is in the public interest to regulate the risk of injury from blood disorders, chromosomal abnormalities and leukemia associated with benzene inhalation under the Consumer Product Safety Act rather than the Federal Hazardous Substances Act because of the desirability of avoiding possibly lengthy, resource-consuming, and inefficient rulemaking proceedings under the FHSA and because of the availability of civil penalties under the CPSA.

The Commission also believes that the complexity and formality of the rulemaking proceedings under the FHSA, in contrast to rulemaking proceedings under the CPSA, may make it difficult for interested persons to participate.

Therefore, consumer products, except gasoline and solvents and reagents for laboratory use, containing benzene as an intentional ingredient or as a contaminant shall be regulated under the CPSA.

(Sec. 30(d), Pub. L. 92-573, 86 Stat. 1231, as amended, 90 Stat. 510 (15 U.S.C. 2079(d)).)

Interested persons are invited to submit, on or before June 30, 1978, written comments regarding this proposal. Written comments and any accompanying data or material should be submitted preferably in five copies, addressed to the Secretary, Consumer Product Safety Commission, Washington, D.C. 20007. Comments may be accompanied by a memorandum or brief in support thereof. Received comments may be seen in the Office of the Secretary, Third Floor, 1111 18th Street NW., Washington, D.C. during working hours Monday through Friday.

Dated: May 15, 1978.

SADYE E. DUNN,
Acting Secretary, Consumer Product Safety Commission.

[FR Doc. 78-13592 Filed 5-18-78; 8:45 am]

[6355-01]

[16 CFR Part 1307]

CONSUMER PRODUCTS CONTAINING BENZENE AS AN INTENTIONAL INGREDIENT OR AS A CONTAMINANT

Proposal to Ban

AGENCY: Consumer Product Safety Commission.

ACTION: Proposed rule.

SUMMARY: The Commission proposes to ban all consumer products, except gasoline and solvents or reagents for laboratory use, containing benzene, as an intentional ingredient or as a contaminant at a level 0.1 percent or greater by volume. The Commission is taking this action because of information indicating that benzene inhalation may cause blood disorders, chromosomal abnormalities, and leukemia or cancer of the white blood cells in consumers exposed to benzene-containing products. The Commission is particularly interested in public comment on the proposed contaminant level and on the issue of health effects of all potential benzene substitutes. In taking this action, the Commission is granting a petition (HP77-12) from the Health Research Group.

DATES: (1) Written comments on the proposal must be received by June 30, 1978. (2) There will be an opportunity for interested persons to orally present data, views, or arguments on June 14, 1978, at 9:30 a.m. Those wishing to make oral presentations should notify the Office of the Secretary by June 7, 1978. Additionally, a summary or copy of testimony is to be submitted to the Office of the Secretary two working days prior to the public meeting. (3) The Commission proposes that consumer products containing benzene as an intentional ingredient manufactured or imported 60 or more days after publication of the final rule in the FEDERAL REGISTER be declared banned hazardous products. Alternatively, the Commission is proposing a shorter effective date with retrospective application for products containing intentionally-added benzene. Under this option, consumer products containing intentionally-added benzene which are initially introduced into interstate commerce 15 or more days after publication of a final rule would be banned hazardous products. All other consumer products containing intentionally-added benzene, no matter when manufactured or initially introduced into commerce, would be banned 30 days after publication of a final ban. As to products that contain benzene as a contaminant, it is expected that the ban will apply to all such products which are manufactured or imported after the date which is 120 days after publication of the final rule. (4) The time during which the Commission must either publish a final rule or withdraw this proposal is extended to October 16, 1978.

ADDRESSES: Written comments should be submitted to the Office of the Secretary, Consumer Product Safety Commission, Washington, D.C. 20207. To the extent practicable, comments on the issue of benzene as an intentional ingredient should be separated from comments on the issue of benzene as a contaminant. Persons wishing to make oral presentations should contact Richard Danca in the Office of the Secretary. These oral presentations will be held in the Commission's 3rd floor hearing room at the address below. All material which the Commission has that is relevant to this proceeding, including any comments that may be received regarding this proposal, may be seen in, and copies obtained from, the Office of the Secretary, Consumer Product Safety Commission, 3rd floor, 1111 18th Street NW., Washington, D.C.

FOR FURTHER INFORMATION CONTACT:

Francine Shacter, Office of Program Management, Consumer Product Safety Commission, Washington, D.C. 20207—202-492-6557. Persons wishing to make oral presentations should contact: Richard Danca, Office of the Secretary, 202-634-7700.

SUPPLEMENTARY INFORMATION:

BACKGROUND

On May 5, 1977 the Commission received a petition (HP 77-12) from the Health Research Group requesting that the Commission declare benzene a banned hazardous substance under the Federal Hazardous Substances Act (FHSA), 15 U.S.C. 1261 et seq., and remove all products containing benzene from the marketplace because of evidence that benzene causes leukemia. The petitioner note that new evidence linking worker exposure to benzene to the induction of leukemia had prompted the Occupational Safety and Health Administration (OSHA) to publish an Emergency Temporary Standard (ETS) for benzene on May 3, 1977 (42 FR 22516). HP 77-12 was later supplemented by a statement in support of the petition from Dr. Michael McCann of the Center for Occupational Hazards, Inc. Dr. McCann expressed particular interest in the banning of benzene-containing art supplies used by consumers.

In gathering information on HP 77-12, Commission staff attended hearings sponsored by OSHA which covered their ETS as well as the OSHA proposed permanent standard for benzene (42 FR 27452; May 27, 1977). The OSHA hearings, which began on July 19, 1977 and were held on successive work days through August 10, 1977, included testimony on health, economic, and environmental issues relevant to the OSHA regulations.

Based on information gathered from the OSHA hearings and from other sources, discussed more fully below, the Commission has granted the petition and proposes to ban all consumer products, except gasoline and solvents or reagents for laboratory use, containing benzene as an intentional ingredient. The Commission also proposes to ban all consumer products containing benzene as a contaminant at a level of 0.1 percent or greater by volume. The Commission is taking this action because of scientific and technical information indicating that benzene inhalation may cause blood disorders, chromosomal abnormalities, and leukemia in consumers exposed to benzene-containing products. While the Commission recognizes that much of the data discussed below concern benzene exposure in an industrial setting, the Commission points out that it does have some information concerning consumer exposure. In addition, the Commission points out that there is general scientific and medical agreement that there is no known safe level of exposure to a carcinogen and that such exposure should, therefore, be reduced to the lowest feasible level.

Although the petition was filed under the FHSA and it appears that the risk of injury from leukemia described by the petitioner is regulatable under the FHSA, the Commission proposes to ban benzene-containing products under the Consumer Product Safety Act (CPSA), 15 U.S.C. 2051, et seq. Section 30(d) of the CPSA, 15 U.S.C. 2079(d), as amended, provides that a risk of injury associated with a consumer product which can be eliminated or reduced to a sufficient extent by action under a transferred act, such as the FHSA, may nonetheless be regulated under the CPSA if the Commission, by rule, determines that regulation under the CPSA is in the public interest. Also published in this issue of the FEDERAL REGISTER is a proposed rule issued in accordance with section 30(d) of the CPSA stating the Commission's finding that it is in the public interest to ban the products described herein under the CPSA.

Current Commission regulations under the FHSA (see 16 CFR § 1500.14 (a)(3) and (b)(3)) require that products containing 5 percent or more by weight of benzene must be specially labeled with the signal word "danger", the word "poison", the skull and crossbones symbol, and the statement of the hazard "vapor harmful." Products containing 10 percent or more by weight of benzene must bear the additional statement of hazard "harmful or fatal if swallowed" as well as the statement "If swallowed, do not induce vomiting. Call physician immediately."

The Commission also has a regulation under the Poison Prevention Packaging Act of 1970 (PPPA) applicable to benzene. (16 CFR § 1700.14(a)(15)). The regulation provides that solvents for paints and other similar surface-coating materials that contain 10 percent or more by weight of benzene and that have a viscosity of less than 100 Saybolt Universal Seconds (SUS) at 100° F must be packaged in child-resistant packaging.

Elsewhere in this issue of the FEDERAL REGISTER, the Commission is proposing to amend the FHSA and PPPA regulations to indicate that they will not be applicable to products that are banned hazardous products under any final CPSA rule. The amendments are proposed to be effective when this Part 1307 becomes effective.

OTHER AGENCY ACTION

The Commission notes that other federal agencies have taken action with respect to benzene. As mentioned above, on May 27 1977, OSHA published its proposed permanent standard for benzene which would limit employee exposure to benzene to 1 ppm as an 8 hour time-weighted average concentration with a ceiling level of 5 ppm for any 15 minute period during an 8 hour day. The proposed standard excluded operations in which the only exposure to benzene is from liquid mixtures which contain no more than 1 percent or less of benzene by volume (0.1 percent after one year from the effective date). (42 FR 27452). On February 10, 1978, OSHA published its final permanent standard for benzene limiting employee exposure to the same levels announced in the proposal. However, the percentage exemption for liquid mixtures containing benzene was dropped because OSHA found that the percentage of benzene in a liquid is not necessarily controlling of an employee's exposure level. Thus, the final standard applies to all liquid mixtures containing benzene regardless of the percent of benzene.[1] The standard also provides for the measurement of employee exposure, engineering controls, work practices, personal protective clothing and equipment, signs and labels, medical surveillance, and employee training (see 42 FR 5918).

In addition to the OSHA action, the Commission notes that the Environmental Protection Agency (EPA) on June 8, 1977 published its decision to list benzene as a hazardous air pollutant under section 112 of the Clean Air Act on the basis of scientific reports which strongly suggest an increased incidence of leukemia in humans exposed to benzene (42 FR 29332).

BENZENE'S PROPERTIES AND USES

Benzene (C_6H_6), also known as Benzol, is a clear, colorless, strong-smelling liquid which is extremely flammable. Benzene has a low boiling point and a high vapor pressure and hence evaporates rapidly under ordinary atmospheric conditions, giving off vapors nearly three times heavier than air. It has significant solvent properties and mixes well with the majority of organic solvents. Benzene occurs naturally in crude oil and also as a product of decomposition of organic material.

Today, benzene is used primarily as an intermediate substance in the production of other chemicals. Cumene, cyclohexane, dichlorobenzene, dodecylbenzene, ethylbenzene, maleic anhydride and nitrobenzene are a few of these chemicals. Additionally, benzene is used as a component of unleaded gasoline to increase the octane rating, as a solvent and reactant in numerous chemical applications in laboratories, and in the manufacture of detergents and pesticides. Benzene is predominantly produced today from petroleum by catalytic reformation.

Benzene's characteristics as a solvent have made it figure prominently over the years in the manufacture of consumer products. It has been used in its pure form in some products and in other products as a component of solvents that are used in manufacturing the final product. Based on Commission staff communications with manufacturers, the consumer products which have been identified as containing benzene as an intentional ingredient are paint removers, rubber cements, gasoline, and solvents and reagents for laboratory use. Other products, notably many products containing petroleum distillates, which include many common solvents such as toluene and naphtha, contain benzene as a contaminant. (For further discussion, see the section of this Preamble entitled ECONOMIC CONSIDERATIONS).

NATURE AND DEGREE OF THE RISK OF INJURY

Benzene has long been recognized as a toxic substance capable of producing acute and chronic effects. Inhalation is the primary route of exposure to a human being. Although present evidence suggests that benzene is not readily absorbed through the skin, significant quantities of benzene could be absorbed in the case of injured skin.

The acute effects of benzene exposure include drowsiness and loss of consciousness at high doses. The chronic effects of benzene exposure, which form the basis for this proposed regulation, are detailed below in a technical report on benzene prepared by a Commission consultant, Michael L. Freedman, M.D. of New York University Medical Center. The chronic effects listed in the report fall into three general categories: (1) pancytopenia (a decrease in all the formed elements of the blood—red cells, white cells, and platelets); (2) chromosomal abnormalities; and (3) leukemia, particularly acute myelogenous (produced in the bone marrow) leukemia.

The Freedman report and accompanying bibliography is set forth in its entirety as an appendix to this document. That report discusses the many studies reported in the scientific literature which have implicated benzene exposure in the development of various blood disorders, including leukemia, and chromosomal abnormalities. The Commission has considered this report along with other available information in proposing the ban and makes the following observations.

(1) Pancytopenia, including aplastic anemia, has been demonstrated in

---

[1] On March 28, 1978 (43 FR 12890), OSHA proposed to amend the final benzene standard to exclude from its coverage work operations where the sole exposure to benzene is from liquid mixtures containing 0.1% or less benzene or the vapors from these liquids. At the same time OSHA stayed the benzene standard to the extent it covers such work operations, pending completion of the rulemaking and proceedings on the amendment.

both humans and animals following exposure to benzene, although it has not been observed subsequent to exposure to other compounds, such as toluene and xylene, which are commonly associated with benzene environmentally. Both pancytopenia and aplastic anemia, although they are non-cancerous diseases, may themselves be fatal. An additional concern is that some or all of the blood disorders induced by benzene may, if allowed to continue, either progress to or represent a preluekemic stage which may eventually evolve into an actual leukemia.

(2) The development of various chromosomal aberrations, both stable and unstable in nature, in peripheral leukocytes and bone marrow cells have been attributed to benzene exposure.

(3) Leukemia is the most serious blood disorder associated with benzene exposure. Once leukemia is diagnosed, there is virtually no chance of recovery. Despite advances in leukemia therapy, the prognosis of acute myelogenous leukemia (the most prevalent subtype of leukemia in adults and the type most commonly associated with benzene) remains poor. OSHA in their permanent benzene standard states that the average life expectancy of even the victims who enter into a remission (generally averaging 6–8 months) ranges from 12–18 months. (43 FR 5925).

Latency periods between exposure to benzene and development of leukemia can vary widely, ranging from two to twenty years. However, leukemia has been observed in workers exposed for as little as four months to benzene or for as much as 15 years or longer. Individual susceptibility to the effects of benzene also varies significantly. In a recent study of workers exposed to benzene from 1940–1949 in two Ohio plants manufacturing the product "Pliofilm", Infante (see reference No. 51 in Freedman bibliography) demonstrated a five-fold excessive risk of death from all leukemias and a tenfold excess of deaths from myeloid and monocytic leukemias in the study population compared with the number expected in a non-exposed cohort. Infante stated that benzene concentrations at the two plants ranged from 0 to 10 or 15 parts per million. This study as well as other occupational studies noted in the Freedman report argue convincingly that benzene is a human leukemogen.

The Commission recognizes that much of the documentation in the consultant's report concerns benzene exposure in an industrial setting. However, as the report indicates, in 1977 the National Institute for Occupational Safety and Health conducted an experiment to measure the amount of benzene exposure to a person using a paint remover containing 52 percent benzene by volume (Reference No. 137). An end table was stripped in a home garage having essentially no air circulation. Five sequential, five minute air samples from the worker's breathing zone were taken. The benzene concentration ranged from 73 to 225 ppm with a mean of 130 ppm. In addition, the Commission notes that values from 1971 exist for exposure levels in one man or husband-wife combination commercial paint stripping operations utilizing paint strippers containing approximately 40 percent benzene. The breathing zone average atmospheric concentration of benzene ranged from 24 to 1216 ppm in 8 such establishments. (Reference No. 138).

In view of the data indicating that blood disorders, including leukemia, have been shown to result from benzene exposure in an industrial setting at levels reported to be as low as 10–15 ppm, the Commission believes that a serious health hazard to consumers may be presented by the use of benzene-containing consumer products.

As an additional matter, the Commission's Health Sciences staff has calculated an assessment of the leukemia risk from consumer exposure to benzene-containing paint strippers. The assessment which uses data from the Infante study discussed above and the NIOSH paint stripper experiment also discussed above is on file at the Commission's Office of the Secretary. For purposes of this assessment, the Commission considered the use of a paint stripper containing 52 percent benzene by a consumer, for 5 hours a day ten times in his life, to be a reasonably foreseeable lifetime exposure. The increased lifetime risk of death from myelogenous and monocytic leukemia induced by this exposure is estimated by the assessment at between 142 and 346 per million exposed.

This is equivalent to a benzene-induced increased risk of between 15.2 percent and 37.1 percent due to the use of such paint strippers. While the Commission is not relying upon a specified percentage of increased risk in issuing this ban, the Commission believes that in view of the seriousness of the risks of injury from benzene exposure, any increased risk is unacceptable.

In addition, the Commission emphasizes that once the carcinogenicity of a substance has been established qualitatively, any exposure must be considered to be attended by risk when considering any given population, because individual susceptibility to a carcinogen varies widely. The Commission, therefore, believes that consumer exposure to benzene at low levels poses a carcinogenic risk to certain individuals. In recognition of the present state of scientific knowledge that there is no known safe level of exposure to a carcinogen, the Commission concludes that it is appropriate that exposure to benzene be reduced to the lowest feasible level.[2] (For further discussion on this point, see the section of the preamble entitled DESCRIPTION OF THE PROPOSED BAN.)

DESCRIPTION OF THE PROPOSED BAN

Proposed Part 1307 would declare that consumer products containing benzene as an intentional ingredient are banned hazardous products under section 8 of the CPSA. Consumer products that contain benzene as an impurity at a level of 0.1 percent or greater by volume would also be declared to be banned hazardous products. For purposes of this Part 1307, "benzene as an intentional ingredient" means benzene which is added deliberately to a product to impart specific characteristics. "Benzene as a contaminant" is any benzene appearing in a final product which is not an intentional ingredient, including any benzene which appears as a component of or impurity in a raw material used in the final product.

The proposed ban applies to products that are customarily produced or distributed for sale to or for the personal use, consumption or enjoyment of consumers in or around a household or residence, a school, in recreation or otherwise. The Commission proposes that consumer products containing benzene as an intentional ingredient manufactured or imported 60 or more days after the publication of the final rule in the FEDERAL REGISTER be declared banned hazardous products. Alternatively, the Commission is proposing a shorter effective date with retrospective application for products containing intentionally-added benzene. Under this option, consumer products containing intentionally-added benzene which are initially introduced into interstate commerce 15 or more days after publication of a final rule would be banned. All other consumer products containing intentionally-added benzene, no matter when manufactured or initially introduced into commerce, would be banned 30 days after publication of a final rule. For

---

[2] The National Cancer Institute's Ad Hoc committee on the Evaluation of Low Levels of Environmental Carcinogens (1970) has stated: "no level exposure to a chemical carcinogen should be considered toxicologically insignificant for man. For carcinogenic agents, a safe level for man cannot be established by application of our present knowledge." (cited in the OSHA permanent benzene standard at 43 FR 5947). In addition, Bernard Goldstein, M.D. and Muzaffer Askoy, M.D., experts in the field of benzene toxicity, both testified before OSHA that there is no known safe level of exposure to benzene and that, therefore, any exposure should be reduced to the lowest feasible level. (See 43 FR 5947).

products containing benzene as an impurity, it is expected that the ban will apply to all such products which are manufactured or imported after the date which is 120 days after publication of the final rule.

Gasoline is exempted from the products banned by the proposal. While most gasoline is specifically for use with motor vehicles and would appear to be considered "motor vehicle equipment" excluded from Commission authority under the CPSA (see section 3(a)(1)(C)), the Commission notes that some gasoline is used by consumers "in or around a permanent or temporary household or residence, a school, in recreation, or otherwise." For instance, gasoline is often sold to consumers for use as a general household solvent or for use with lawn mowers, snowmobiles, and so forth, and the Commission possibly could take regulatory action under the CPSA as to this gasoline. However, since this gasoline generally comes from the same source (i.e. the retail pumps) as motor vehicle gasoline and may be subject to regulation by OSHA, EPA, and/or the National Highway Traffic Safety Administration (NHTSA), the Commission is of the opinion that exposure to benzene from gasoline is a complex interagency problem which would probably best be resolved on an interagency basis.

In addition, the Commission believes that any attempt to eliminate benzene from gasoline would have a major economic impact, necessitating thorough feasibility studies which are beyond the scope of this proposal. In this regard the Commission notes that benzene is present in gasoline as a contaminant and in unleaded gasoline, as an intentional ingredient used to increase the octane rating. The use of benzene and its closely related aromatic distillates, xylene and toluene, has increased in the past few years because the distillates are the principal substitutes for tetraethyl lead. The Commission has information indicating that the removal of benzene from gasoline motor fuel would entail major adjustments in the petroleum refining industry in terms of capital equipment and performance and price and availability of gasoline as well as other petrochemicals and has, therefore, concluded that the use of benzene in consumer product gasoline should not be restricted at this time. The Commission emphasizes, however, that the complexity of the gasoline question need not delay Commission action on other benzene-containing products.

Besides exempting gasoline, the proposed ban contains an exemption for benzene manufactured and distributed for use in laboratories as a solvent or a reagent. Because benzene is the basic unit on which aromatic hydrocarbon chemistry is built and can be expected to be found in nearly pure form in school laboratories at the high school, college, and graduate levels, the Commission has decided to propose an exemption for such laboratory benzene. The Commission notes that it has information indicating that benzene is essential for certain experiments at the junior and senior college level as well as at the graduate level. The Commission believes that where careful laboratory procedures are followed, such as the use of properly functioning hoods and other appropriate practices, exposure to benzene vapors should be very low. (See OSHA's permanent standard for the regulation of benzene, 43 FR 5918, at 5938.)

In this regard, the Commission points out that OSHA's permanent benzene standard covers benzene exposure in research facilities in both industrial and academic laboratories, where there is at least one employee. All containers of benzene in such laboratories are required to be bear specified labeling, warning of the cancer hazard. In addition, laboratories must comply with all other provisions of the performance standard, which, as is explained in the preamble to the standard, could be met by laboratories in many instances through the use of properly designed hoods. Therefore, the Commission believes that an absolute exemption for laboratory benzene is appropriate because students in educational laboratories using benzene will be protected by the requirements of the OSHA standard.

It should be noted that the proposed ban not only affects products which contain benzene as an intentional ingredient, but also products which contain benzene as a contaminant. In recognition of the medical and scientific agreement that there exists no known safe level of exposure to a carcinogen, the Commission believes that consumer exposure to carcinogens such as benzene must be reduced to the lowest feasible level. The Commission, further, believes that with respect to benzene, the lowest feasible level is not zero, but some low level of benzene contamination in products. (Note: With consumer exposure, unlike occupational exposure, standards based on air monitoring procedures or a requirement for sophisticated protective equipment are impractical. The Commission, therefore, has not attempted to set a permissible air concentration level for benzene, but proposes to rely instead on a percentage limitation on benzene in products.)

The reason that benzene cannot be eliminated entirely from all products is because of the physical properties of petrochemical refining process. As has been mentioned above, benzene occurs naturally in crude oil. It is also formed in the petrochemical refining process as part of a product called reformate from which gasoline and other petroleum products are distilled. These products are distilled into various petroleum distillate fractions according to boiling point range. The boiling point ranges tend to be fairly crude in normal refinery practice so that any particular distillate may have residual or contaminant fractions with boiling points both above and below it. The purity of the distillate depends on the age and technical sophistication of each refinery operation. Additional refining steps may be performed on the distillate in order to strip away the contaminating fractions.

The Commission recognizes that without resorting to extremely expensive special techniques, these refining processes do not result in chemically pure products. The Commission is also aware that reduction of contaminants below the lowest levels generally available in commerce would entail very expensive changes in refinery equipment throughout the industry. Therefore, the Commission proposes to set as the lowest feasible level of benzene contamination, the lowest level generally available in commerce.

Economic information available to the Commission indicates that most common solvents or petroleum distillates which may show traces of benzene contamination are generally available with less than 0.1 percent benzene contamination. Manufacturers, furthermore, substitute one distillate for another or use various combinations of distillates, with a high degree of flexibility. (See "Economic Assessment of a Potential CPSC Benzene Regulation" on file at the Commission's Office of the Secretary). While some benzene substitutes are available with benzene contamination considerably lower than 0.1 percent and others have a higher contamination level (see Table 1 below), the Commission has selected the 0.1 percent level as representative of a feasible impurity level. In so doing, the Commission has preliminarily concluded that it is not now technically feasible to avoid a level of benzene contamination in consumer products below 0.1 percent and has, therefore, decided to propose a ban of products containing benzene traces at or above the level.

While the Commission's current information indicates that contamination below 0.1 percent is the lowest feasible impurity level for benzene in products, the Commission, in recognition of the limited information available on benzene as a contaminant, particularly wishes to solicit public comment on the issue of an appropriate level and the costs of achieving various levels. Based on the comments and other relevant material which may be received in response to the proposal, the Commission may revise the allowable impurity level when the ban is finalized.

ECONOMIC CONSIDERATIONS

Information available to the Commission indicates that the only consumer products which currently contain benzene as an intentional ingredient are paint removers, rubber cements, gasoline, and solvents and reagents for laboratory use. It appears that the OSHA regulations have resulted in a reduction of the use of benzene in consumer products. In fact, only a few firms are using benzene in their consumer products. Since benzene is produced and sold predominantly for industrial use, it is expected that CPSC action will not have a major impact on producers of benzene.

The paint remover industry is one of two industries likely to be affected by a CPSC ban. (As noted above, gasoline and laboratory benzene are proposed to be exempted from the ban.) Generally, there has been a widespread movement away from the use of benzene in paint removers. Commission staff contact with 49 firms revealed only 2 firms presently using benzene in their paint removers. Commission staff estimates that less than 1 percent of the paint removers currently on the market contain benzene as an intentional ingredient.

The reformulation of removers involves the substitution for benzene of various combinations of methylene chloride, toluene, chlorinated hydrocarbons, mineral spirits, or other solvents. Most manufacturers agree that methylene chloride, usually used in conjunction with other solvents, makes a better remover than those currently produced with benzene. However, methylene chloride is a good deal more expensive than benzene. Other solvents which may be substituted for benzene—toluene, for example—will yield a less effective product for approximately the same price. In general, the economic impact on consumers of a ban will be in the form of price increases or utility losses, depending upon which solvent is substituted for benzene.

A ban on benzene-containing products is expected to have a minimal impact on the manufacturers of paint removers. For the few firms that still use benzene in the removers, there may be additional costs associated with reformulation, but these are expected to be small.

The rubber cement industry is another industry likely to be affected by a CPSC ban. In recent years there has been a steady decline in shipments of rubber cement, both in absolute terms and as a proportion of total adhesives shipments. From 1963 to 1973, shipments fell from $87 million to $55 million. During the same period, shipments of adhesives rose from $353 million to $820 million. This implies an absolute decrease in demand for rubber cement, probably in response to the substitution of other adhesives for comparable applications.

Of 52 rubber cement producers contacted, ony one is currently using benzene in his product. Commission staff estimates that less than 0.1 percent of the rubber cements on the market contain benzene. Reformulation costs are, furthermore, expected to be negligible.

Other consumer products such as varnish, wood stains, and household cleaners were investigated by Commission staff to determine whether benzene was present as an intentional ingredient. The products were found not to contain benzene, except possibly as a solvent impurity. Benzene, of course, is also used in almost a pure form in scientific laboratories as a solvent or a reagent. Such benzene is proposed to be exempt from Part 1307.

In addition to products containing intentionally added benzene, products which contain benzene as a contaminant at a level of 0.1 percent or greater are affected by the proposed ban. Many common solvents show traces of brenzene contamination. In fact, any petroleum distillate, even a highly refined solvent, will contain traces of benzene, because of the physical properties of the petroleum refining process. Since there appears to be a wide range of petroleum distillates or solvents generally available with benzene contamination of less than 0.1 percent, it is not expected that the ban on products containing benzene as an impurity will be burdensome to manufacturers or consumers. (See Table 1 below.)

TABLE 1.—*Typical benzene impurities*

	Percent	PPM
Heptane	0.0025	25
Hexane	.5	5,000
Mineral spirits	.1	1,000
Methyl ethyl ketone		0
Methylene chloride		0
Naphtha	.1	1,000
Toluene	.02	200
Xylene	.0025	25

Source: Various industry sources.

As noted above, a ban on consumer products containing benzene will necessitate the use of alternate solvents in manufacturing these products, many of which are other petroleum distillates. Possible substitutes are noted in the chart above. These solvents, with the exception of methylene chloride, are generally not as efficient as benzene, but are accceptable substitutes in most cases. The potential health effects of the use of these substitutes, however, is an issue which the Commission believes merits further study. While the Commission has some information on the health effects of potential substitutes which is summarized below, the Commission believes further information is necessary and specifically requests comment on the subject.

The Commission notes that although conflicting evidence exists, it appears that toluene and xylene may be slightly more acutely toxic than benzene. However, because toluene and xylene are not as volatile as benzene, less of these chemicals may be available in the atmosphere for exposure. (See International Labour Office. "Benzene users, toxic effects, substitutes." Geneva, May 16–19, 1967, pp. 48–89.) Neither toluene nor xylene appear to present the same chronic hazards as benzene. The "Benzene Health Effects Assessment", published as an external review draft in 1977 by the Environmental Protection Agency, states:

Hematoxicity, particularly pancytopenia, has been observed in both humans and animals, following exposure to benzene. The toxicity does not follow exposure to other compounds such as toluene and xylene commonly associated with benzene environmentally. (p. 2).

The Commission notes that a review of the scientific literature on benzene indicates that outbreaks of pancytopenia have followed the introduction of benzene to the workplace and have ceased when benzene has been substituted for by other solvents.

Methylene chloride has been for many years the standard active ingredient in paint removers. It is also a likely substitute for benzene in paint remover reformulations. However, it has been noted that methylene chloride, when inhaled, forms carbon monoxide in the body which reduces the blood's potential to carry oxygen and causes stress on the heart. Therefore, the use of methylene chloride-containing products by heart patients or heavy smokers may cause severe illness or even death. (Stewart and Hake, "Paint-Remover Hazard" in JAMA 235: 398–401, 1976). (In a separate matter, the Commission is considering the possibility of requiring special labeling on paint removers containing methylene chloride to warn of this hazard.)

Significant questions have also been raised concerning the potential chronic toxicity of methylene chloride. Long-term animal testing is now being conducted by the Dow Chemical Company and by the National Cancer Institute in an attempt to ascertain whether methylene chloride poses a potential carcinogenic hazard. Results from one of these laboratories should be available shortly. In addition, methylene chloride has been shown to be mutagenic in two types of short-term tests. (See, Proceedings of the 2nd International Conference on Environmental Mutagens: Progress in Genetic Toxicology—Scott, D., Bridges, B. A., Sobells, FH., eds., Elsevier, North Holland (1977), pp. 249-258.) Further, the Toxic Substances Control Act (TSCA) Interagency Testing Committee has recently recommended to the Administrator of the Environmental Protection Agency that methylene chloride be added to the list of chemicals recommended for priority testing under TSCA. Because of these serious health concerns, the Commission particularly solicits comment on the toxicity of methylene chloride.

The Commission staff economic analysis and supporting materials are available from the Office of the Secretary.

ENVIRONMENTAL CONSIDERATIONS

An assessment has been made of the potential environmental impact of the proposed ban, as required by the National Environmental Policy Act. Based on this assessment, the Commission concludes that there are no significant environmental effects associated with the proposed and that no environmental impact statement is necessary.

The use of benzene as an ingredient in consumer products other than in gasoline and solvents and reagents for laboratory use has decreased over the years to the point where only a few small firms continue to use it. The reasons for this decline are a recognition of the health effects of benzene, the declining maximum allowable exposure limits imposed by State and federal health authorities, and the increased use of benzene as a valuable petro-chemical feedstock that is used to make other chemicals.

Because the use of benzene as an ingredient in consumer products has declined significantly, and since other chemicals have become the standard of the industries that produce such consumer products, no significant environmental effect is expected to result from the ban on benzene as an intentional ingredient. In addition, no significant environmental effect is expected as a result of limiting the level of benzene contamination in products. The level that is proposed appears to be a typical level of contamination in common petroleum distillates, the class of products that is likely to contain benzene as a contaminant.

All of the substitutes for benzene are acutely toxic to humans when inhaled. However, benzene alone has been shown to cause chronic blood disorders, including leukemia, in persons who are exposed to it. While questions have been raised concerning the chronic toxicity of methylene chloride, methylene chloride has long been the standard ingredient in paint strippers. Therefore, a ban on benzene will not substantially increase an already significant use of methylene chloride in these products. The ban, if finalized, is expected to have a beneficial effect on the health of consumers.

The method of disposal of rubber cement and paint strippers that do not contain benzene is the same as for those that do. In nearly all cases of use, the active ingredient in the product is volatile and evaporates into the air. Following use, the container and its residue are disposed of along with other household refuse. The ban is, therefore, expected to have no significant impact on land fills or other disposal media, or on air or water quality.

The environmental assessment is available for review in the Office of the Secretary of the Commission.

ANTICIPATED EFFECTIVE DATES

The Commission has preliminarily assessed the possible effects on the relevant industries of banning consumer products, except gasoline and solvents and reagents for laboratory use, containing benzene as an intentional ingredient or as a contaminant at a level of 0.1 percent or greater by volume. The Commission notes that the few firms currently using benzene as an intentional ingredient will have to reformulate their products, test the new formulations, and acquire different chemical supplies. Based on discussion with the affected companies and experience with other industries with similar problems, the Commission believes that since few reformulation problems should be presented, 60 days is a sufficient period of time for these firms to bring their products into compliance. Therefore, the Commission proposes that consumer products containing intentionally added benzene manufactured or imported 60 or more days after publication by the Commission of a final rule be declared banned hazardous products.

Alternatively, because of the severe nature of the hazard associated with consumer products containing high levels of benzene, the Commission is also proposing a shorter effective date with retrospective application to all products still in the channels of commerce containing intentionally added benzene. The Commission believes it may be necessary to halt all distribution in commerce of products containing intentionally added benzene shortly after the final ban is published so that consumers will not be able to purchase them and thus, continue to be exposed to benzene while supplies are exhausted. Shortening an effective date to less than 30 days from the date of publication of a final rule is authorized by section 553(d)(3) of the Administrative Procedure Act if the Commission so provides for good cause found and published with the rule.

Therefore, the Commission may make the final ban applicable to consumer products containing intentionally added benzene which are initially introduced into commerce 15 days after publication of a final ban. In addition under this option, all other consumer products containing intentionally added benzene, no matter when manufactured or initially introduced into commerce, would be banned 30 days after the publication of a final ban. This means that 15 days after publication of a final rule, manufacturers would be prohibited from shipping the product to distributors, retailers, consumers, or to others. Further, 30 days after publication of a final rule all manufacturing, selling, offering for sale or distributing in commerce of the described products would be prohibited, no matter when initially introduced into commerce.

The Commission does not believe that this alternative proposed effective date will have a major economic impact because so few firms are currently making products containing intentionally added benzene. The Commission staff is continuing to investigate the potential economic and environmental effects associated with the selection of this alternative date. The Commission is particularly interested in public comment on both proposed effective dates.

As to firms that manufacture products containing benzene as a contaminant at a level that is higher than the proposed limit, the Commission notes that these firms must consider alternative chemicals or new sources of supply and may also have to work down or otherwise dispose of inventories of contaminated ingredients. While little information exists on the difficulties that various firms may actually have in dealing with these problems, the Commission staff estimates that 120 days would be a reasonable time period for firms to bring their products into compliance. On this basis, the Commission proposes that consumer products containing benzene as an impurity at levels of 0.1 percent or greater manufactured or imported 120 days or more after publication of a final rule be banned.

The Commission has decided that the ban on benzene as a contaminant in products need only apply to products manufactured or imported after the effective date, not to those held in inventory after the effective date. The Commission notes that these products contain significantly smaller amounts of benzene than those for which the earlier effective date is proposed. The Commission also believes that banning products containing benzene as an impurity manufactured or imported prior to the effective date of this Part 1307 might result in a chaotic situation since the products were manufactured prior to a significant amount of testing; and therefore, extensive testing would be needed to determine which products contain 0.1 percent or more of benzene.

SECTION 8—FINDINGS

Section 8 of the CPSA requires that, in order to propose a ban, the Commission must find that a consumer product is being or will be distributed in commerce, that it presents an unreasonable risk of injury and that no feasible consumer product safety standard under the CPSA would adequately protect the public from that unreasonable risk.

The Commission finds that the products banned by the proposal are being and will be distributed in commerce.

A. UNREASONABLE RISK OF INJURY

In determining whether a specific risk of injury is "unreasonable", the Commission generally balances the probability that the risk will result in harm and the gravity of the harm against a rule's effect on the product's utility, cost, and availability to the consumer. In this instance, the Commission notes that the risk of injury from benzene-containing products is great and includes leukemia as well as other blood disorders. In addition, the Commission staff risk assessment suggests an increased lifetime risk of death from myelogenous and monocytic leukemia of between 15.2 percent and 37.1 percent due to the use of benzene-containing paint strippers for 5 hours a day, ten times in a lifetime.

The Commission has considered the effect on the cost, utility, and availability of benzene-containing products and it is noted that this rule would eliminate the few such consumer products, except gasoline and laboratory benzene, currently on the market. However, the Commission also notes, as is discussed under Economic Considerations, that there are readily available substitute products at approximately the same cost which will yield a slightly less effective product.

After considering the nature and degree of the hazard from benzene-containing products and the availability of substitutes, the Commission concludes that the risk of injury clearly outweighs the effects of the proposed rule on the cost, utility, and availability of the products, and finds that benzene-containing products present an unreasonable risk of injury to the public.

B. FEASIBILITY OF A STANDARD

The Commission has considered all available information concerning the risk of injury associated with consumer products containing benzene as an intentional ingredient and as a contaminant and believes that the risk of injury is inherent in the products as long as they contain benzene.

The Commission believes that it is not possible to manufacture safe rubber cements or paint removers containing benzene because a safe level of exposure to benzene is not known. Furthermore, a standard such as OSHA's which sets a permissible air concentration level for benzene and requires protective clothing and equipment is not feasible for the consumer due to the cost of equipment and the knowledge required for its use.

In addition, except for laboratory use, the Commission notes that it does not believe that any required labeling for benzene-containing consumer products no matter how explicit and graphic would be adequate to address the risk of injury. (Current regulations under the FHSA at 16 CFR § 1500.14 (a)(3) and (b)(3) require cautionary labeling for products which contain 5 percent or more by weight of benzene including the words "danger", "vapor harmful", and "poison".) Since there is no known safe level of exposure to a carcinogen, the Commission believes that exposure to benzene must be reduced to the lowest possible level. The cautionary labeling, which would be necessary to approach that level, such as suggesting the use of a laboratory ventilation hood or the wearing of an adequate facemask when using benzene products, is impractical for the average consumer, outside of a supervised laboratory situation. The Commission notes in this regard that the type of respirator (chemical cartridge respirator) that would mitigate the effects of benzene-containing products is not generally available to households and even if it was, the specialized maintenance of this device would be burdensome to consumers. In addition, there is the possibility that any recommended protective measures may not be followed in practice or may be followed by the user but not by bystanders or visitors.

Therefore, the Commission concludes that a standard, including a labeling standard, for consumer products containing benzene as an intentional ingredient or as a contaminant is not feasible at this time and that only banning these products can adequately protect the public from the unreasonable risk of injury associated with them.

EXTENSION OF TIME TO PUBLISH FINAL RULE

Section 9(a)(2) of the CPSA (15 U.S.C. 2058(a)(2)) requires the Commission in declaring a product to be a "banned hazardous product" to publish a notice of proposed rulemaking in the FEDERAL REGISTER and to provide interested persons with an opportunity to participate in the rulemaking through submission of written data, views, or arguments. In addition to these written comments, interested persons must also be given an opportunity for the oral presentation of data, views, or arguments.

Section 9(a)(1) of the CPSA provides that within 60 days after publication of a proposed rule, the Commission shall promulgate a final rule or withdraw the proposal. The Commission may extend the 60 day period for good cause and must publish its reasons for extension in the FEDERAL REGISTER.

In the case of benzene-containing consumer products, the Commission has expressed a particular interest in obtaining public comment on the proposed contaminant level, on the issue of toxicity of substitutes, on the proposed effective dates, and on the various health and economic findings required by section 9(c) of the CPSA (see below). to afford an ample opportunity for obtaining such comment, the Commission has decided that, in addition to holding a public meeting in Washington, D.C., the comment period, in which the public is invited to submit written comments on the proposed ban, is to be longer than the 30 day period usually provided. Written comments may be submitted until June 30, 1978.

It is anticipated that because of the extended comment period and the need for careful analysis of the public's oral and written views, more time will be required than the 60 day period provided in section 9(a) of the CPSA between publication of a proposal and publication of any final rule.

Therefore, the Commission has decided that this 60 day period should be extended. In order to conserve the resources that would be needed to issue such an extension of time during the comment and evaluation period, the Commission announces in this proposal that good cause is found to extend for 90 days, until October 16, 1978, the time in which the Commission must either issue a final rule concerning benzene-containing consumer products or withdraw the proposal to ban. This period may be further extended for good cause and announced in the FEDERAL REGISTER.

CONCLUSION AND PROPOSAL

Based on the foregoing information, the Commission proposes to declare that all comsumer products, except gasoline and solvents or reagents for laboratory use, containing intentionally added benzene or containing benzene as a contaminant at levels of 0.1 percent or greater are banned hazardous products.

Accordingly, pursuant to provisions of the CPSA (sections 8, 9, and 30(d), 86 Stat. 1215-17, 1231, as amended 90 Stat. 506, 510; 15 U.S.C. 2057, 2058, 2079(d)), the commission proposes that Title 16, Chapter II, be amended by adding to Subchapter B the following new Part 1307:

Part 1307—Ban on Consumer Products Containing Benzene as an Intentional Ingredient or as Contaminant

Sec.
1307.1 Scope and application.
1307.2 Purpose.

Sec.
1307.3 Definitions.
1307.4 Exemptions.
1307.5 Consumer products containing benzene as an intentional ingredient as banned hazardous products.
1307.6 Consumer products containing benzene as a contaminant as banned hazardous products.

AUTHORITY: Secs. 8, 9, 30(d), Pub. L. 89-573, as amended, Pub. L. 94-284, 86 Stat. 1215-17, 1231 as amended 90 Stat. 506, 510 (15 U.S.C. 2057, 2058, 2079(d)).

§ Scope and application.

(a) In this Part 1307 the Consumer Product Safety Commission declares that all consumer products, except gasoline and solvents and reagents for laboratory use, containing intentionally added benzene are banned hazardous products under sections 8 and 9 of the Consumer Product Safety Act (CPSA) (15 U.S.C. 2057 and 2058). The Commission also declares that consumer products, except gasoline and solvents and reagents for laboratory use, containing benzene as a contaminant at levels of 0.1 percent or greater by volume are banned hazardous products under sections 8 and 9. This Part applies to products which are customarily produced or distributed for sale to or for the personal use, consumption, or enjoyment of consumers in or around a permanent or temporary household or residence, a school, in recreation or otherwise.

(b) The Commission has found (1) that the products described in subsection (a) are being or will be distributed in commerce, (2) that they present an unreasonable risk of injury, and (3) that no feasible consumer product safety standard under the CPSA would adequately protect the public from the unreasonable risk of injury associated with these products.

§ 1307.2 Purpose.

The purpose of this rule is to ban consumer products containing benzene as an intentional ingredient or as a contaminant. These products present an unreasonable risk of injury because exposure to the fumes increases the risk of developing blood disorders, chromosomal abnormalities, and leukemia.

§ 1307.3 Definitions.

(a) The definitions in section 3 of the Consumer Product Safety Act (15 U.S.C. 1052) apply to this Part 1307.

(b) "Benzene as an intentional ingredient" is benzene which is added deliberately to a product to impart specific characteristics.

(c) "Benzene as a contaminant" is any benzene appearing in a final product, which benzene is not an intentional ingredient. This includes any benzene which appears as a component of or impurity in a raw material used in the final product.

(d) "Imported" is the first entry of a product within a U.S. port of entry.

§ 1307.4 Exemptions.

The following categories of products are exempt from the ban established by this Part 1307:
(a) Gasoline.
(b) Benzene manufactured and distributed for use in laboratories as a solvent or reagent.

§ 1307.5 Consumer products containing benzene as an intentional ingredient as banned hazardous products.

On the basis that consumer products containing benzene as an intentional ingredient present the hazards of blood disorders, chromosomal abnormalities, and leukemia to the public, such products, except gasoline and solvents or reagents for laboratory use, manufactured or imported after the date which is 59 days after publication of the final rule in the FEDERAL REGISTER are banned hazardous products. Alternatively, such products, except gasoline and solvents or reagents for laboratory use, which are initially introduced into commerce 15 or more days after publication of a final ban are banned hazardous products. In addition, all other consumer products containing benzene as an intentional ingredient, no matter when manufactured or initially introduced into commerce, are banned hazardous products 30 days after the publication of a final ban.

§ 1307.6 Consumer products containing benzene as a contaminant as banned hazardous products.

On the basis that consumer products containing benzene as a contaminant at levels of 0.1 percent or greater by volume present the hazards of blood disorders, chromosomal abnormalities, and leukemia to the public, such products, except gasoline and solvents and reagents for laboratory use, manufactured or imported after the date which is 119 days after the publication of the final rule in the FEDERAL REGISTER, are banned hazardous products.

WRITTEN COMMENTS

Interested persons are invited to submit written comments before the close of business on June 30, 1978. Comments may be accompanied by written data, views, and arguments and should be submitted preferably in five copies, addressed to Secretary, Consumer Product Safety Commission, Washington, D.C. 20207. Received comments and other related material may be seen in the Office of the Secretary, Third Floor, 1111 18th Street NW., Washington, D.C., during working hours Monday through Friday.

ORAL PRESENTATIONS

Interested persons will be afforded an opportunity to make an oral presentation of data, views, or arguments in addition to the opportunity to make written comments, or any aspect of the proposed ban on June 14, 1978. The proceedings for the oral presentation will be held at 9:30 a.m. at the Commission's hearing room at 1111 18 Street NW., Washington, D.C. The procedural regulations for oral presentation, 16 CFR Part 1109, promulgated October 14, 1975 (40 FR 49122), shall govern this proceeding.

All persons wishing to make an oral presentation should notify Richard Danca of the Office of the Secretary, 202-634-7700, no later than close of business June 7, 1978, for scheduling purposes. A summary or outline of each oral presentation and an estimate of the length of time it will require should be filed with the Office of the Secretary at least business days prior to the oral presentation.

SECTION 9 FINDINGS

Section 9(c) of the CPSA (15 U.S.C. 2058(c)) requires the Commission, prior to promulgating a consumer product safety rule, to consider and make certain findings for inclusion on that rule. The Commission has discussed information relevant to these findings in the preamble of this document. However, in order to aid the Commission in obtaining the most complete and accurate information before making these findings, the Commission is particularly interested in receiving, either by written comment or through the oral presentation, additional data, views, and arguments as to the following:

1. The degree and nature of the risk of injury the rule is designed to eliminate or reduce;
2. The approximate number of consumer products, or types of classes thereof, subject to the rule;
3. The need of the public for consumer products subject to such rule and the probable effect of such rule upon the utility, cost, or availability of such products to meet that need;
4. Any means of achieving the objective of the rule while minimizing adverse effects on competition or disruption or dislocation of manufacturing and other commercial practices consistent with the public health and safety;
5. The necessity of the rule to eliminate or reduce the unreasonable risk of injury associated with the consumer products subject to the rule;
6. Whether the rule is in the public interest;
7. The feasibility of a consumer product safety standard under the CPSA to protect the public adequately from the unreasonable risk of injury associated with the products which are the subject of this proposed ban;

8. The potential environmental effects of the rule;

9. Available toxicity data regarding acute and chronic effects of all potential substitutes for benzene in products;

10. The appropriateness of the proposed contamination level for benzene in consumer products and the cost of achieving this level or any other suggested level;

11. The proposed effective dates;

12. Any adverse effects that this proposed ban will have on elderly and handicapped persons; and

13. The need and appropriateness of the gasoline and laboratory benzene exemptions.

Dated: May 15, 1978.

SADYE E. DUNN,
Acting Secretary, Consumer Product Safety Commission.

APPENDIX

Benzene is a clear, colorless, non-corrosive, flammable liquid with a rather pleasant odor. Benzene has a low boiling point and a high vapor pressure which causes it to evaporate rapidly, giving off vapors nearly three times heavier than air. These properties are responsible for the fact that considerable concentrations of benzene will rapidly appear in the air when benzene is used in open systems.

In the U.S. approximately 11 billion pounds of benzene were produced in 1976 (*1*). About 86 percent of this benzene was used as an intermediate in the production of other organic chemicals. The remainder was used in the manufacture of detergents, pesticides, solvents and paint removers. Benzene is also present in motor fuels and raises the octane levels of gasoline: In United States gasoline, the average benzene content is roughly one percent but ranges from 0.2-4.3 percent (*2*).

In consumer products benzene may be present in such products as rubber cements, paints, varnishes, stain removers and adhesives. Benzene has been recognized as being a toxic substance for over 75 years. It causes both acute and chronic effects. Considerable evidence has been collected that the chronic toxic effects of benzene include leukemia, bone marrow suppression and chromosomal abnormalities. Since benzene is implicated as a carcinogen and that for any carcinogen there is no zero risk level for a given population, exposure to this compound must be limited.

A. TOXIC EFFECTS OF BENZENE

1. *Metabolism.* The major route of entry of benzene into the body is via inhalation. Benzene rapidly reaches equilibrium with the air in the lungs and rapidly diffuses into the blood (*3*). In experiments with rats, the concentration of benzene in blood was shown to parallel the concentration in the inspired air (*4*). Benzene is retained in the body for many hours after exposure is terminated, and elimination through the lungs accounts for only 12 percent of the elimination during the first six hours after exposure.

Exposure to benzene can also be by direct contact, although benzene absorption through non-damaged human skin is not very great (*5-7*). When direct contact occurs, obviously there usually is inhalation of the vapors.

Benzene is highly lipid soluble and accumulates in various body organs in proportion to their fat content. It accumulates particularly in fat, bone marrow and the central nervous system. Elimination of benzene from the body takes place by respiration and by metabolism. Elimination of unchanged benzene from respiration ranges from 12-50 percent of the total amount absorbed, while very small amounts are excreted in the urine (*8*).

Benzene appears to be metabolized mainly by the liver. The first step appears to be a reaction that forms benzene oxide. This intermediate has been suggested as a possible candidate for the bone marrow toxicity of benzene (*9*). The principal metabolites are the phenols, formed by enzymatic hydroxylation. The phenols can then further processed by sulfation or glucuronization to give the conjugated end products catechol, hydroquinone and phenyl-mercapturic acid which are also found in the urine in small amounts as well as phenol (*7*). It has been reported that both catechol and hydroquinone –depress bone marrow activity, and therefore are also candidates for the chronic toxic effects of benzene (*8*).

2. *Acute toxicity.* Benzene concentrations of about 20,000 ppm are fatal within minutes, with death occurring from acute circulatory failure or coma. The symptoms of acute toxicity may include convulsions, muscle tremors, salivation, nystagmus and phenomena of very intense asphyxiation due to paralysis of the medullary respiratory center (*9*). Sub-lethal exposures produce nervous excitation, euphoria, headache, nausea and vomiting, weakness and abdominal pains. This will be followed by a period of depression which can lead to cardiovascular collapse and unconsciousness. Exposures of about 250-500 ppm gives signs and symptoms of mild poisoning, such as dizziness, drowsiness, headache and nausea. These usually disappear rapidly after removal from exposure.

Direct skin contact with liquid benzene also may cause redness and blistering. Chronic skin dryness and scaling have also been found.

3. *Chronic toxicity in man.* The major site of benzene's chronic toxicity is on the bone marrow, the blood-forming organ. In this section representative studies from the medical literature will be cited, demonstrating the diversity and seriousness of benzene exposure on hematopoiesis. This is not meant to be an all-inclusive review of this subject, rather an indication of the diseases that have been implicated to occur with chronic benzene exposure.

The emphasis in this section is primarily on studies that have evaluated substantial numbers of occupationally exposed people, particularly where measurement of dose has been attempted.

In most reported cases of benzene hematoxicity, exposure has been in a working situation where benzene was used as a solvent or manufactured as a product. One problem is that exposure usually occurs to a mixture of volatile compounds, rather than to benzene alone. Despite this, the overwhelming evidence points to an association of hematopoietic disease with benzene. The evidence includes the fact that benzene has been the common-denominator in various settings where exposure to other compounds has varied extensively; there are good epidemiological studies showing that outbreaks of hematoxicity followed the introduction of benzene to the workplace and responded to replacement of benzene with other solvents; and that bone marrow toxicity occurs in animals exposed only to benzene.

It is possible that compounds inhaled along with benzene alter the expression of hematotoxicity. Both other aromatic hydrocarbons, as well as chemically unrelated chemicals could conceivably modify benzene metabolism in man.

(a) *Pancytopenia and aplastic anemia.* Pancytopenia refers to a decrease in all of the formed elements of the blood, red blood cells (erythrocytes), white blood cells (leukocytes), and platelets (thrombocytes). In benzene exposures these elements decrease primarily due to a decrease in production. The term aplastic anemia is often used to describe benzene toxicity. However, this term classically denotes a condition in which pancytopenia is accompanied by a marked decrease in the number of hematopoietic precursor cells within the bone marrow (hypoplasia). In benzene toxicity manifested by pancytopenia, the bone marrow studies have varied, with hypoplasia (*11-13*), normal cellularity (*11, 14*), and even hyperplasia (*11, 14*), being found. For the purpose of this discussion, pancytopenia will be discussed as a single entity, regardless of the bone marrow findings (if performed).

Among the earliest investigations of pancytopenia in benzene exposure were those of workers in rotogravure printing plants (*15-17*). In the studies, the workers had been exposed to benzene for a period of 3-5 years. The benzene contents of the ink solvents and thinners were between 10-80 percent. The air concentrations were analyzed and ranged from 11-1,060 ppm with a median level of 132 ppm. The workers complained of fatigue, lethargy, dizziness, headache, shortness of breath and dryness of the mucuous membranes. Hematological studies were performed in 332 exposed male workers and 81 controls. Sixty-five of these people had some degree of hematological toxicity while 23 were severely affected. Six of these people were hospitalized; the remainder continued working and, following discontinuation of benzene use, recovered (*17*). These investigators found anemia, macrocytosis (large red cells) and thrombocytopenia (low platelet count). They rarely observed neutropenia (low granulocytic white cell count), but a decrease in lymphocytes appeared more commonly. They also found in a few cases prolongation of the bleeding or clotting times, an increase in capillary fragility and an increase in serum bilirubin and reticulocytes.

Wilson (*10*) in 1942 reported a study of 1,104 workers in an American rubber factory exposed to a maximum level of 500 ppm benzene, with an average of approximately 100 ppm. Of these, 83 were found to have mild hematological abnormalities and 25 more had severe pancytopenia. Nine of these 25 were hospitalized and 3 died as a result of the pancytopenia.

A study in England of 200 women exposed to benzene in 13 aircraft factories in England was reported in 1944 (*18*). This was in the manufacture of rubber and the women had been exposed in solvents containing 5-20 percent aromatic hydrocarbons including benzene. In this study, neutropenia was the most consistent indicator of benzene toxicity. No significant differences in red cells or lymphocyte counts was found.

In a rubber raincoat factory in Sweden (*19*), 189 people were studied. The benzene levels were measured and found to be 137-218 ppm. Hematological toxicity was found in 60 people. After 16 months of no further benzene exposure, 46 had recovered, 12 still had significant abnormalities and 2 had died. Prominent in this group was the finding of thrombocytopenia.

In an Australian air force workshop study (20) of 87 exposed people there was one case of fatal pancytopenia. The benzene peak concentration ranged between 10-1,400 ppm and in most areas the benzene concentration was well above 100 ppm. The average concentration was 10-35 ppm. The solvents contained up to 53 percent benzene.

In 1961 (21) a study was performed in rubber coating plants where petroleum naphtha containing 1.5-9.3 percent benzene was in use. Benzene levels were usually less than 25 ppm but did go as high as 125 ppm. In one plant, 5 of 32 men had anemia; in a second plant, 1 of 9 was anemic and in a third plant, all workers studied were normal.

In the Soviet Union, Doskin (22) reported an evaluation of 365 workers exposed to benzene for 1-3 years. The benzene levels ranged between 10-40 ppm. It was reported that 40 percent of the workers had hematological changes during the first year. The most common early sign was thrombocytopenia, followed by anemia. An initial increase in white blood cell count was followed in some cases by leukopenia.

In another study, Kozlova and Volkova (23) studied 252 workers exposed to benzene over a 5-year period. Initially the benzene concentrations ranged from 47-310 ppm, but by the end of the 5 years had decreased to 25-47 ppm. The workers were classified into 3 groups according to exposure levels, the lowest being 24-39 ppm. In all 3 groups there was a decrease in all cell counts, with the most severe cases being in the highest exposure group. These investigators found leukopenia (mainly neutropenia) and thrombocytopenia as the most prominent early findings. Red cell changes were noted as a later phenomenon.

Savilakti et al. (24, 25) evaluated people exposed to benzene in a shoe factory in Finland. In this group of 147 persons, 107 were found to have abnormal blood counts, the most common being thrombocytopenia. Ten people required hospitalization and one died of severe pancytopenia. In 1964 (9 years after cessation of benzene exposure) 125 of these workers were re-examined. Combined assessment of the cell counts revealed the difference between the exposed and the controls was significant for males, but not for females. There was a greater tendency towards recovery of leukocytes rather than erythrocytes or platelets.

Girard and colleagues in France (26, 27) questioned all of their patients with hematological diseases concerning benzene exposure and reported the percentage of individuals with a given hematological disease and known exposure to benzene. Pancytopenia had the highest association. Of all of their patients with the diagnosis of aplastic anemia, 20.8 percent had exposure to benzene or toluene. In contrast, only 4 percent of controls (non-hematological disease) had exposure to these agents.

Aksoy et al. (28) in Turkey examined 217 apparently healthy men who had worked for 3 months to 17 years in shoe factories using an adhesive containing benzene. These workers were exposed to 30-210 ppm benzene. In 51 of these people some form of hematological abnormality was found. Anemia was the most common finding, but since it was correctable by iron, it was not ascribed to benzene. However, the volume of the cells (mean corpuscular volume) was high normal to slightly elevated range which is unusual for iron deficiency and is more consistent with what has been seen with benzene (macrocytosis). Leukopenia with or without thrombocytopenia was the most common finding attributed to benzene. Pancytopenia was seen in 6 people.

In 1977 Aksoy et al. (29) reported they had previously studied 46 patients with aplastic anemia who had been exposed to benzene. Of these patients 14 had died of aplastic anemia, 5 had developed acute leukemia, 1 had developed myeloid metaplasia, 22 were in complete remission, 2 were still under treatment and 2 were lost to follow up. All of these people had been exposed to adhesives containing 9-88 percent benzene and exposure was generally in small shops with poor ventilation. Where measured, the benzene concentration was 150-650 ppm.

In addition to this data in humans there is strong support from the medical literature, using animal models, that benzene causes pancytopenia. The route of administration to animals has been both by inhalation and injection (usually subcutaneously).

In the inhalation studies, Weiskatten et al. (30) produced leukopenia, slight anemia and hemorrhages in rabbits exposed to approximately 240 ppm benzene for 10 hours a day. The effects were observed within 2 weeks of exposure. This study found that "small mononuclear white cells" (probably lymphocytes) were the most depressed.

Svirbely et al. (31) studied rats and dogs exposed to 1000 ppm benzene 5 days a week, 7 hours a day for 28 weeks. The rats had intermittent leukopenia and lymphocytopenia, while the dogs had lymphocytopenia throughout the study.

Li et al. (32) found varying results in dogs exposed to 600 ppm benzene for 42 hours per week over periods of time dependent on their response to benzene. Some animals (least affected) could be exposed to up to one year, while others became moribund in 5-6 weeks. These investigators monitored leukocytes and platelets and found a decrease in the number of both. Protein deficient diets produced greater reductions, which were exacerbated by high fat diets.

Wolff et al. (6) exposed rats, guinea pigs and rabbits to benzene for 7 hours periods for various numbers of days. They stated that they saw leukipenia at levels as low as 80 ppm.

Deichmann et al. (4) exposed rats to benzene at various doses for 5 hours per day, 4 days a week. They report leukopenia at a benzene concentration as low as 44 ppm for 2-8 weeks. They did not observe leukopenia at concentrations of 31, 29 or 15 ppm. Female animals were more sensitive than males.

Nau et al. (33) found leukopenia in rats with exposure to 1000, 200, and 50 ppm benzene. The reported toxicity occurred more rapidly with the higher doses, but did occur at 50 ppm after long exposures.

Ikeda (34) exposed rats to 1000 ppm benzene for 7 hours a day, 5 days a week for 60 days. They noted leukopenia which decreased in the following order: adult males, young males, adult females, young females.

Baje et al. (35) noted leukopenia in rats exposed to 400 ppm benzene 7 hours a day for 13 weeks.

Benzene-induced hematological toxicity in animals has also been produced by administering benzene by injection. Santession (36) in 1897 was the first person to suggest benzene as a cause of hemorrhagic phenomena by subcutaneous injection of rabbits. Selling (37) was able to demonstrate severe leukopenia in rabbits by injection of 1 ml/kg benzene (as a 50% solution in olive oil) per day in 4-9 days. In Selling's study complete aplasia of the bone marrow was first noted.

Gerarde (38) performed similar experiments in rats treated daily for 2 weeks with 1 ml/kg of benzene (50 percent solution in olive oil). Leukopenia was noted during this time which was reversible.

Latta and Davies (39) gave rats 50 percent benzene in olive oil at 2, 3, and 4 ml/kg daily for up to 60 days. They noted first a temporary stimulation of white cell production. After this the leukocyte count fell with lymphocyte production being more severely impaired. With the high dose they found disappearance of bone marrow myelocytes after 24 days and complete aplasia after 60 days.

Speck et al (11-13) gave benzene to rabbits at doses of 0.2 and 0.3 ml/kg per day and produced severe leukopenia within 1-9 weeks. They also noted a decrease in red cells, hemoglobin, reticulocytes and platelets. Pancytopenia was a common finding. The bone marrows from the pancytopenic animals were interpreted as very hypoplastic (21 percent), hypoplastic (32 percent), normal (26 percent) and hypercellular (21 percent).

Lee et al. (40) treated mice with benzene in corn oil at 400 mg/kg and 2200 mg/kg as single doses and demonstrated that benzene depresses the incorporation of radioactive iron into circulating red cells. Their results are compatible with an effect of benzene at the early bone marrow precursors of red cells.

Summary of Benzene-Associated Pancytopenia

Benzene exposure has been clearly demonstrated to produce hematotoxicity in man. The most commonly reported effect is a decrease in one or more of the formed elements of the blood. In more severe cases this will take the form of pancytopenia, including in some cases an aplastic bone marrow. The evidence that benzene is causally related to pancytopenia includes studies showing that outbreaks of pancytopenia throughout the world have followed the introduction of benzene to the workplace and that these outbreaks will stop when benzene is substituted for by other solvents.

As confirmatory evidence, it is possible to induce similar hematological effects in animals treated with benzene. At the present time it is not possible to establish a definitive, safe dose or time exposure, nor is it possible to state what (if any) age, sex, social or ethnic groups are more or less susceptible to benzene toxicity.

(b) *Leukemia*: Leukemia is a term for a group of diseases with an abnormal proliferation of white blood cells or their precursors in the peripheral blood and bone marrow, often with the appearance of abnormal immature white cell precursors in the blood. Leukemia is considered to be a malignant neoplasm. Evidence in man industrially exposed to benzene has strongly implied that benzene is directly related to the development of acute myeloblastic leukemia and its variants such as erythroleukemia and acute myelomonocytic leukemia. There have been numerous case reports in the medical literature suggesting that there is this association. However, in this report the more recent larger epidemiological studies will be reviewed.

In the past 10 years, Aksoy and his co-workers in Turkey have performed careful studies strongly suggesting that benzene exposure leads to acute leukemia (29, 41-44). Their evidence includes reports of workers with aplastic anemia which progressed through a "pre-leukemic" phase to a typical acute myeloblastic leukemia or erythroleukemia. The outbreak of leukemia occurred in shoemakers when benzene use in an adhesive was introduced. Leukemia cases subsided after benzene use was discontinued.

Aksoy et al. (27, 41, 42) observed 26 patients with acute leukemia in this group of shoemakers. The benzene concentration in air reached a maximum of 210-650 ppm when the benzene-containing adhesives were in use. The cases included 14 cases of acute myeloblastic leukemia, 3 of acute lymphoblastic leukemia, and 1 each of acute monocytic and promyelocytic leukemia. These investigators had evaluated 28,500 shoe workers in Istanbul and calculated a leukemia risk of 13 per 100,000 in this group (29). This is compared to an incidence of leukemia in Turkey of 2.5-3.0 per 100,000. In order to be extremely careful in these studies, Askoy et al compared their results to the incidence of leukemia in developed nations (6 per 100,000). The incidence of leukemia in shoeworkers in Turkey was statistically higher ($p < 0.02$) than this number. These studies, therefore, if anything, underestimate the risk of leukemia on benzene exposure. Furthermore, these investigators did not age-adjust their findings, which also underestimates the risk of leukemia.

Recently, Aksoy (29) summarized his experiences with this group of patients. He reported that between 1967-1971 there was a notable incidence of pancytopenia followed by a peak incidence of leukemia between 1971-1973. Since 1969 there was a gradual decline in the use of benzene in this occupation, and there was a decline in leukemia after 1973, with no cases in 1976. Aksoy also reported that he was able to identify pancytopenia in 27.5 percent of the cases 6 months to 6 years before the onset of acute leukemia. The pancytopenia appeared to be recovering before the onset of acute leukemia. Aksoy (29) has stated that he is of the opinion that pancytopenia is not a necessary precursor to acute leukemia.

Another series of studies have been reported by Vigliani and his colleagues in Northern Italy. In recent reviews Vigliani (46) and Vigliani and Forni (47) summarized their experiences from 1942-1975. In Pavia they reported 135 patients with benzene hematotoxicity of which 13 died of acute myeloblastic leukemia (or its variants) and 3 of aplastic anemia. In Pavia the patients were workers in the shoe industry and were exposed to adhesives containing benzene. Benzene concentrations ranged from 25-600 ppm, but usually were 200-500 ppm. In Milan they saw 66 cases of benzene hematotoxicity, of which 11 were cases of acute myeloblastic leukemia (or its variants). In addition to these 11 fatal cases of leukemia, 7 additional patients died of aplastic anemia. In Milan many of the patients were also in the shoe industry, but other occupational groups were also affected. They reported an outbreak of benzene hematotoxicity in the rotogravure industry which occurred when inks and solvents containing benzene were introduced to the workplace. Benzene concentrations in the air were 200-1500 ppm. Vigliani has reported that he knows of at least 150 cases of benzene-induced leukemia in Italy (46), and has also reported a 20-fold higher risk of acute leukemia in workers exposed to benzene (48).

In France, Girard and his colleagues (49, 50) have evaluated the frequency of a benzene exposure in 401 patients hospitalized with hematological diseases as compared to a control group of 124 patients hospitalized for nonhematological diseases. They found a statistically significant increase in benzene exposure in patients with aplastic anemia, acute leukemia, and chronic lymphocytic leukemia as compared to the control.

Recently in the USA, Infante et al. (51) have performed an epidemiological evaluation of benzene-exposed workers in two "Pliofilm" factories. These investigators report a five-fold increased risk of total leukemia and a 10-fold risk of acute myeloblastic leukemia (or its variant, myelomonocytic leukemia) in these workers. Since they were unable to determine the status of 25 percent of the workers and assumed that they were still living, if anything, this study would underestimate the risk of leukemia.

In contrast to the pancytopenic effects of benzene where there are abundant animal models, there has only been one study reported in 1932 (52) claiming that animals receiving benzene developed leukemia. The absence of controls and the uncertainty as to the strain of mice in this study makes it impossible to evaluate this study. While at least one study is presently being carried out in animals, it is too premature to draw conclusions from it (53).

Summary of Benzene-Associated Leukemia

Benzene exposure has been associated with an increased risk of development of acute myeloblastic leukemia and its variants. Most of this evidence has been collected from epidemiological studies of workers exposed to benzene, showing an increased number of cases of leukemia in these groups. Substitution of other solvents for benzene results in a decreased number of cases. There are no animal model systems at the present time. It is not yet possible to establish a safe dose or time exposure, nor is it possible to identify any group of people more susceptible to a leukemogenic effect of benzene. It is not known whether pancytopenia must precede leukemia.

(c) *Other abnormalities possibly associated with benzene exposure:* While pancytopenia and acute myeloblastic leukemia are the diseases most commonly considered with benzene exposure, there are other abnormalities which have been reported to occur.

Abnormalities in leukocyte function have been reported by a number of investigators. These changes could be important, as these cells are necessary to fight infection. Among the granulocyte abnormalities are a decrease in phagocytic function (54), a decrease in leukocyte alkaline phosphatase activity (55, 56), an alteration in osmotic fragility (57), a change in fluorescent characteristics of the nuclei (58) and the presence of the Pelger-Huet anomaly of the nuclei (28, 59-62). The Pelger-Huet anomaly, when it is acquired, has been associated with leukemias.

Lymphocytopenia has been reported to be a common finding in benzene toxicity (28) and Goldwater suggested that this is even more common than granulocytopenia (16). Other investigators have failed to observe this, and there are some reports of lymphocytosis (62, 63) in man. In animals, several groups have shown that benzene decreases lymphocytes and lymphoid tissues (64, 66). Snyder et al. (65) have recently shown that lymphocytopenia is a very early manifestation of toxicity in rats and mice exposed to 100-300 ppm benzene in the air. Further studies have suggested that there is an altered immune function in benzene-exposed people. Included in these have been findings of leukocyte agglutinins (66), decreased serum immunoglobinulins (67, 68) and complement levels (69) and antibodies against red cells, platelets, and leukocytes (70).

The "immune system" of the body includes lymphocytes, lymphoid tissue, immunoglobulins, and complement. This system is involved in protecting people from substances "foreign" to the body, such as infectious agents. It may also be involved in protection against development of malignancies. The noted effects of benzene on the immune system then raise the question of whether the association with leukemia might arise from an altered immune surveillance. Another possibility is that benzene somehow alters the bone marrow so that it is recognized as "foreign" and is attacked by the so-called "suppressor lymphocytes" which have recently been postulated to be involved in aplastic anemia in people (71, 72).

Still other leukocyte findings have been reported which include an increased number of another type of white cell, the eosinophil (24, 25), as well as the monocyte, (73) with abnormal-appearing monocytes present.

Abnormal platelet function has also been reported in people exposed to benzene independent of a decrease in the number of platelets (74-83). Morphdogical abnormalities in peripheral blood platelets, as well as the bone marrow precursor, the megakaryocyte (28, 80, 81, 82, 83) have been reported. This also suggests that the coagulation factors are abnormal (75, 79).

Macrocytosis (large volume of the red cells) is the most frequently reported abnormality of the red cell associated with benzene toxicity (16, 19, 28, 85-88). Large red cell precursors (in the bone marrow, macroerythroblasts) have also been reported (89, 90). It has been suggested that macrocytosis is of prognostic importance and might be a warning signal of benzene hematotoxicity (91, 92).

Significant destruction of red cells (hemolysis) has also been reported in some people exposed to benzene (84, 93-97). There have also been a few cases of paroxysmal nocturnal hemoglobinuria (98), a rare hemolytic disorder due to an acquired defect of the red cell membrane. Of note is that this rare disorder has been associated with the development of both aplastic anemia and acute myeloblastic leukemia.

In recent years studies have been performed both in people exposed to benzene (99, 100) and in an in vitro rabbit reticulocyte system (101-105) showing that benzene affects heme synthesis by inhibiting the enzymatic system necessary to synthesize this compound. Heme is a porphyrin which is a portion of the hemoglobin molecule, but is also a part of many other proteins. It is found in all cells, but it is in greatest amounts in red cells and then liver cells. Recent work has shown that heme is necessary for maximal protein synthesis in most if not all cells (106). When heme synthesis in red cell precursors is decreased due to a deficiency of iron (iron deficiency anemia and the anemia of chronic disease) less protein (including hemoglobin) and less red cells, will be produced. When the enzymes necessary to synthesize heme are inhibited protein snythesis and red cell production are also inhibited, but iron accumulates in the mitochondria of these cells. The accumulation of iron in the mitochondria gives a characteristic morphological picture and is termed a "sideroblastic anemia" (107). Of note is that "sideroblastic anemia" is a common presentation of a "preleukemia". It has been postulated (108-109) that a deficiency of heme in bone marrow pluripotential stem cells (precursors of all the marrow elements) could conceivably result in an inability to make protein and thus an inability to mature, divide, and make any of the formed cells of the blood.

While the association of benzene exposure and acute myeloblastic leukemia appears definite, the association with other types of leukemia and malignancies is still unclear. In France it has been noted that there is a relatively high incidence of chronic lymphocytic leukemia, often greater than that of acute myeloblastic leukemia in people exposed to benzene (49, 50, 110). Furthermore, three cases of chronic lymphocytic leukemia have been reported from the Soviet Union (110). In contrast, Aksoy in Turkey, Vigliani in Italy, and Infante in the United States have not reported this association. The reason for this difference is not yet clear.

There is also inconclusive evidence suggesting that benzene might be associated with other diseases, including acute lymphoblastic leukemia, lymphomas (110-118), chronic myelogenous leukemia, and myelofibrosis and myeloid metaplasia (98, 119-121).

Evidence has been presented that benzene causes chromosomal aberrations. Since most (if not all) carcinogens are mutagens, the observation that benzene causes chromosomal damage in man suggests that it is also a mutagen. Even though there is a rapid, convenient, accurate in vitro system ("The Ames Test") for mutagen testing (122), apparently benzene is too toxic for use in this test system (123). In any event the assessment of damage to mammalian chromosomes is probably much more directly relevant to estimation of human health hazards from a mutagen, than any in vitro test system.

Chromosomal damage can lead to cell death if it interferes with cell division (mitotic poison). If it does not interfere with cell division, the lesion is replicated and a mutation may result. Not all mutations will lead to malignant changes, but there is an increased likelihood of a malignancy when mutational events occur.

The mitotic poison effect of benzene has been well demonstrated. There is a decrease in DNA synthesis in in vitro cultures of human (124, 125) and animal cells after in vivo exposure to benzene (11-13, 38, 126, 127). A defect in RNA synthesis has also been reported (11) in the basophilic normoblast, but not in earlier red cell precursors. In these studies (11) it was implied that marrow from benzene-treated mice, injected into other mice (lethally irradiated), could still form colonies of bone marrow cells in the spleens. This is evidence that the pluripotential stem cells are not affected by benzene. These studies in animals used very high doses of benzene (0.2-0.3 ml/kg/day at least 3 times per week for 6-12 weeks). The studies were performed only after the animals had become pancytopenic. Of note is that toluene and xylene have not been toxic in these animal systems. Indeed, a recent report by Andrews et al. (128) has shown that toulene can protect against benzene-induced decreased incorporation of ^{59}Fe into red cells.

There have been several studies attempting to locate a nuclear lesion within cells of humans exposed to benzene. In healthy workers with a history of exposure to benzene, Tough and Court-Brown (129-130) have noted unstable chromosome damage such as breaks and gaps in cultured lymphocytes. Chromosomal abnormalities in people exposed to benzene without any clinical abnormalities have also been noted in other studies (55, 131, 132). These changes were not noted in workers exposed to toluene (133).

In addition there have been numerous reports of chromosomal abnormalities in patients with hematological disease and a history of benzene exposure (44, 133-136). However, it is impossible to state whether or not these abnormalities are indeed due to benzene or a result of the underlying disease.

Summary. Benzene exposure has been implicated as causing abnormalities in the function of leukocytes, platelets, red cells, and the immune system. In addition, abnormalities in prophyrin and heme synthesis have been reported. It is possible, but not yet proven, that there is an increased risk of developing chronic leukemias and lymphomas. The abnormalities in chromosomes seen with benzene exposure are consistent with a mutagenic potential of this agent. However, these changes could result from the proven ability of benzene to act as a mitotic poison.

B. CONCLUSIONS

(1) Chronic benzene exposure has been shown to be associated with an increased risk of developing hematological disorders including pancytopenia, acute myeloblastic leukemia (and its variants, erythroleukemia and myelomonocytic leukemia), and chromosomal abnormalities. There is also evidence that benzene exposure causes abnormalities in porphyrin and heme synthesis and is associated with qualitative abnormalities in leukocytes, platelets, red cells, and the immune system. It is possible that there is also an increased risk of developing other leukemias, lymphomas, and hematological disorders.

(2) Neither the dose of benzene nor the time of exposure required to develop these complications is known. Neither is it known if benzene or a metabolite of benzene is the toxic agent. No particular groups have been definitively shown to be more susceptible to benzene.

(3) In consumer products there is no possible control as to the method of use, extent of exposure or length of exposure. An experiment to determine the amount of benzene exposure from a consumer product was recently performed by NIOSH (137). They used a paint remover purchased over the counter which was found by gas chromatography to contain 52 percent benzene by volume. To simulate a typical exposure, an end table was stripped in a home garage. Five samples of air were collected and found to contain 73-225 ppm benzene, with a mean of 130 ppm. These results were similar to that reported previously by Otterson measuring air concentrations of benzene in commercial furniture stripping factories. Therefore, the presence of benzene in consumer products constitutes a health hazard.

(4) There is no evidence at present that toluene or xylene which have been used as substitutes for benzene have the same chronic effects as benzene.

REFERENCES

1. Wrenn, G. Testimony of, to occupational Safety and Health Administration, U.S. Dept. of Labor, July 19, 1977.
2. Sexton, J. Testimony of, to Occupational Safety and Health Administration, U.S. Dept. of Labor, July 27, 1977.
3. Hunter, C. G. and Blair, D. Benzene: Pharmacokinetic studies in man. Ann. Occup. Hyg. 15:193-199, 1972.
4. Deichmann, W. B., MacDonald, W. E. and Bernal, E. The hemopoietic tissue toxicity of benzene vapors. Toxicol. and Applied Pharmacol. 5:201-224, 1963.
5. Bowditch, M. and Elkins, H. B. Chronic exposure to benzene (benzol). I. The industrial aspects. J. Indust. Hyg. and Toxicol. 21:321-330, 1939.
6. Wolff, M. A., Rowe, V. K., McCollister, D. D., Hollingsworth, R. L., and Oyen. F. Toxicological studies of certain alkylated benzenes and benzene. AMA Arch. Indust. Health, 14:387-398, 1956.
7. Conca, G. L. and Maltagliata, A. Studies on the absorption of benzene through the skin. Med. Lavora 46:194-198, 1955. Abstracted Indust. Hyg. Digest 20:13, 1956.
8. Rusch, G. M., Leong, B. K. S. and Laskin, S. "Benzene Metabolism" in: *A Critical Review of Benzene Toxicity* (S. Laskin and B. D. Goldstein, Eds) in press, 1977.
9. Jerina, D. M. and Daly, J. W. Arene oxides: A new aspect of drug metabolism. Science 185:573-582, 1974.
10. Wilson, R. H. Benzene poisoning in industry. J. Lab. Clin. Med. 27:1517-1521, 1942.
11. Kissling, M. and Speck, B. Further studies in experimental benzene-induced aplastic anemia. Blut. 25:97-103, 1972.
12. Moeschlin, S. and Speck, B. Experimental studies on the mechanism of action of benzene in the bone marrow (radioautographic studies using ^{3}H-thymidine). Acta. Haematol 38:104-111, 1967.
13. Kissling, M. and Speck, B. Chromosome aberrations in experimental benzene intoxication. Helv. Med. Acta. 36:59-66, 1972.
14. Hunter, F. T. Chronic exposure to benzene (benzol). II. The clinical effects. J. Indust. Hyg. and Toxicol. 21:331-354, 1939.
15. Greenberg, L., Mayers, M. R., Goldwater, L. and Smith, A. R. Benzene (benzol) poisoning in the rotogravure printing industry in New York City. J. Indust. Hyg. and Toxicol. 21:395-420, 1939.
16. Goldwater, L. J. Disturbances in the blood following exposure to benzol. J. Lab. Clin. Med. 26:957-973, 1941.
17. Goldwater, L. J., Tewksbury, M. P. Recovery following exposure to benzene (benzol) J. Indust. Hyg. 23:217-231, 1941.
18. Hamilton-Peterson, J. L., Browning, E. Toxic effects in women exposed to industrial rubber solutions. Br. Med. J. 1:349-352, 1944.
19. Helmer, K. J. Accumulated cases of chronic benzene poisoning in the rubber industry. Acta. Medica. Scand. 118:354-375, 1944.
20. Hutchings, M., Frescher, S., McGovern, F. B., and Coombs, F. A. Investigation of benzol and toluol poisoning in Royal Australian Air Force Workshop. Med. J. Australia 2(23):681-693, 1947.
21. Pagnotto, L. D., Elkins, H. B., Brugsch, H. G. and Walkley, E. J. Industrial benzene exposure from petroleum naptha-I in the rubber coating industry. Am. Ind. Hyg. Assoc. J. 22:417-421, 1961.

22. Doskin, T. A. Effect of age on the reaction to a combination of hydrocarbons. Hygiene and Sanitation, 36:379–384, 1971.
23. Kaslova, R. A. and Volkova, A. P. Blood picture and phagocytic activity of leukocytes in workers having contact with benzene. Gig. Sanit. 25:29–34, 19??
24. Savilahti, M. Over 100 cases of benzene poisoning in a shoe factory. Archiv fur Gewerbepathologie und Gewerbehygiene 15:147–157, 1956.
25. Hernberg, S., Savilahti, M., Ahlman, K., and Asp, S. Prognostic aspects of benzene poisoning. Br. J. Indust. Med. 23:204–209, 1966.
26. Girard, R., Prost, G., and Tolot, F. Comments on indemnification for benzene-induced leukemia and aplasia. Arch. Mol. Prof. 32(9) 581–583, 1971.
27. Girard, R., Tolot, F., and Bourret, J. Malignant hemopathies and benzene poisoning. Medicina del Lavoro 62:71–76, 1971.
28. Aksoy, M., Dincol, K., Akgun, T., Erdem, S., and Dincol, G. Haematological effects of chronic benzene poisoning in 217 workers. Br. J. Indust. Med. 28:296, 302, 1971.
29. Aksoy, M. Testimony of, to the Occupational Safety and Health Administration. U.S. Dept. of Labor, July 13, 1977.
30. Weiskotten, H. C., Schwartz, S. C., and Steensland, H. S. J. Med. Res. 41:425, 1920.
31. Svirbely, J. L., Dunn, R. C., and von Oettingen, W. F. The chronic toxicity of moderate concentrations of benzene and its homologues for rats and dogs. J. Indust. Hyg. and Toxicol. 26:47–46, 1944.
32. Li, T. W., Freeman, S. The effect of protein and fat content of the diet upon the toxicity of benzene for rats. Am. J. Physiol. 145:158–165, 1945.
33. Nau, C. A., Neal, J., Thornton, M. C_5–C_{12} fractions obtained from petroleum distillates. An evaluation of their potential toxicity. Arch. Env. Health 12:382–393, 1966.
34. Ikeda, M. Enzymatic studies on benzene intoxication. J. Biochem. 55:231–243, 1964.
35. Baje, H., Benkel, W., and Heiniger, H. J. Untersuchungen zur leukipoese im knochenmark der ratte nach chronischer, benzol-inhalation. Blut 21:250–257, 1970.
36. Santresson, C. G. Ueber chronische vergiftungen mit steinkohlen theerbenzin: vier to desfalle. Arch. Hyg. Berlin. 31:336–376, 1897.
37. Selling, L. Benzol as a leucotoxin. Studies on the degeneration and regeneration of the blood and hematopoietic tissue. Johns Hopkins Hospital Reports 17:83–142, 1916.
38. Gerarde, H. W. and Ahlstrom, D. B. Toxicologic studies on hydrocarbons. XI. Influence of dose on the metabolism of mono-n-alkyl derivatives of benzene. Toxicol. and Applied Pharmacol. 9:185–190, 1966.
39. Latta, J. S. And Davies, L. T. Effects on the blood and hematopoietic organs of the albino rat of repeated administration of benzene. Arch. Pathol. 31:55–67, 1941.
40. Lee, E. W., Kocsis, J. J. and Snyder, R. Acute effects of benzene on ^{59}Fe incorporation into circulating erythrocytes. Toxicol and Applied Pharmacol 27:431–433, 1974.
41. Aksoy, M., Dincol, K., Erdem, S. and Dincol, G. Acute leukemia due to chronic exposure to benzene. Am. J. Med. 52:160–166, 1972.
42. Aksoy, M., Erdem, S. and Dincol, G. Types of leukemia in chronic benzene poisoning. A study in thirty-four patients. Acta Haematologica 55:65–72, 1976.

43. Aksoy, M., Erdem, S. and Dincol, G. Leukemia in shoe workers exposed chronically to benzene. Blood 44:837–841, 1974.
44. Aksoy, M., Erdem, S., Erdogan, G. and Dincol, G. Acute leukaemia in two generations following chronic exposure to benzene. Human Heredity 24:70–74, 1974.
45. Aksoy, M., Erdem, S., Erdogan, G. and Dincol, G. Combination of genetic factors and chronic exposure to benzene in the aetiology of leukaemia. Human Heredity 26:149–153, 1976.
46. Vigliani, E. C. Leukemia associated with benzene exposure. Ann. N. Y. Acad. Sci. U.S.A. 271:143–151, 1976.
47. Vigliani, E. C., Forni, A. Benzene and leukemia. Environ. Res. 11:122–127, 1976.
48. Vigliani, E. C. and Saita, G. Benzene and leukemia. New Engl. J. Med. 271:872–876, 1964.
49. Girard, R., Rigaut, P., Bertholon, J., Tolot, F. and Bourret, J. Les expositions benzeniques meconnues. Leur recherche systematique au cours des hemopathie hospitalises. Arch. Mol. Prof. 29:723–726, 1968.
50. Girard, R., Reval, L. La frequence d'une exposition benzinique au cours des hemopathies graves. Nouv Revue Fr. Hemat. 10:477–484, 1970.
51. Infante, P. F., Rinsky, R. A., Wagoner, J. K. and Young, R. J. Leukaemia in benzene workers. Lancet 2:76–78, 1977.
52. Lignac, G. O. E. Benzene leukemia in humans and white mice. Krankheitsforsch 91:403–453, 1932.
53. Goldstein, B. Testimony of, to the Occupational Safety and Health Administration. U.S. Dept. of Labor, Aug. 10, 1977.
54. Kaslova, T. A. and Volkova, A. P. Blood picture and phatocytic activity of leukocytes in workers having contact with benzene. Gig. Sanit. 25:29–34, 1960.
55. Girard, R., Mallein, M. L., Bertholon, J., Coeur, P. and Cievreux, J. Etude de la phosphatase alcaline leucocytaire et du caryotype des ouvriers exposes au benzene. Arch. Mol. Prof. 31(1-2):31–38, 1970.
56. Girard, R., Mallein, M. L., Bertholon, J., Coeur, P. and Tolot, F. Leukocyte alkaline phosphatase and benzene exposure. Med. Lavora. 61:502–508, 1970.
57. Pollini, G. and Colombi, R. Changes in the osmotic resistance of the leukocyte in persons exposed to benzene. Lavora Umano 16:177–184, 1964.
58. Kolesar, D. and Ballog, O. Studies by fluorescent microscopy of the effect of occupational benzene exposure on leucocyte nuclei changes. Bratislavske Lekarske Listy 40 II (4):212–219, 1965.
59. Saita, G. and Moreo, L. A case of chronic benzene poisoning with a Pelger-Huet type leucocyte anomaly. Med. Lavoro 57(5):331–335, 1966.
60. Zini, C. and Alessandri, M. Anomalia leucocitoria pseudopelgeriana in un caso di emopatia benzolica con leucosi acuta terminale. Haematol 52:258–266, 1967.
61. Sellyei, M. and Kelemen, E. Chromosome study in a case of granulocytic leukemia with "pelgerisation" 7 years after benzene pancytopenia. Eur. J. Cancer 7:83–85, 1971.
62. Bernard, J. La lymphocytose benzenique. Sangre 15:501–505, 1942.
63. Doskin, T. A. Effect of age on the reaction to a combination of hydrocarbons. Hyg. and Sanit. 36:379–384, 1971.
64. Latta, J. S. and Davies, L. T. Effects on the blood and hemopoietic organs of the albino rat of repeated administration of benzene. Arch. Pathol. 31:55–67, 1941.
65. Synder, C. A., Goldstein, B. D. and Laskin, S. Chronic inhalation studies of benzene with rats and AKR mice. Am. Indust. Hyg. Conf., May 1976, p. 75, Abstract.
66. Lange, A., Smolik, R., Zatonski, W. and Glazman, H. Leukocyte agglutinins in workers exposed to benzene, toluene, and xylene. Intl. Arch. Arbeitsmed. 31:45–50, 1973.
67. Lange, A., Smolik, R., Zatonski, W. and Szymanska, J. Serum immunoglobulin levels in workers exposed to benzene, toluene, and xylene. Intl. Arch. Arbeitsmed. 31:37–44, 1973.
68. Roth, L., Serban, V., Turcanu, P. and Moise, G. Qualitative Veranderungen der Lymphozyten nach Benzol absorption. Zschr. Inn. Med. Jahrg. 25:932–34, 1970.
69. Smolik, R., Grzybek-Nryncewica, K., Lange, A. and Zatonski, W. Serum complement level in workers exposed to benzene, toluene, and xylene. Intl. Arch. Arbeitsmed. 31:243–247, 1973.
70. Revnova, N. V. Concerning auto-immunity shifts in chronic occupational benzene poisoning. Gig. Tr. Prof. Zobol 6(7):38–42, 1962.
71. Kagan, W. A., Ascensao, J. A., Pahwa, R. N., Hansen, J. A., Goldstein, G., Valera, E. B., Incefy, G. S., Moore, M. A. S., and Good, R. A. Aplastic anemia: Presence in human bone marrow of cells that suppress myelopoiesis. Proc. Natl. Acad. Sci. U.S.A. 73:2890–2894, 1976.
72. Hoffman, R., Zanjani, E. S., Lutton, J. D., Zalusky, R. and Wasserman, L. R. Suppression of erythroid-colony formation by lymphocytes from patients with aplastic anemia. New Engl. J. Med. 296:10–13, 1977.
73. Roth, L., Turcanu, P., Dinu, I. and Moise, G. Monocytosis in those who work with benzene and chronic benzene poisoning. Folia Haematol. 100:213–224, 1973.
74. Binet, L., Conte, M. and Bourliere, F. Intoxication benzolique mortelle chex une femme vendant des sacs en cuir synthetique. Presence de benzene dans le sang. Bull. Mem. Soc. Med. Hop. Paris 61:118, 1945.
75. Craveli, A. Fibrinolysis, blood platelets, fibringen and other blood coagulation tests in clinical benzene poisoning. Med. Lavoro 53(11)722–727, 1962.
76. Favre-Gilly, M., and Bruel, M. Syndrome hemarragique benzolique avec simple alteration morpologique des plaquettes. De l'utilite de'examen systematique des plaquettes sur lame chez les ouvriers exposes au benzol. Arch. Mol. Prof. 91:274–277, 1948.
77. Inceman, S. and Tangun, Y. Impaired platelet-collagen reaction in a case of acute myeloblastic leukemia due to chronic benzene intoxication. Turk. Tip. Cemig. Mecm. 35:417–424, 1969.
78. Kliche, N., Meubrink, H. and Wohl, F. Chronische benzolschaden bu dachdeckern Z. Ges. Hyg. 15:310–316, 1969.
79. Monteverde, A., Grazioli, C. and Fumagalli, E. The effect of fibrinogen and platelets on the thromboelastogram in benzene poisoning. Med. Lavoro 54: (2)95–102, 1963.
80. Saita, G. and Sbertoli, G. L'agglutins gramma nul intossicazione cronica da benzolo. Med. Lavoro 45:250–253, 1954.
81. Saita, G., Sbertoli, G. and Farina, G. F. Thromboelastographic investigations in benzene haemopathy. Med. Lavoro 55(11):655–664, 1964.
82. Sroczynski, J., Kossmann, S. and Wegiel, A. Coagulation system in experimental poisoning with benzene vapors. Bull Slezby Sanit Epidemiol Vojewodztwa Katowickiego 15:131–134, 1971.
83. Solov'eva, E. A. Morphological changes of blood platelets and of megakaryocytes in chronic benzene poisoning. Gig. Tr. Prof. Zobol 7(11):57–60, 1963.

84. Erf, L. A. and Rhoads, C. P. The hematological effects of benzene (benzol) poisoning. J. Indust. Hyg. and Toxicol. 21:421-435, 1939.
85. Aksoy, M., Dincol, K., Erdem, S., Akgun, T. and Dincol, G. Details of blood changes in 32 patients with pancytopenia associated with long-term exposure to benzene. Br. J. Indust. Med. 29:56-64, 1972.
86. Hutchings, M., Drescher, S., McGovern, F. B. and Coombs, F. A. Investigations of benzol and toluol poisoning in Royal Austrialian Air Force Workshops. Med. J. Australia 2(23):681-693, 1947.
87. Kauppila, O. and Setala, A. Chronic benzene poisoning: Report of 5 cases. Ann. Med Int. Fenniae 45:49-51, 1956.
88. Vilter, R. W., Janold, T., Will, J. J., Mueller, J. F., Freedman, B. I. and Hawkins, V. Refractory anemia with hyperplastic bone marrow. Blood 15:1-29, 1960.
89. Gorini, P., Colombi, R. and Pecorari, D. Investigation of the origin of the macrocytosis in the cytoplasmic anemia from chronic benzene poisoning. Lavoro Umano 11:121-133, 1959.
90. Curletto, R. and Ciconali, M. Haematological disorders in benzene poisoning. Med. Lavoro 53(8, 9):505-546, 1962.
91. Corsico, R., biscaldi, G. P. and Lalli, M. Benzene-induced changes in the size of haemopoietic and haematic cells. Lavoro Umano 19(1):16-27, 1967.
92. Esteban, J. M. Benzene poisoning—Review of some cases—The warning signal Med. y Sequridad de Trabago 12(45):11-23, 1964.
93. Andre, R. and Dreyfus, B. Anemie hemolytique grave associee a un purpura hemorragique thrombopenique. Role probable due benzol transfusion massive splenectomie guerison. Sangre 22:57-65, 1951.
94. Ferrara, A. and Balbo, W. Malattia hemlitica in soggetto esposto a rischio prof di tipo benzolico. Riv. Infart. Mol. Prof. 44:713, 1957.
95. Marchal, G. and Duhamel, G. L'anemie hemolytique dans les leucemies. Sangre 21:254-261, 1950.
96. Nissen, N. I. and Ohlsen, A. S. Erythromyelosis. Oversigt og et tilfuelde hos en benzolarbegder. Ugeskr Laeg 114:737-742, 1952.
97. Saita, G. and Moreo, L. Haemolytic attack following inhalation of a single massive dose of benzol. Med. Lavoro 52(11):713-716, 1961.
98. Aksoy, M., Erdem, S. and Dincol, G. Two rare complications of chronic benzene poisoning: Myeloid metaplasia and paroxysmal nocturnal hemoglobinuria. Report of two cases. Blut 30:255-260, 1975.
99. Kahn, H. and Muzyka, V. I. The effect of benzene on the -aminolevulinic acid and porphyrin content in the cerebral cortex and in the blood. Ind. Hyg. and Prof. Assoc. Disorders 3:59-60, 1970.
100. Kahn, H. and Muzyka, V. I. The chronic effect of benzene on porphyrin metabolism Work—Environ. Health 10:140-143, 1973.
101. Forte, F. J., Cohen, H. S., Rosman, J. and Freedman, M. L. Hemin Reversal of benzene-induced inhibition of reticulocyte protein synthesis. Blood 47:145-154, 1976.
102. Freedman, M. L. Wildman, J. M., Rosman, J., Eisen, J. and Greenblatt, D. R. Benzene inhibition of in vitro rabbit reticulocyte haem synthesis at delta-aminolevulinic acid synthetase: Reversal of benzene toxicity by pyridoxine. Br. J. Haematol 35:49-60, 1977.
103. Freedman, M. L., Spieler, P. J., Rosman, J. and Wildman, J. M. Cyclic AMP maintenance of rabbit reticulocyte haem and protein synthesis in the presence of ethanol and benzene. Br. J. Haematol, in press.
104. Wildman, J. M., Freedman, M. L., Rosman, J. and Goldstein, B. Benzene and lead inhibition of rabbit reticulocyte heme and protein synthesis: Evidence for additive toxicity of these two components of commercial gasoline. Res. Commun. Chem. Pathol. and Pharmacol 13:473-487, 1976.
105. Greenblatt, D. R., Rosman, J. and Freedman, M. L. Benzene and ethanol additive inhibition of rabbit reticulocyte heme and protein synthesis. Environ. Res. 13:425-431, 1977.
106. Freedman, M. L., Geraghty, M. and Rosman, J. Hemin control of globin synthesis. Isolation of a hemin-reversible translational repressor from human mature erythrocytes. J. Biol. Chem. 249:7290-7294, 1974.
107. Freedman, M. L. and Rosman, J. A. A rabbit reticulocyte model for the role of hemin-controlled repressor in hypochromic anemias. J. Clin. Invest. 57:594-603, 1976.
108. Cohen, H. S., Freedman, M. L. and Goldstein, B. D. The problem of benzene in our environment. Clinical and molecular considerations. Am. J. Med. Sci., in press.
109. Freedman, M. L. The molecular site of benzene toxicity. IN: *A Critical Review of Benzene Toxicity* (S. Laskin and B. D. Goldstein, Eds), in press.
110. Girard, R., Mollen, M., Fourel, R. and Tolot, F. Lymphose et intoxication benzolique professionelle chronique. Arch. Mol. Prof. 22(10-11):781-786, 1960.
111. Goguel, A., Covigneaux, A. and Bernard, J. Benzene leukemias in the Paris region from 1950 to 1965.
112. Tareeff, E. M., Kontchalovskaya, N. M. and Zorina, L. A. Benzene leukemias. Acta Unio Intl Contra Can. Crum 19:751-755, 1963.
113. Aksoy, M., Erdem, S., Dincol, K., Hepyuksel, T. and Dincol, G. Chronic exposure to benzene as a possible contributory factor in Hodgkin's Disease. Blut 38:293-298, 1974.
114. Boussér, J., Neyde, R. and Fabre, A. Un cas d'hemopathie benzolique tres retardee a type de lymphosarcome. Arch. Mol. Prof. 9:130, 1948.
115. Casirola, G. and Santagati, G. Considerazioni su due casi, clinicamente primitivi, di linfomi maligni a sede splenica. Haematologica 54:85-96, 1969.
116. Mallory, T. B., Gall, E. A. and Brickley, W. J. Chronic exposure to benzene (benzol) III. The pathologic results. J. Ind. Hyg. and Toxicol. 21(8):355-377, 1939.
117. Paterni, L. and Sarnari, V. Involutional myelopathy due to benzene poisoning with the appearance of a mediastinal reticulosarcoma at an advanced stage. Securitas 50(10):55-59, 1965.
118. Torres, A., Giralt, M. and Raichs, A. Coexistencia de antecedentes benzolicos cronicas y plasmocitoma multiple. Presentacion de los casos. Sangre 15:275-279, 1970.
119. Gall, E. A. Benzene poisoning with bizarre extra-medullary hematopoiesis. Arch. Pathol. 25:315-326, 1938.
120. Rawson, R., Parker, F. and Jackson, H. Industrial solvents as possible etiologic agents in myeloid metaplasia. Science 6:541-542, 1941.
121. McLean, J. A., Blood dyscrasia after contact with petrol containing benzol. Med. J. Australia 47(11):845-849, 1960.
122. Ames, B. N. A bacterial system for detecting mutagens and carcinogens. IN: *Mutagenic Effects of Environmental Contaminants*, p. 57-63, H. E. Sutton and M. I. Harris, New York, Academic Press, 1972.
123. Infante, P. F. Testimony of, to the Occupational Safety and Health Administration, U.S. Dept. of Labor, July 26, 1977.
124. Koizumi, A., Dobashi, Y., Tachibana, Y., Tsuda, K. and Katsunuma, H. Cytokinet and cytogenetic changes in cultured human leucocytes and Hela cells induced by benzene. Ind. Health 12:23-29, 1974.
125. Dobashi, Y. Influence of benzene and its metabolites on mitosis of cultured human cells. Jap. J. Ind. Health. 16:453-461, 1974.
126. Speck, B., Schnider, T. H., Gerber, U. and Moeschlin, S. Experimentalle untersuchungen, uber den wirkungs mechanismus des benzols auf das knochenmark Schweizerische. Med. Woch 38:1274-1276, 1966.
127. Speck, B., and Moeschlin, S. Die workung von toluol, xylol, chloramphenicol und thiouracil auf dos knoshen. Experimentelle autoradiographische studien mit ^{1}-H-thymidin. Schweizerische Med. Woch 98:1684-1686, 1968.
128. Andrews, L. S., Lee, E. W., Witmer, C. M., Kocsis, J. L. and Snyder, R. Effects of toluene on the metabolism, disposition and hemopoietic toxicity of (^{3}-H) benzene. Biochem Pharmacol 26:293-300, 1977.
129. Tough, I. and Court-Brown, W. M. Chromosome aberrations and exposure to ambient benzene. Lancet 1:684, 1965.
130. Tough, I. M., Smith, P. G., Court-Brown, W. M. and Harnden, D. G. Chromosome studies on workers exposed to atmospheric benzene. Eur. J. Cancer 6:49-55, 1970.
131. Khan, H. and Khan, M. H. Cytogenetic studies following chronic exposure to benzene. Arch. Toxikol 31:39-49, 1973.
132. Hartwich, G. and Schwanity, G. Chromosomenuntersuchungin nach chronischer benzol-exposition. Deustche Med. Woch 97:15-49, 1972.
133. Forni, A., Pacifico, D. and Limonta, A. Chromosome studies in workers exposed to benzene or toluene or both. Arch. Environ. Health 22:373-378, 1971.
134. Forni, A., Moreo, L. Chromosome studies in a case of benzene-induced erythro-leukemia. Eur. J. Cancer 5:459-463, 1969.
135. Forni, A., Vigiani, E. C. Chemical Leukemogenesis in Man Ser Maemat 7(2):211-23, 1974.
136. Wurster-Hill, D. H., Cornwell, G. G. and McIntyre, O. R. Chromosomal aberrations and neoplasm—a family study. Cancer 33(1):72-81, 1974.
137. Young, R. J., Rinsky, P. A., Infante, P. F. and Wagoner, J. K. Benzene in Consumer Products. Science 199:248, 1977.
138. Otterson, E. J. Furniture stripping with benzene-based solvents, a small plant health problem. Transactions of the Thirty-third Annual Meeting of the American Conference of Governmental Industrial Hygienists. Toronto, Canada, May 24-28, 1971.

[FR Doc. 78-13663 Filed 5-18-78; 8:45 am]

[6355-01]

[16 CFR Part 1500]

PRODUCTS REQUIRING SPECIAL LABELING UNDER SECTION 3(b) OF THE FHSA; BENZENE

Proposed Amendment

AGENCY: Consumer Product Safety Commission.

ACTION: Proposed amendment to rule.

SUMMARY: The Commission proposes to amend a regulation under the Federal Hazardous Substances Act (FHSA) which mandates special cautionary labeling for products containing 5 percent or more by weight of benzene. The proposed amendment provides that the FHSA regulation will not be applicable to consumer products, except gasoline, which are subject to a new regulation under the Consumer Product Safety Act (CPSA) banning benzene-containing consumer products, except gasoline and solvents and reagents for laboratory use. Elsewhere in this issue of the FEDERAL REGISTER the Commission proposes the new CPSA banning regulation.

DATES: Comments concerning this proposal must be received by June 30, 1978. It is expected that this amendment will become effective when any final CPSA ban concerning benzene is effective.

ADDRESS: Comments should be sent to Office of the Secretary, Consumer Product Safety Commission, Washington, D.C. 20207.

FOR FURTHER INFORMATION CONTACT:

Francine Shacter, Office of Program Management, Consumer Product Safety Commission, Washington, D.C. 20207, 301-492-6557.

SUPPLEMENTARY INFORMATION: In this issue of the FEDERAL REGISTER the Commission has proposed a rule (16 CFR Part 1307) under section 8 of the Consumer Product Safety Act (CPSA), declaring all consumer products, except gasoline and solvents or reagents for laboratory use, containing benzene as an intentional ingredient or as a contaminant at levels of 0.1 percent or greater to be banned hazardous products.

The Commission notes that there is currently in existence a labeling regulation under the Federal Hazardous Substances Act (FHSA) applicable to benzene which is intended to address, among other things, the risk of blood dyscrasias which may result from benzene inhalation. This regulation, at 16 CFR 1500.14 (a)(3) and (b)(3), requires that products containing 5 percent or more by weight of benzene must be specially labeled with the signal word "danger", the word "poison", the skull and crossbones symbol, and the statement of the hazard "vapor harmful." Products containing 10 percent or more by weight of benzene must bear the additional statement of hazard "harmful or fatal if swallowed" as well as the statement "If swallowed, do not induce vomiting. Call physician immediately."

The Commission now proposes to amend the FHSA regulation to indicate that it will not be applicable to products covered by the FHSA regulation that are also banned hazardous products under any final CPSA rule. Since proposed Part 1307 is applicable only to consumer products and specifically exempts gasoline, the FHSA labeling requirements are proposed to be retained for benzene-containing gasoline and for other hazardous substances containing benzene which are not consumer products—for example, certain automotive products which are excluded from Commission jurisdiction under the CPSA, but not under the FHSA, to the extent they are distributed for use in or around a household. In addition, the Commission concludes that any consumer products containing benzene that might not be covered by any final ban because of the date they are manufactured, imported or introduced into commerce, should continue to be subject to the FHSA labeling requirements.

Since the purpose of this proposed amendment is to eliminate the possibility of confusion concerning labeling products that are banned, it is proposed that the amendment be conditional upon Part 1307 being issued in final form and its continuing in full force and effect. If, at any time, any requirement of Part 1307 relating to products within the scope of § 1500.14 (a)(3) and (b)(3) is stayed, revoked, or set aside by judicial or other action, the amendment restricting the application of the FHSA labeling requirement is to be withdrawn, and a FEDERAL REGISTER notice will be issued reinstating full applicability of the FHSA regulation. It is proposed that this amendment will become effective at the same time any final CPSA ban concerning benzene is effective.

AMENDMENT

Accordingly, pursuant to provisions of the Federal Hazardous Substances Act (sec. 2(f)(1)(A), 3(b), 74 Stat. 372, 374, as amended 80 Stat. 1303-1304, 15 U.S.C. 1261(f)(1)(A), 1262(b); sec. 10(a), 74 Stat. 378, 15 U.S.C. 1269); and under authority vested in the Commission by the Consumer Product Safety Act (sec. 30(a), 86 Stat. 1231, 15 U.S.C. 2079(a)), the Commission proposes to amend 16 CFR 1500.14(b)(3) by adding a new subparagraph (iv) as follows:

§ 1500.14 Products requiring special labeling under section 3(b) of the act.

(a) * * *
(b) * * *
(1) * * *
(2) * * *
(3) *Benzene, toluene, xylene, petroleum distillates.*
(i) * * *
(ii) * * *
(iii) * * *
(iv) Since the Commission has issued a regulation under the Consumer Product Safety Act (CPSA) banning benzene-containing consumer products, except gasoline, covered by this labeling regulation (see 16 CFR Part 1307), subparagraph (i) of this subsection is inapplicable to consumer products containing benzene as an intentional ingredient, except gasoline, which are covered by Part 1307.

Subparagraph (i) continues to apply to gasoline and other hazardous substances containing benzene, intended or packaged in a form suitable for use in or around the household, which are not "consumer products" as that term is defined in the CPSA.

This amendment is conditional upon Part 1307 continuing in full force and effect.

Dated: May 15, 1978.

SADYE E. DUNN,
Acting Secretary, Consumer Product Safety Commission.

[FR Doc. 78-13664 Filed 5-18-78; 8:45 am]

[6355-01]

[16 CFR Part 1700]

SUBSTANCES REQUIRING SPECIAL PACKAGING; BENZENE

Proposed Amendment

AGENCY: Consumer Product Safety Commission.

ACTION: Proposed amendment to rule.

SUMMARY: The Commission proposes to amend a regulation under the Poison Prevention Packaging Act of 1970 (PPPA), which requires child-resistant packaging for certain benzene-containing solvents for paint or other similar surface coating materials. The proposed amendment provides that the PPPA regulation will not be applicable to consumer products covered by the PPPA regulation which are subject to a new regulation under the Consumer Product Safety Act (CPSA) banning benzene-containing consumer products, except gasoline and solvents and reagents for laboratory use. Elsewhere in this issue of the FEDERAL REGISTER the Commission proposes the new CPSA regulation.

DATES: Comments concerning this proposal must be received by June 30, 1978. It is expected that this amendment will become effective when any final CPSA rule concerning benzene is effective.

ADDRESS: Comments should be sent to: Office of the Secretary, Consumer Product Safety Commission, Washington, D.C. 20207.

FOR FURTHER INFORMATION CONTACT:

Francine Shacter, Office of Program Management, Consumer Product Safety Commission, Washington, D.C. 20207, 301-492-6557.

SUPPLEMENTARY INFORMATION: In this issue of the FEDERAL REGISTER the Commission has proposed a rule (16 CFR Part 1307) under section 8 of the Consumer Product Safety Act (CPSA), declaring all consumer products, except gasoline and solvents or reagents for laboratory use, containing benzene as an intentional ingredient or as a contaminant at levels of 0.1 percent or greater to be banned hazardous products.

There is currently in existence a regulation under the Poison Prevention Packaging Act of 1970 (PPPA) at 16 CFR §1700.14(a)(15) requiring child-resistant packaging for prepackaged liquid solvents (such as removers, thinners, brush cleaners, etc.) for paints or other similar surface-coating materials (such as varnishes and lacquers) that contain 10 percent or more by weight of benzene, toluene, xylene, petroleum distillates, or combinations thereof, and that have a viscosity of less than 100 Saybolt Universal Seconds at 100° F.

It is apparent that many of the products covered by §1700.14(a)(15) will become banned hazardous products when proposed Part 1307, applicable to benzene-containing consumer products, except gasoline and laboratory benzene, is issued in its final form. Therefore, to eliminate any possible confusion which may arise from retaining a PPPA special packaging regulation for certain products which are now proposed to be banned hazardous products under the CPSA, the Commission proposes to amend §1700.14(a)(15) to indicate that it will not be applicable to products covered by the PPPA regulation that are also banned hazardous products under any final CPSA rule.

Since the purpose of this proposed amendment is to eliminate the possibility of confusion concerning requiring special packaging for products that are banned, the amendment is conditional upon Part 1307 being issued in final form and continuing in full force and effect. If, at any time, any requirement of Part 1307 relating to products within the scope of §1700.14(a)(15) is stayed, revoked, or set aside by judicial or other action, the amendment restricting the application of the PPPA special packaging requirement is to be withdrawn, and a FEDERAL REGISTER notice will be issued reinstating full applicability of the PPPA regulation. The Commission proposes that this amendment will become effective when any final ban under the CPSA concerning benzene becomes effective.

AMENDMENT

Accordingly, pursuant to provisions of the Poison Prevention Packaging Act of 1970 (Pub. L. 9-601; Secs. 2(4), 3, 5, 84 Stat. 1670-72, 15 U.S.C. 1471(4), 1472, 1474) and under authority vested in the Commission by the Consumer Product Safety Act (sec. 30(a), 86 Stat. 1231, 15 U.S.C. 2079(a)), the Commission proposes to amend 16 CFR §1700.14(a)(15) by adding the following language at the end of paragraph 15:

§1700.14 Substances requiring special packaging.

(a) Substances.

* * * * *

(15) * * * Since the Commission has issued a regulation under the Consumer Product Safety Act (CPSA) banning consumer products containing benzene as an intentional ingredient (see 16 CFR 1307), this paragraph is inapplicable to those benzene-containing consumer products which are covered by Part 1307. (This amendment is conditional upon Part 1307 continuing in full force and effect.)

Dated: May 15, 1978.

SADYE E. DUNN,
Acting Secretary, Consumer Product Safety Commission.

[FR Doc. 78-13655 Filed 5-18-78; 8:45 am]

Index

Abbreviations, 99
Absorber design, 140
Absorption equipment, 140
Access to records, 223
Accuracy, 89
Activated carbon, 131
Acute effects, 177
  exposure, 94, 95
Adiabatic oxidation, 143, 144
Adsorption
  isosteres, 133
  isotherm, 131, 132
  potential, 135
  rate, 137
Air/benzene mixture, 5
  respirator, 50
  samples, 70, 82, 84
  shipments, 63, 64
Airborne concentration, 75, 165, 166
Aircraft, 63
Ambient air, 166
Analysis, 80, 82, 83
Analytical, 81, 88
  procedures, 84
  techniques, 84
Anemia, 95
Anesthetic effect, 94
Aniline, 118, 120
Aplastic anemia, 95, 96
Aquatic toxicity rating, 105
Aromatic solvents, 26

Atmospheric sampling, 85
Autoignition, 4, 8
Automotive gasolines, 26
Aviation gasolines, 26
  turbine fuels, 26

Benzene
  absorption, 141
  -containing products, 26
  contamination, 30
  derivatives, 24, 25
  emissions, 129
  ethylene, 3
  exposure, 71, 78, 97, 119
    concentration, 79
  fate, 117
  free substitute, 167
  from light oil, 18
  hexachloride, 23
  isosteres, 134
  molecule, 2
  oxidation rate, 150
  poisoning, 77
  production, 12, 13, 119
  reaction, 3
  recovery, 130
  regulated areas, 52
  spill, 47
  substitute, 112
Benzenesulfonic acid, 30
Biological monitoring, 71, 73, 77

Biphenyl, 24
Blood, 73, 75, 94
  alteration, 95
  analysis, 76
  count, 71
  tests, 77
Boiling point, 1, 4
Bone marrow, 94, 95
Brain lipid cells, 94
Breathanalysis, 74
Breathing apparatus, 50
Bromobenzene, 22
Bubblers, 83
Bulk sample concentration, 121
  terminals, 193
By-product manufacturing, 120

Car evaporation losses, 118
  exhausts, 118
Carbon adsorption, 129, 130, 154
Carriers, 64
CAS number, 4
Catalyst deactivation, 152
  poisoning, 153
Catalytic alkylation, 32
  oxidation, 129, 151, 153
  reforming, 9, 11
Categories, 48
Caution labels, 53
Certification, 63, 64
Charcoal, 83, 90, 92,
  tube, 91
Chemical cartridge respirator, 50
Chlorination, 3
Chlorobenzene, 43, 45
Chromosomal damage, 97
Chronic effects, 177
Circulatory problems, 94, 95
Claims, 61
Closed systems, 199
Coal-derived benzene, 11, 13
  chemical plants, 124, 126
  tar, 26
Coke, 119
  oven, 119
    batteries, 199
    tar, 26

[Coke]
  plants, 193
  -producing plants, 17
Colorimetric indicator tubes, 83
Combustible liquid code, 46
Combustion, 8
  process, 145
Compliance cost, 190
Compressibility, 6
Consumer Product Safety Commission, 120, 227
Consumer product, 229, 236
Containers, 52, 58, 66
Contaminated air, 129
  inerts, 129
Contamination, 62
Control, 128
Critical density, 5
  pressure, 5
  temperature, 5
Cumene, 20, 29, 33, 35, 118
Cyclohexane, 20, 27, 33, 38, 39, 120
  production, 3

Damaged containers, 62
DDT, 120
Dehydrogenation, 3
Delivery drivers, 60
Density, 5
Department of Transportation (DOT), 53, 58
  requirement, 59
Dermal absorption, 94
Dermatitis, 94
Desorption, 91
  cycle, 133
  efficiency, 92
  process, 138
Detector tubes, 83, 84
Detergent alkylate, 120
Determination of benzene, 84
Dichlorobenzene, 21, 44, 118, 120
Diesel fuel oils, 26
Dispatchers, 59
Dodecylbenzene, 20, 23, 42, 43
Drums, 54

Index

Economic consideration, 190, 233
  impact, 195
Emission, 117, 118, 119
  control, 129, 143
Environmental consideration, 234
Environmental Protection Agency (EPA), 47
Equipment cleaning, 90
Erythema, 94
Ethylbenzene, 20, 21, 32, 33, 34, 120
Ethylene, 32
Evacuated bomb, 82
Evaporation rate, 5
Exposure, 47, 94, 102
  concentration, 123
  level, 123
  routes, 102
Extinguishing media, 46

Farm tractor fuels, 26
Fire
  safety, 46
  hazard properties, 46
First aid, 49, 51
Flame
  burner, 147
  speed, 8
Flammability limit, 8
Flammable liquid label, 54
Flash point, 4
Flowmeter calibration, 86
Foreign shipments, 65
Freight
  containers, 67, 68
  handlers, 61
Fumaric acid, 24

Gas
  analyzer, 85
  detector system, 87
  grab sample technique, 82
  mask, 50
  turbine fuel oil, 26
Gasoline, 30, 117, 118, 121
  distribution, 118

Handling and storage, 223
  practices, 49
Hazard classes, 53
Hazardous materials shipment, 58
Health effects, 176
  hazard data, 222
Heat
  of adsorption, 135, 136
  of combustion, 5, 7
  of fusion, 5
  of vaporization, 5, 8, 131
  recovery, 129, 146
    system, 146, 148
Hematological values, 72
Hematology guidelines, 224
Heptane, 112
Hexane, 26, 112
  exposure, 113, 114
History of regulation, 162
Human studies, 188

Ignition source, 46, 47
Impact
  on energy, 195
  on prices, 195
Infrared portable analyzer, 83
Ingestion, 94
Inhalation, 94
Instruments, 83

Kerosene, 26, 154
  absorption, 142

Label, 54
  and sign, 52
Labeling, 56, 66
Laboratories, 194, 199
Laboratory equipment, 90
Lethal
  concentration, 101, 105
  dose, 101, 103
    fifty (LD_{50}), 103
Leukemia, 94, 95, 96, 181, 239
Leukocyte function, 97
Leukopenia, 95

Light oil, 13
  recovery, 12
Lightning protection code, 46
Line notation, 4
Liquid
  label, 54
  mixtures, 167
Liquefied petroleum gases, 26
Lubricating oil, 26
Lymphatic system, 97

Macroeconomic variables, 195
Maleic anhydride, 20, 25, 38, 41, 118
Marking, 56, 66
Mass transfer coefficient, 141, 142
Medical
  examination, 71
  monitoring, 71
  requirement, 222
  surveillance, 191, 221
Melting point, 4, 5
Metabolic action, 73
Metadichlorobenzene, 44
Methyl ethyl ketone, 112
Methylene chloride, 112
  inhalation, 115
Modes of transport, 65
Molecular
  formula, 4
  weight, 4
Monitoring
  biological, 71
  measurement procedures, 223
  medical, 71
Monochlorobenzene, 118
Multicomponent contaminants, 138, 139

National Electric Code, 46
National Fire Prevention Association Code, 46
Nature of operations, 166
Nervous system, 94
National Institute for Occupational Safety and Health (NIOSH), 83, 92, 121

Nitrobenzene, 20, 24, 28, 38, 40, 118
Nongasoline petroleum products, 198

Observation of monitoring, 223
Occupational exposure, 84
  to benzene, 174
Oil
  absorption, 129, 141
  cracking, 10
  gas production, 193
  spill, 117, 119
Orthodichlorbenzene, 21
Occupational Safety and Health Administration (OSHA), 46, 52, 161, 174
Oxidation
  reaction, 149
  temperature, 146

Packaging, 56, 243
Paint
  removers, 26, 30
  thinners, 26
Pancytopenia, 96, 97, 238
Paradichlorbenzene, 22
Partition coefficient, 5
Personnel, 58
Permissible airborne exposure, 200
Pertinent legal authority, 176
Petrochemical industry, 9, 125, 193
Petroleum
  -derived benzene, 9, 10
  refineries, 192
Phenol, 31, 36, 118, 120
  conjugates, 81
  production, 3
Phenylstearic acid, 22
Physical
  properties, 4
  state, 4
Placard, 62, 67, 68, 69
Plants, producing cumene, 36
Platelet function, 97
Polymer manufacturing, 27
Production
  capacity, 12, 36, 41

Index

[Production]
  plants, 12
Propylene, 33
Protective
  clothing, 48, 222
  equipment, 221
Pyrolysis gasoline, 9, 10

Rail
  cars, 68, 69
  shipment, 63
Rate of adsorption, 136
Reaction, 1
Reaction kinetics, 149
Reactivity hazard data, 223
Reagents, 90
Red blood cells, 97
Refractive index, 5
Regulated areas, 52
Regulations, 128
Regulatory analysis, 163
Requirements
  for carriers, 59
  for shippers, 54
Residence time, 146
Respirator, 48, 49
  type, 50
  usage, 50
Road drivers, 60
Rubber
  industry, 20
  manufacturing, 97
  products, 27, 194
  tire manufacturing, 27

Safety, 46
Sales personnel, 59
Sample
  collection, 90
  types, 84
Sampling, 70, 79, 80, 89, 90
  bag, 82, 90
  instrument, 86
  procedure, 88
Sealed containers, 200
Service station, 118

Shipment
  of flammable liquids, 54, 56
  violation, 66
Shipping
  document requirement, 62
  papers, 63, 64, 66
  personnel, 58
Silica gel, 83
Signs
  and labels, 221
  and symptoms, 223
Skin absorption, 94
Solubility, 1
  in water, 5
Solvent
  naphthas, 26
  properties, 1
Sources, 117
Specific gravity, 5
  heat, 7
Spectrophotometry, 83
Spill, 47, 117
  and leak procedure, 223
Spray tower, 141
Standards preparation, 92
Static electricity, 46
Storage, 46, 47, 122
Structural formula, 2
Studies, human, 181
Styrene, 32, 34, 118
  production, 3
Surface tension, 4
Surveillance, 224
Switching ticket, 63
Synonyms, 4

Tank, cargo, 68
  portable, 68
Thermal
  conductivity, 5
  oxidation, 129, 143, 154
  vortex burner, 146
Thrombocytopenia, 95
Toluene, 20, 112
  disproportionation, 11
Toxic
  concentration, 101, 103

[Toxic]
  dose, 98
    concentration, 106
Toxicity, 112
  data, 98
  ranges, 48
Toxicology, 223
  notations, 104
Traffic, 61
Training programs, 51, 221
Transfer station, 122
Transportation, 65, 194

Units of measurement, 48, 79, 78
Urinary
  phenol analysis, 71, 74, 80
  sulfate analysis, 74
Urine sample, 80, 81

Vapor
  density, 4
  pressure, 1, 4, 6, 134
Vehicle placarding, 67
Viscosity, 4

Wash oil
  absorption, 140, 154
Waste disposal, 48
Water shipments, 64
Work
  factors, 166
  practices, 47
Workers, exposed, 76

Xylenes, 26, 112